Battery Management Systems

Philips Research

VOLUME 1

Editor-in-Chief
Dr. Frank Toolenaar
Philips Research Laboratories, Eindhoven, The Netherlands

SCOPE TO THE *'PHILIPS RESEARCH BOOK SERIES'*

As one of the largest private sector research establishments in the world, Philips Research is shaping the future with technology inventions that meet peoples' needs and desires in the digital age. While the ultimate user benefits of these inventions end up on the high-street shelves, the often pioneering scientific and technological basis usually remains less visible.

This 'Philips Research Book Series' has been set up as a way for Philips researchers to contribute to the scientific community by publishing their comprehensive results and theories in book form.

Ad Huijser

Battery Management Systems
Design by Modelling

by

Henk Jan Bergveld
Wanda S. Kruijt
Peter H.L. Notten

Philips Research Laboratories,
Eindhoven, The Netherlands

KLUWER ACADEMIC PUBLISHERS
DORDRECHT / BOSTON / LONDON

A C.I.P. Catalogue record for this book is available from the Library of Congress.

ISBN 1-4020-0832-5

Published by Kluwer Academic Publishers,
P.O. Box 17, 3300 AA Dordrecht, The Netherlands.

Sold and distributed in North, Central and South America
by Kluwer Academic Publishers,
101 Philip Drive, Norwell, MA 02061, U.S.A.

In all other countries, sold and distributed
by Kluwer Academic Publishers, Distribution Center,
P.O. Box 322, 3300 AH Dordrecht, The Netherlands.

Printed on acid-free paper

Explanation of cover:
The cover shows a transparent battery as an illustration of the use
of battery models for the design of Battery Management Systems
designed by Hennie Alblas

Printed in the Netherlands.

To Peggy, Bert and Pascalle

Table of contents

4. Battery modelling   55

7. Optimum supply strategies for Power Amplifiers in cellular phones   241

List of abbreviations

ACPI	Advanced Configuration and Power Interface
ACPR	Adjacent Channel Power Ratio
ADC	Analogue-to-Digital Converter
AM	Amplitude Modulation
BER	Bit Error Rate
BIOS	Basic Input Output System
BMS	Battery Management System
CAC	Compensated Available Charge
CC	Constant Current
CDMA	Code-Division Multiple Access
$Cd(OH)_2$	Cadmium hydroxide
CHC	Charging Control
CV	Constant Voltage
DAC	Digital-to-Analogue Converter
DCR	Discharge Count Register
DCS	Digital Cellular System
DMC	Dimethyl carbonate
DQPSK	Differential Quadrature Phase Shift Keying
DSP	Digital Signal Processor
DTC	DeskTop Charger
e	Electron
EC	Ethylene carbonate
ECC	Energy Conversion Control
EDGE	Enhanced Data rates for GSM Evolution
EM	Electro-Magnetic
EMC	Ethyl methyl carbonate
EMF	Electro-Motive Force
EMI	Electro-Magnetic Interference
ESR	Equivalent Series Resistance
FDD	Frequency Division Duplex
FDMA	Frequency-Division Multiple Access
FM	Frequency Modulation
FSK	Frequency Shift Keying
GMSK	Gaussian Minimum Shift Keying
GSM	Global System for Mobile communication
H^+	Proton
H_2O	Water
H_2SO_4	Sulphuric acid
HVIC	High-Voltage IC
ID	Identification
IEC	International Electrotechnical Commission
IIC	Interface IC
KOH	Potassium hydroxide
LED	Light-Emitting Diode
LCD	Liquid-Crystal Display
LR	Linear Regulator
Li-ion	Lithium-ion
$LiCoO_2$	Lithium cobalt oxide

$LiMn_2O_4$	Lithium manganese oxide
$LiNiO_2$	Lithium nickel oxide
$LiPF_6$	Lithium hexafluorophosphate
LiC_6	Lithium graphite
LMD	Last Measured Discharge
LSE	Least-Square Error
MSK	Minimum Shift Keying
NAC	Nominal Available Charge
NADC	North American Digital Cellular
NiCd	Nickel-cadmium
NiMH	Nickel-metalhydride
$Ni(OH)_2/NiOOH$	Nickel hydroxide/nickel oxyhydroxide
NTC	Negative-Temperature Coefficient resistor
O_2	Oxygen
OQPSK	Offset Quadrature Phase Shift Keying
Ox	Oxidized species
PA	Power Amplifier
Pb	Lead
PbO_2	Lead dioxide
PCB	Printed-Circuit Board
PEO	Polyethylene oxide
PFC	Programmed Full Count
PM	Power Module
PTC	Positive-Temperature Coefficient resistor
PWM	Pulse-Width Modulation
QAM	Quadrature Amplitude Modulation
QPSK	Quadrature Phase Shift Keying
Red	Reduced species
RF	Radio-Frequency
Rx	Receive
SBC	Smart Battery Charger
SBD	Smart Battery Data
SBS	Smart Battery System
SEI	Solid Electrolyte Interface
SHE	Standard Hydrogen reference Electrode
SLA	Sealed lead-acid
SMBus	System Management Bus
SMPS	Switched-Mode Power Supply
SoC	State-of-Charge
SoH	State-of-Health
TCH	Traffic Channel
TCM	Timer-Control Module
TDD	Time Division Duplex
TDMA	Time-Division Multiple Access
Tx	Transmit
UMTS	Universal Mobile Telecommunication System
UPS	Uninterruptible Power Supply
VRLA	Valve-regulated lead-acid
$Zn-MnO_2$	Zinc-manganese dioxide
3G	Third Generation

List of symbols

Symbol	Meaning	Value	Unit
A	Electrode surface area		m^2
A	Area of spatial element		m^2
A_{bat}	Battery surface area		m^2
A_{Cd}^{max}	Surface area of cadmium electrode		m^2
a_i	Activity of species i		mol/m^3
a_i^{ref}	Activity of species i in the reference state		mol/m^3
a_i^b	Bulk activity of species i		mol/m^3
a_i^s	Surface activity of species i		mol/m^3
c_i	Concentration of species i		mol/m^3
c_i^b	Bulk concentration of species i		mol/m^3
c_i^s	Surface concentration of species i		mol/m^3
C_{ch}	Chemical capacitance		mol^2/J
C_{el}	Electrical capacitance		F
C_{th}	Thermal capacitance		J/K
C^{dl}	Double-layer capacitance		F
C_o^{dl}	Double-layer capacitance per unit area		F/m^2
C_H	Helmholtz capacitance		F
C_{G-C}	Gouy-Chapman capacitance		F
C_{para}	Parasitic capacitance in DC/DC converter		F
Cap_{rem}	Remaining battery capacity		Ah
Cap_{max}	Maximum possible capacity that can be obtained from a battery		Ah
CF_i	Cost function for output variable i		-
D_i	Diffusion coefficient of species i		m^2/s
d_i	Diffusion layer thickness of species i		m
D_j	Anti-parallel diodes that model Butler-Volmer relation for reaction j		-
dV/dt_{lim}	Change of battery voltage in time, used as parameter in proposed SoC indication system		V/s
E	Error in 'battery empty' prediction of an SoC indication system		%
E_i	Potential of electrode i		V
E_i^{eq}	Equilibrium potential of electrode i		V
E_{bat}^{eq}	Equilibrium potential of battery, or EMF		V
E^{eq*}	Apparent equilibrium potential		V
E_i^o	Standard redox potential of electrode i		V
E_{bat}^o	Standard redox potential of battery		V
E_{par}^a	Activation energy of parameter *par*		J/mol
E_{ch}	Chemical energy		J
E_{el}	Electrical energy		J
E_{th}	"Thermal energy"		JK
E_{max}	Maximum energy stored in capacitor or coil		J

Symbol	Meaning	Value	Unit
E_q	Energy term, normalized to current, used in simple overpotential description of (Eq. 6.4)		J/A
F	Faraday's constant	96485	C/mol
f_c	Channel frequency		Hz
f_{RF}	RF frequency		Hz
f_{switch}	Switching frequency in DC/DC converter		Hz
f_T^d	Relation between measured battery parameter and SoC in direct-measurement SoC system		%
$f^{bk}{}_{V,T,I}$	Function that translates coulomb counter contents into SoC based on the basis of V, T and I measurements in a book-keeping system		%
I_j^o	Exchange current for reaction j		A
I_a	Anodic current		A
I_c	Cathodic current		A
I_{dl}	Double-layer current		A
I_{lim}	Current level that determines state in proposed SoC indiction system		A
I_{sup}	Supply current		A
J_i	Diffusion flux of species i		mol/(m^2.s)
J_{ch}	Chemical flow		mol/s
J_{th}	Heat flow		W
$k_{a,j}$	Reaction rate constant for oxidation (anodic) reaction j		Unit depends on reaction
$k_{c,j}$	Reaction rate constant for reduction (cathodic) reaction j		Unit depends on reaction
k_b	Backward reaction rate constant		Unit depends on reaction
k_f	Forward reaction rate constant		Unit depends on reaction
K_{O_2}	Oxygen solubility constant		mol/(m^3.Pa)
l_{Ni}	Thickness of nickel electrode, grain size		m
l_{pos}	Thickness of positive electrode, grain size		m
l_{neg}	Thickness of negative electrode, grain size		m
l_{elyt}	Thickness of electrolyte in Li-ion model		m
m	Number of electrons in reaction		-

Symbol	Meaning	Value	Unit
m	Number of spatial elements in the electrolyte in Li-ion model		-
M	Number of parameter sets in optimization process		-
m_i	Molar amount of species i		mol
m^{ref}	Molar amount in the reference state		mol
m^o_{Cd}	Molar amount of cadmium nuclei at $t=t_o$		mol
M_{Cd}	Molecular weight of cadmium		kg/mol
n	Number of electrons in reaction		-
n	Number of capacitors in capacitive voltage converter		-
N	Number of nuclei		-
N	Number of points taken into account in optimization process		-
p	Number of spatial elements in a system		-
P	Pressure		Pa
P_{out}	Output power of PA in cellular phone		dBm
P_{sup}	Supply power for PA		W
P_{ch}	Chemical power		W
P_{el}	Electrical power		W
P_{th}	"Thermal power"		WK
$par(T)$	Temperature-dependent parameter		Unit depends on parameter
par^o	Pre-exponential factor for parameter par		Unit depends on parameter
Q	Charge		C
Q_{th}	Heat		J
$Q_{Cd,Max}$	Maximum capacity of cadmium electrode		C
$Q_{Ni,Max}$	Maximum capacity of nickel electrode		C
$Q_{Cd(OH)_2,Ni}$	Overdischarge reserve of $Cd(OH)_2$ at the nickel electrode		C
$Q_{Cd,Cd}$	Overdischarge reserve of cadmium at the cadmium electrode		C
$Q^{max}_{LiCoO_2}$	Maximum capacity of $LiCoO_2$ electrode		C
$Q^{max}_{LiC_6}$	Maximum capacity of LiC_6 electrode		C
$r(t)$	Radius of hemispherical particles		m
R	Gas constant	8.314	J/(mol.K)
R_{ch}	Chemical resistance		Js/mol^2
R_e	Electrolyte resistance		Ω
R_{leak}	Resistance that models self-discharge in Li-ion model		Ω
R_{bypass}	On-resistance of bypass switch in DC/DC converter		Ω
R_{coil}	ESR of coil in DC/DC converter		Ω

Symbol	Meaning	Value	Unit
R_{loss}	Ohmic-loss resistance in DC/DC converter		Ω
R_{switch}	On-resistance of switch in DC/DC converter		Ω
R_{el}	Electrical resistance		Ω
R_{load}	Transformed antenna impedance seen at collector/drain of final PA stage		Ω
R_{th}	Thermal resistance		K/W
R_S	Series resistance of switch S		Ω
$R_{\Omega k}$	Ohmic and kinetic resistance		Ω
R_d	Diffusion resistance		Ω
$R_d C_d$	Time constant diffusion overpotential		s
$R_k C_k$	Time constant kinetic overpotential		s
S_i	Switch i in DC/DC converter		-
SoC_E	First SoC in equilibrium state, just after re-entry from transitional state		%
SoC_S	First SoC when discharge state is entered		%
SoC_t	Last SoC in transitional state just before equilibrium state is entered		%
t	Time		s
t_{rem}	Remaining time of use		s
$t_{1,actual}$	Experimental remaining time of use from the moment a discharge current is applied until the battery is empty		s
T	Switching period in DC/DC converter		s
T	Temperature		K
T_{amb}	Ambient temperature		K
T_{oper}	Operating time		s
T_{period}	Period time of burst drawn by PA model		s
$U_{pos,i}$	Interaction energy coefficient for $LiCoO_2$ electrode in phase i		-
$U_{neg,i}$	Interaction energy coefficient for LiC_6 electrode in phase i		-
$U_q(I)$	Inverse step function	1 for I≤0 and 0 for I>0	-
V	Voltage		V
V_{bat}	Battery voltage		V
V_{con}	Control voltage		V
V_{EoD}	End-of-Discharge voltage		V
V_{error}	Error voltage in behavourial PA model	V_{error}=0 or V_{error}=1	V
V_{nom}	Nominal supply voltage for PA		V
V_{sup}	Supply voltage		V
$V_{sup,opt}$	Optimum supply voltage, as applied in efficiency control		V
$V_{sup,min}$	Minimum PA supply voltage at which linearity specification can still be met		V

Symbol	Meaning	Value	Unit
V_{switch}	Switch drive voltage in DC/DC converter		V
$V(t)$	Volume of deposited material at time t		m^3
V^o	Initial volume of hemispherical particles at time t_o		m^3
V_g	Free gas volume inside battery		m^3
Var_i	Output variable in optimization process		Depends on Var_i
$W_{i,j}$	Normalizing and weighing factor for output variable Var_i		Depends on Var_i
x	Distance		m
w_j, x_j, y_j, z_j	Reaction order of species in reaction j	1^1	-
x_i	Mol fraction of species i		-
$x^{pos}_{phasetransition}$	Mol fraction of Li^+ ions at which phase transition occurs in positive electrode		-
$x^{neg}_{phasetransition}$	Mol fraction of Li^+ ions at which phase transition occurs in negative electrode		-
z_i	Valence of ionic species i		-
α	Transfer coefficient	$0<\alpha<1$	-
α_{th}	Heat transfer coefficient		$W/(K.m^2)$
γ	Activity coefficient	1^1	-
γ_g	Fugacity coefficient		$mol/(m^3.Pa)$
δ	Diffusion layer thickness		m
δ	Duty cycle in DC/DC converter	$0<\delta<1$	-
δ_H	Helmholtz layer thickness		m
δ_{PA}	Duty cycle of PA = ON/OFF ratio	$0<\delta_{PA}<1$	-
ΔG^o	Gibbs free energy change under standard conditions		J/mol
$\Delta G^{o'}$	Gibbs free energy change under standard conditions, taking constant activity terms for e.g. OH^- or H_2O species into account		J/mol
ΔG_O	Change in (Gibbs) free energy in an oxidation reaction		J/mol
ΔG_R	Change in (Gibbs) free energy in a reduction reaction		J/mol
$\Delta \bar{G}_O$	Change in free electrochemical energy in an oxidation reaction		J/mol
$\Delta \bar{G}_R$	Change in free electrochemical energy in a reduction reaction		J/mol
ΔH	Change in a reaction's enthalpy		J/mol
ΔS	Change in a reaction's entropy		J/(mol.K)
Δx	Thickness of a spatial element		m
ε_o	Permittivity in free space	$8.85.10^{-12}$	$C^2/(N.m^2)$
ε_r	Dielectric constant		-
ϕ	Electrostatic or Galvani potential		V
ϕ_e	Electrostatic electrode potential		V
ϕ^s	Electrostatic electrode potential		V

Symbol	Meaning	Value	Unit
ϕ	Electrostatic electrolyte potential		V
ϕ_{Ox}	Electrostatic potential of Ox ions in electrolyte		V
Φ_i	Conduction period for switch in capacitive voltage converter		s
2Φ	Conduction angle for PA		rad
μ_i	Chemical potential of species i		J/mol
$\overline{\mu}_i$	Electrochemical potential of species i		J/mol
μ_i^o	Standard chemical potential of species i		J/mol
ρ_{Cd}	Gravimetric density of cadmium		kg/m^3
v_i	Stoichiometric factors for species i in a reaction equation		-
τ	Time constant		s
τ_d	Diffusion time constant		s
τ_q	Time constant associated with increase in overpotential in an almost empty battery		s
ω_{RF}	RF angular frequency		rad/s
η^{ct}	Overpotential of charge transfer reaction		V
η^d	Diffusion overpotential		V
η^k	Kinetic overpotential		V
η^{Ω}	Ohmic overpotential		V
$\eta_{\Omega k}$	Combined ohmic and kinetic overpotential		V
η_q	Increase in overpotential when the battery becomes empty		V
$\eta_{DC/DC}$	DC/DC converter efficiency		%
η_{max}	Maximum attainable PA efficiency		%
$\eta_{max,theory}$	Maximum theoretical PA efficiency		%
η_{PA}	PA efficiency		%
θ_{ex}	Extended surface area		m^2
$\zeta_{pos,i}$	Constant for LiCoO$_2$ electrode in phase i		-
$\zeta_{neg,i}$	Constant for LiC$_6$ electrode in phase i		-

Note 1: Value of 1 has been assumed in this book

Series preface

As one of the largest private sector research establishments in the world, Philips Research is shaping the future with technology inventions that meet peoples' needs and desires in the digital age. While the ultimate user benefits of these inventions end up on the high-street shelves, the often pioneering scientific and technological basis usually remains less visible.

This 'Philips Research Book Series' has been set up as a way for our researchers to contribute to the scientific community by publishing their comprehensive results and theories in a convenient book form.

Dr Ad Huijser
Chief Executive Officer of Philips Research

Preface

This book covers the interesting and multidisciplinary subject of Battery Management Systems (BMS), in particular the design of BMS with the aid of simulation models. The basic tasks of BMS are to ensure optimum use of the energy stored in the battery (pack) that powers a portable device and to prevent damage inflicted on the battery (pack). This becomes increasingly important due to the larger power consumption associated with added features to portable devices on the one hand and the demand for longer run times on the other hand. In addition to explaining the general principles of BMS tasks such as charging algorithms and State-of-Charge (SoC) indication methods, the book also covers real-life examples of BMS functionality of practical portable devices such as shavers and cellular phones.

Simulations offer the advantage over measurements that less time is needed to gain knowledge of a battery's behaviour in interaction with other parts in a portable device under a wide variety of conditions. This knowledge can be used to improve the design of a BMS, even before a prototype of the portable device has been built. Although the advantages are obvious, the application of simulation to the design of BMS is relatively new, mainly due to the absence of useable battery models so far. Therefore, a large part of this book is devoted to the construction of such simulation models for rechargeable batteries. With the aid of several illustrations it is shown that design improvements can indeed be realized with the battery models constructed in this book. The battery models can be seen as 'transparent batteries' in which the course of all reactions and the origin of certain externally visible battery behaviour can be traced. This is also illustrated in the cover illustration of this book. Apart from the battery models, models of other parts of a portable device are also used. A battery's electrical and thermal environments can be modelled in this way.

After the formal definition of a BMS in chapter 1, chapter 2 provides more detailed information on BMS. It is shown that the complexity and sophistication of a BMS depend not only on the portable device in which it is included, but certainly also on the battery type. The characteristics of the various parts that constitute a BMS are explained and examples of BMS in different types of portable products are given. The emphasis in these examples is on illustrating the influence of several factors on the BMS complexity.

Chapter 3 starts with a brief historical overview that illustrates the development of batteries from 250 BC until the present day. Commonly encountered battery-related terms are defined and the main characteristics of the most common battery types are listed. The operational mechanism of batteries is described in simple terms, offering people without an (electro)chemical background some essential information on the main processes that are responsible for battery behaviour. The several contributions to the overall battery voltage are also defined.

Chapter 4 forms the core of this book and describes the subject of battery modelling. This chapter should be read by those that would like to obtain detailed knowledge on the occurrence of chemical and physical processes inside batteries and how these processes can be modelled. A new method of battery modelling is introduced that is based on the concept of physical system dynamics. The essence of this concept is the analogy between energy storage and energy dissipation definitions in various physical domains, such as the electrical domain, the chemical domain and the thermal domain. As an advantage of this method, these domains are coupled enabling the transfer of energy from one domain to another. As a result, the battery voltage, internal gas pressure and temperature can be simulated coherently under a wide variety of conditions. The general principles of modelling are described first in a step by step approach. The models are built up in the form of network models, making them more suitable for use in an electronic environment with electronic-circuit simulation tools such as SPICE. The general principles of modelling are then successively applied to a NiCd battery and a Li-ion battery. For both battery types the mathematical description of all the relevant processes and reactions that take place inside that particular battery type are given. Chapter 4 also covers the subject of finding the correct parameter values for the models. This is done by means of a new mathematical method for the NiCd model. The simulation examples at the end of the chapter illustrate the use of the constructed battery models as 'transparent batteries'.

Chapter 5, 6 and 7 give more detailed information on the most important BMS tasks. Simulation results obtained with the models described in chapter 4 serve as an example of the use of these models to improve the design of BMS. Chapter 5 discusses the charging algorithms for NiCd, NiMH and Li-ion batteries encountered in practice. Moreover, charging simulations are performed with both the NiCd and Li-ion model discussed in chapter 4. A new charging algorithm based on a maximum allowed battery temperature under the name *thermostatic charging* is introduced based on simulation results. Measurement results are presented to confirm the advantages revealed by the simulations. Possible causes for a decrease in cycle life of Li-ion batteries when they are charged faster with increased current and voltage values are investigated by means of simulations.

The subject of State-of-Charge (SoC) indication is discussed in chapter 6. Three possible methods are identified and described. Experiments performed with a commercially available SoC indication system are described and the shortcomings of this system are identified. A new set-up for an SoC system for Li-ion batteries is proposed in an attempt to overcome these shortcomings. The method is based on a combination of the methods mentioned at the beginning of the chapter. The method was developed using the battery models described in chapter 4. The system gives an estimation of the remaining time of use during the valid conditions during discharge of the battery. The chapter ends with a discussion of experimental results obtained with the proposed system. A comparison with the results obtained with the commercially available system shows that many more experiments will have to be carried out to further investigate the new system. Several suggestions for improvements are given.

Chapter 7 compares several ways of powering a Power Amplifies (PA) in a cellular phone with respect to efficiency. On the basis of trends in the cellular telephony market it is shown that reducing the power consumption of the PA in talk mode is becoming increasingly important. This can be achieved by increasing the PA efficiency, which is defined as the ratio of the transmitted power and the power drawn at the supply pin. The concept of efficiency control is introduced, which implies that the PA is always powered with the minimum needed supply voltage. It is shown that inductive DC/DC converters are most suitable in terms of efficiency to generate this minimum supply voltage from the battery voltage. The benefits of efficiency control are illustrated by means of simulations using the battery models of chapter 4 and by means of measurements of practical PAs.

The various application examples of battery models discussed in this book illustrate that these models are indeed beneficial for the design of BMS. The authors would like to encourage the co-operation of people with different backgrounds such as electrochemistry and electrical engineering to improve the design of BMS. As illustrated in this book, a lot can be gained from this co-operation. This book looks at the exiting field of battery management from both viewpoints. Therefore, the authors hope to have laid a foundation for many more successful research and development projects dedicated to getting the most out of our batteries. The future is wireless!

Acknowledgements

The authors would like to thank the management of Philips Research Laboratories for the opportunity to publish their research results in this book. Paul Regtien, Frans Schoofs, Peter Blanken and Hans Feil are acknowledged for their substantial contributions to the content of this book. The authors are also indebted to René van Beek for performing many battery-related measurements described in this book. The following persons are acknowledged for various contributions: Bram van den Akker, André van Bezooijen, Jan Reinder de Boer, Ralf Burdenski, Vincent Cachard, Kok-how Chong, René du Cloux, Valère Delong, Robert Einerhand, Wouter Groeneveld, Saskia Hanneman, Anton Hoogstraate, Henk Jan Pranger, Siep Onneweer, William Rey, Jurgen Rusch and Wouter de Windt. The authors would like to thank their colleagues at Matsushita, Japan, for the co-operation during a joint battery project. Domine Leenaerts, Gerard van der Weide and the Philips Translation Service are acknowledged for proofreading of the original manuscript. Hennie Alblas is acknowledged for drawing all figures in this book. Finally, the authors would like to thank their families for the continuous support and understanding.

HENK JAN BERGVELD
WANDA KRUIJT
PETER NOTTEN

Eindhoven, April 2002

Chapter 1
Introduction

1.1 The energy chain

The demand for portable electronic consumer products is rapidly increasing. Examples of fast-growing markets of portable products are notebook computers, cellular and cordless phones and camcorders. Figure 1.1 shows the expected world-wide shipment of cellular handsets until the year 2004 [1], with the total number of handsets sold increasing dramatically. At present, almost half of the shipments of cellular handsets are replacement sales, which means that people who already have a cellular phone buy them. By 2004, almost all cellular handsets sold will be replacement sales.

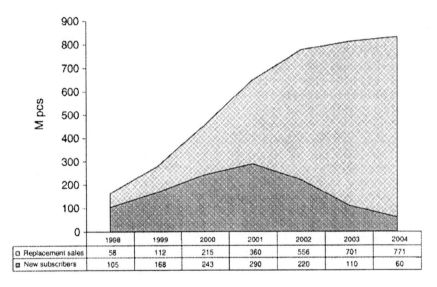

	1998	1999	2000	2001	2002	2003	2004
□ Replacement sales	58	112	215	360	556	701	771
▨ New subscribers	105	168	243	290	220	110	60

Figure 1.1: Expected world-wide shipments of cellular handsets in millions per type of sales [1]

In view of battery costs and environmental considerations, the batteries used to power these portable devices should preferably be rechargeable. The demand for rechargeable batteries is increasing in line with that for portable products. Table 1.1 shows the sales of the most important rechargeable battery types in Japan in 1998 [2]. The majority of the rechargeable batteries are manufactured in Japan. Table 1.1 shows that the number of batteries sold increased dramatically in 1998 relative to 1997. More information on rechargeable battery types will be given in chapter 3.

The energy yielded by a portable device in the form of, for example, sound or motion, ultimately derives from the electrical energy supplied by the mains. The conversion of energy from the mains to the eventual load inside a portable product can be described as an energy chain [3], which is shown in Figure 1.2. The links of the energy chain are a charger, a battery, a DC/DC converter and a load.

Table 1.1: Sales of the most important rechargeable battery types in Japan in millions in 1998 [2]

Battery type	Sales in 1998 (1M pieces)	Growth with respect to 1997
NiCd	588	85%
NiMH	640	112%
Li-ion	275	141%

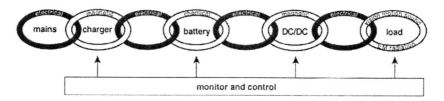

Figure 1.2: The energy chain symbolizing the energy transfers from mains to load in a portable product

Electrical energy from the mains is fed to the battery through the charger during charging. The charger uses electromagnetic components like a transformer or an inductor. Here, electrical energy from the mains is first transformed into magnetic energy and then back into electrical energy. The battery stores the electrical energy in the form of chemical energy. During discharge of the battery, chemical energy is converted back into electrical energy.

The DC/DC converter is an optional link. It is not to be found in every portable product. There are two reasons for its presence. First of all, the battery might deliver a voltage which is not suitable for operation of the load. Secondly, each circuit part of the load should be operated from the lowest possible supply voltage for efficiency reasons, because a surplus in supply voltage is often dissipated in the form of heat. In both cases, the DC/DC converter powers the load with the lowest possible supply voltage, irrespective of the battery voltage. The DC/DC converter uses an inductor, which translates the electrical energy from the battery into magnetic energy and back into electrical energy again. In the load, the electrical energy from the DC/DC converter is converted into sound, light, Electro-Magnetic (EM) radiation or mechanical energy.

In order to make optimum use of the energy inside the battery, all conversions of energy in the energy chain should be well understood and made as efficient as possible. This is especially true when miniaturization of the portable product is desired. Miniaturization is an important trend for many portable devices. Portable phones, for example, are becoming smaller and smaller. When the volume of a portable device decreases, the amount of dissipated power must also decrease. The reason is that a smaller volume will yield a higher temperature when the amount of dissipated power remains the same. There is also a trend towards increasing complexity and functionality of portable products. An example is e-mail and internet facilities added to the functionality of cellular phones. Adding complexity to the total system adds to the total power consumption of the load. So assuring efficient energy conversions becomes even more important. This can be achieved by monitoring and controlling all the links in the energy chain. This is schematically shown in Figure 1.2.

There are algorithms that check and control the links in the energy chain. A first example is a charging algorithm, which keeps track of the battery status and controls the charger by interrupting the charging current when the battery is full. Charging should not continue once the battery is full, because otherwise the battery temperature will rise substantially and/or the battery might be damaged. This decreases its capacity and usable number of cycles. Therefore, a proper charging algorithm leads to a more efficient use of the battery and its energy.

A second example is an algorithm that determines the battery's State-of-Charge (SoC). This information can be used to make more efficient use of the battery energy. For example, it can be used as input for charge control, indicating that the battery is full. Also, it is more likely that the user will wait longer before recharging the battery when an accurate and reliable SoC indication is available on a portable device. Less frequent recharging is beneficial for the cycle life of the battery.

A third example is an algorithm that controls a DC/DC converter to power the load with the minimum required supply voltage, dependent on the activity of the load. An example of such a load is a Power Amplifier (PA) in a cellular phone. In the case of a PA, the supply voltage may be lower for lower output power. This leads to better efficiency. This example will be elaborated in chapter 7.

1.2   Definition of a Battery Management System

Three terms apply to the implementation of monitor and control functions in the energy chain. These terms are *battery management, power management* and *energy management*. As a rough indication, battery management involves implementing functions that ensure optimum use of the battery in a portable device. Examples of such functions are proper charging handling and protecting the battery from misuse. Power management involves the implementation of functions that ensure a proper distribution of power through the system and minimum power consumption by each system part. Examples are active hardware and software design changes for minimizing power consumption, such as reducing clock rates in digital system parts and powering down system parts that are not in use. Energy management involves implementing functions that ensure that energy conversions in a system are made as efficient as possible. It also involves handling the storage of energy in a system. An example is applying zero-voltage and zero-current switching to reduce switching losses in a Switched-Mode Power Supply (SMPS). This increases the efficiency of energy transfer from the mains to the battery.

It should be noted that the implementation of a certain function may involve more than one of the three management terms simultaneously. This book will focus on battery management and its inclusion in a system. A definition of the basic task of a Battery Management System can be given as follows:

The basic task of a Battery Management System (BMS) is to ensure that optimum use is made of the energy inside the battery powering the portable product and that the risk of damage inflicted upon the battery is minimized. This is achieved by monitoring and controlling the battery's charging and discharging process.

Keeping in mind the examples of algorithms given in section 1.1, this basic task of a BMS can be achieved by performing the following functions:

- Control charging of the battery, with practically no overcharging, to ensure a long cycle life of the battery.

- Monitor the discharge of the battery to prevent damage inflicted on the battery by interrupting the discharge current when the battery is empty.

- Keep track of the battery's SoC and use the determined value to control charging and discharging of the battery and signal the value to the user of the portable device.

- Power the load with a minimum supply voltage, irrespective of the battery voltage, using DC/DC conversion to achieve a longer run time of the portable device.

1.3 Motivation of the research described in this book

As described in section 1.1, more efficient use of the energy inside a battery is becoming increasingly important in the rapidly growing market for portable products. Manufacturers of portable devices are consequently paying even more attention to battery management. This is reflected in many commercial electronics magazines, such as *Electronic Design* and *EDN*, containing examples of implementations of battery management functions in a system. Many examples are to be found of ICs that implement certain charging algorithms [4]-[7]. Adding intelligence to batteries in portable products to enable e.g. SoC monitoring is also receiving a great deal of attention. The term 'smart battery' is a general buzzword that pops up in many articles [5],[8]-[10]. However, no explanation of battery behaviour is given in this kind of magazines. Therefore, the reason why one battery management IC performs better than another is often not understood. Moreover, it is hard to determine how the functionality of a BMS can be improved.

Besides in the magazines mentioned above, a lot of information on battery management can also be found in the literature. For example, much attention is paid to finding ways of accurately determining a battery's SoC [11]-[13]. The battery management functions described in these articles are derived from extensive battery measurements, for example measurements of battery discharge curves under various conditions. Most of these measurements are very time-consuming. The conclusions are moreover often empirical.

In practice, battery management functions are implemented in portable devices by electrical engineers. These engineers usually treat the battery as a black box. It is usually assumed that the battery is a voltage source with some series resistance. However, in order to improve the functionality of a BMS, at least some understanding of battery behaviour in the system is needed. A prototype of (part of) the portable device is needed for measurements of actual battery charge and discharge behaviour. On the other hand, simulation is a helpful tool in obtaining a better understanding of the behaviour of complex systems under a wide variety of conditions. Simulations take less time than measurements and no prototype is needed. Therefore, the availability of simulation models for batteries would be very helpful for the development of BMS.

Simulation models of batteries are given a lot of attention in the literature. However, some models have been constructed by researchers with an electrochemical background and are very complex and based on many mathematical equations [14],[15]. Consequently, such models are usually not very suitable for electrical engineers who have to simulate a battery in a system. Other models have been constructed by electrical engineers [16]. These models can be readily used in system simulations, but do not lead to a better understanding of battery behaviour. Therefore, simulations with these models do usually not lead to better views on battery management.

This book presents the results of research into battery modelling obtained by combining the expertise of electrical engineering with that of electrochemistry. The result is a method for modelling batteries that can be readily applied to all kinds of batteries [17]-[24]. The models result from translating (electro)chemical theory into equivalent-network models using the principle of physical system dynamics [25]. This enables the use of conventional electronic-circuit simulators that electrical engineers are accustomed to. In addition to the battery voltage and current, the internal gas pressure and battery temperature can also be simulated integrally and coherently under a wide variety of conditions. Apart from the electrical interaction with the battery's environment, it is also possible to simulate the thermal interaction. The modelling method even allows the simulation of effects such as venting under high-pressure build-up inside the battery.

Obviously, close quantitative agreement between the results of simulations using a battery model and measurements of battery behaviour is important. Part of the research described in this book is devoted to optimizing this quantitative agreement. The results of this research will be described in chapter 4.

As is revealed by the title of this book, the derived battery models will be applied in the design of Battery Management Systems. The battery models can be viewed as 'transparent', in which the course of the various reactions can be investigated. For example, the charging efficiency can be easily plotted. Charging efficiencies of different charging algorithms can be compared for optimization. An example of an optimized charging algorithm that was found in simulations with a battery model will be described in chapter 5 [24].

Internal battery behaviour that is normally hard to measure can be visualized with the models. For example, overpotentials of all reactions inside a battery and electrode equilibrium potentials can be easily plotted during the battery's operation. This makes the model a very useful tool in the quest for more accurate SoC indication algorithms. The model can be used to gain insight into the combined action of the various processes taking place inside a battery. Based on this insight, compact descriptions of battery behaviour can be derived. The results of research efforts in this field will be described in chapter 6.

Sometimes a designer is not interested in what goes on inside a battery, for example when simulating the run time of a portable device. In such cases the designer is merely interested in the battery's discharge behaviour under various load conditions. Part of the research described in this book is aimed at finding an optimum method for powering a PA inside a cellular phone. As will be shown in chapter 7, the battery models described in this book offer a simple way of comparing the run times of various PA supply strategies in a cellular phone. In addition to battery models, use will be made of simulation models of a DC/DC converter and a PA to design the BMS.

1.4   Scope of this book

Chapters 2 and 3 provide general information required as a background in the remaining chapters of this book. Chapter 2 describes the various parts of a BMS and their functionality in more detail. Some examples of BMS in different portable products are given to clarify the influence of several factors on the complexity.

Chapter 3 deals with the central part of a BMS, which is the battery itself. Some general information is given on the construction, types, operational mechanism and behaviour of batteries.

The research results are described in chapters 4 to 7. Battery modelling forms the core of this book. The construction of simulation models for rechargeable batteries is described in chapter 4. Those who want to have a thorough understanding of the background and construction of the models should read this chapter. The adopted modelling approach is explained in detail and the model equations for both a NiCd and a Li-ion battery model are derived. Further, the efforts to improve the quantitative agreement between the results of simulations and measurements are thoroughly discussed for the NiCd model. Chapters 5, 6 and 7 deal with the design of BMS. These chapters describe the use of the battery models of chapter 4 and other models to find improved BMS schemes. One does not have to read chapter 4 first in order to understand the content of these chapters.

Battery charging algorithms are discussed in chapter 5. It is shown that battery models can readily be used in the development of new, more efficient charging algorithms. Chapter 6 deals with the determination of a battery's SoC. Several possible methods are compared. In addition, a new SoC indication system is proposed and tested, based on simulations using a battery model and knowledge obtained in battery measurements. An optimum method for efficiently powering PAs in cellular phones is described in chapter 7. This strategy is named *efficiency control* [26]. As a DC/DC converter is necessary to implement efficiency control, some basic information on voltage conversion techniques is given. Measurement results are discussed to define the benefits of efficiency control in practice and to make a comparison with simulation results. Conclusions are drawn in chapter 8 and recommendations are made for further research in the exciting field of BMS in general and battery modelling in particular.

1.5   References

[1]   From slide of *Market Research Team Cellular*, Philips Components, June 2000
[2]   P.H.L. Notten, "Batterij Modellering & Batterij Management", *EET project proposal*, March 2000
[3]   The energy chain concept originates from Frans Schoofs
[4]   R. Cates, R. Richey, "Charge NiCd and NiMH Batteries Properly", *Electronic Design*, pp. 118-122, June 10, 1996
[5]   L.J. Curran, "Charger ICs Reflect Shift to Smart Batteries", *EDN*, pp. S.10-S.11, July 3, 1997
[6]   R. Schweber, "Supervisory ICs Empower Batteries to Take Charge", *EDN*, pp. 61-72, September 1, 1997
[7]   G. Cummings, D. Brotto, J. Goodhart, "Charge Batteries Safely in 15 Minutes by Detecting Voltage Inflection Points", *EDN*, pp. 89-94, September 1, 1994
[8]   A. Watson Swager, "Smart-Battery Technology: Power Management's Missing Link", *EDN*, pp. 47-64, March 2, 1995
[9]   D. Freeman, "Integrated Pack Management Addresses Smart-Battery Architecture", *Begleittexte zum Design & Elektronik Entwicklerforum*, Batterien, Ladekonzepte & Stromversorgungsdesign, pp. 82-86, München, March 31, 1998
[10]   C.H. Small, "ICs Put the Smarts in Smart Batteries", *Computer Design*, pp. 35-40, August 1997
[11]   J. Alzieu, H. Smimite, C. Glaize, "Improvement of Intelligent Battery Controller: State-of-Charge Indicator and Associated Functions", *J. Power Sources*, vol. 67, no. 1&2, pp. 157-161, July/August 1997

[12] J.H. Aylor, A. Thieme, B.W. Johnson, "A Battery State-of-Charge Indicator for Electric Wheelchairs", *IEEE Trans. on industrial electronics*, vol. 39, no. 5, pp. 398-409, October 1992

[13] O. Gerard, J.N. Patillon, F. d'Alche-Buc, "Neural Network Adaptive Modelling of Battery Discharge", *Lecture Notes in Computer Science*, vol. 1327, pp. 1095-1100, 1997

[14] D. Fan, R.E. White, "A Mathematical Model of a Sealed Nickel-Cadmium Battery", *J. Electrochem. Soc.*, vol. 138, no. 1, pp. 17-25, January 1991

[15] T.F. Fuller, M. Doyle, J. Newman, "Simulation and Optimization of the Dual Lithium Insertion Cell", *J. Electrochem. Soc.*, vol. 141, no. 1, pp. 1-10, January 1994

[16] S.C. Hageman, "Simple Pspice Models Let You Simulate Common Battery Types", *EDN*, pp. 117-132, October 28, 1993

[17] W.S. Kruijt, H.J. Bergveld, P.H.L. Notten, *Electronic-Network Modelling of Rechargeable NiCd batteries*, Nat.Lab. Technical Note 211/96, Philips Internal Report, 1996

[18] P.H.L. Notten, W.S. Kruijt, H.J. Bergveld, "Electronic-Network Modelling of Rechargeable NiCd Batteries", Ext. Abstract 96, *191st ECS Meeting*, Montreal, Canada, 1997

[19] W.S. Kruijt, E. de Beer, P.H.L. Notten, H.J. Bergveld, "Electronic-Network Model of a Rechargeable Li-ion Battery", *The Electrochemical Society Meeting Abstracts*, vol. 97-1, Abstract 104, p. 114, Paris, France, August 31-September 5, 1997

[20] W.S. Kruijt, H.J. Bergveld, P.H.L. Notten, *Principles of Electronic-Network Modelling of Chemical Systems*, Nat.Lab. Technical Note 261/98, Philips Internal Report, October 1998

[21] W.S. Kruijt, H.J. Bergveld, P.H.L. Notten, *Electronic-Network Model of a Rechargeable Li-ion Battery*, Nat.Lab. Technical Note 265/98, Philips Internal Report, October 1998

[22] W.S. Kruijt, H.J. Bergveld, P.H.L. Notten, "Electronic-Network Modelling of Rechargeable Batteries: Part I: The Nickel and Cadmium Electrodes", *J. Electrochem. Soc.*, vol. 145, no. 11, pp. 3764-3773, November 1998

[23] P.H.L. Notten, W.S. Kruijt, H.J. Bergveld, "Electronic-Network Modelling of Rechargeable Batteries: Part II: The NiCd System", *J. Electrochem. Soc.*, vol. 145, no. 11, pp. 3774-3783, November 1998

[24] H.J. Bergveld, W.S. Kruijt, P.H.L. Notten, "Electronic-Network Modelling of Rechargeable NiCd Cells and its Application to the Design of Battery Management Systems", *J. Power Sources*, vol. 77, no. 2, pp. 143-158, February 1999

[25] R.C. Rosenberg, D.C. Karnopp, *Introduction to Physical System Dynamics*, McGraw-Hill Series in Mechanical Engineering, New York, 1983

[26] The name *efficiency control* originates from Ralf Burdenski

Chapter 2
Battery Management Systems

This chapter gives general information on Battery Management Systems (BMS) required as a background in later chapters. Section 2.1 starts with the factors that determine the complexity of a BMS and shows a general block diagram. The function of each part in a BMS is discussed in more detail in section 2.2 and examples of adding BMS intelligence are given. The BMS aspects of two types of portable devices are discussed in section 2.3. This serves to illustrate the theory presented in sections 2.1 and 2.2.

2.1 A general Battery Management System

The concept of the energy chain was explained in chapter 1. Essentially, the links in this energy chain already reflect the basic parts of a BMS. In more general terms, the charger can be called a Power Module (PM). This PM is capable of charging the battery, but can also power the load directly. A general BMS consists of a PM, a battery, a DC/DC converter and a load.

The intelligence in the BMS is included in monitor and control functions. As described in chapter 1, the monitor functions involve the measurement of, for example, battery voltage, charger status or load activity. The control functions act on the charging and discharging of the battery on the basis of these measured variables. Implementation of these monitor and control functions should ensure optimum use of the battery and should prevent the risk of any damage being inflicted on the battery.

The degree of sophistication of the BMS will depend on the functionality of the monitor and control functions. In general, the higher this functionality, the better care will be taken of the battery and the longer its life will be. The functionality depends on several aspects:

- *The cost of the portable product*: In general, the additional cost of a BMS should be kept low relative to the cost of the portable product. Hence, the functionality of the monitor and control functions of a relatively cheap product will generally be relatively low. As a consequence, the BMS will be relatively simple. An example of the difference in BMS between a cheap and an expensive shaver will be given in section 2.3.

- *The features of the portable product*: This is closely related to the product's cost. A high-end product will have more features than a low-end product. For example, a high-end shaver with a 'Minutes Left' indication needs more BMS intelligence than a low-end shaver with no signalling at all.

- *The type of battery*: Some types of batteries need more care than others. An example of the influence on the complexity of the BMS when moving from one battery technology to a combination of two battery technologies will be given in section 2.3.

- *The type of portable product*: In some products, a battery will be charged and discharged more often than in others. For example, a cellular phone might be charged every day, whereas a shaver is charged only once every two or three weeks. The number of times a battery can be charged and discharged before it wears out, together with the average time between subsequent charge cycles, determines the lifetime of a battery in a device. So it is more important for this number to be high in a cellular phone than in a shaver. This can be achieved by making the BMS more intelligent. Therefore, a more sophisticated BMS is more important in a cellular phone than in a shaver.

The intelligence needed for the BMS can be divided between the various parts. This *partitioning of intelligence* is an important aspect in designing a BMS. The main determining parameter in this respect is cost. Dedicated battery management ICs can implement intelligence. Useful background information with many examples can be found in [1]-[15]. Measured variable and parameter values and control commands are communicated between the parts of the BMS via a *communication channel*. This channel can be anything from a single wire that controls a Pulse-Width Modulation (PWM) switch to a bus that is controlled by a dedicated protocol [16]-[18].

The structure of a general BMS is shown in Figure 2.1. The partitioning of intelligence is symbolized by placing a 'Monitor and Control' block in every system part. The BMS shown in Figure 2.1 also controls a Battery Status Display. An example is a single Light-Emitting Diode (LED) that indicates the 'battery low' status. It can also be a string of LEDs indicating the battery's State-of-Charge (SoC) or a Liquid-Crystal Display (LCD) that indicates the battery status, including the SoC and the battery condition.

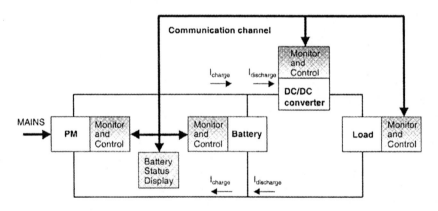

Figure 2.1: A general Battery Management System (BMS)

2.2   Battery Management System parts

2.2.1   The Power Module (PM)

The basic task of the PM is to charge the battery by converting electrical energy from the mains into electrical energy suitable for use in the battery. An alternative for the mains might be other energy sources, such as a car battery or solar cells. In many cases, the PM can also be used to power the portable device directly, for example when the battery is low. The PM can either be a separate device, such as a travel charger, or be integrated within the portable device, as in for example shavers. Especially in the latter case, the efficiency of the energy conversion process has to be sufficiently high, because the poorer the efficiency, the higher the internal temperature of the portable device and hence that of the battery will be. Long periods at elevated temperatures will decrease the battery capacity [19].

The monitor and control functions for the PM can be divided into two types. First of all, the energy conversion process itself has to be controlled and, secondly, the battery's charging process has to be controlled. The Energy Conversion Control (ECC) involves measuring the output voltage and/or current of the PM and controlling them to a desired value. This desired value is determined by the CHarging Control (CHC) on the basis of measurement of battery variables such as voltage and temperature. Moreover, the current flowing into the battery can be input for the CHC function.

The ECC function is situated within the PM itself, when the PM is a separate device. The CHC function can now be divided between the PM, the portable device and the battery, assuming that this battery can be removed from the portable device. The partitioning of the CHC function will depend on cost, but also on the employed charging algorithm.

A removable battery can be charged separately on a so-called DeskTop Charger (DTC). In addition to the standard battery that comes with a product, some users also buy a spare battery. Cellular phone users who make many phone calls during the day sometimes charge the standard battery and the spare battery. It is often possible to simultaneously connect a standard battery in a portable product and a separate spare battery to a DTC. The priority of charging must then be fixed in the system. When the CHC function is incorporated inside the portable device, it has to be present inside the DTC as well. The reason is that the user of the portable product should be able to charge only the spare battery in the DTC. The CHC function inside the portable product can then not be used.

Three simple examples of different ways of partitioning the CHC function are given below. A separate PM and a detachable battery are assumed. In the first example, the CHC function is incorporated in the PM. It is incorporated in the portable product in the second example and it is incorporated with the battery in the third example. A dashed box in the examples denotes the DTC indicating that it is optional. For simplicity, the intelligence needed in the DTC to determine charging priority is ignored.

Partitioning of the CHC function: example 1
The first example is illustrated in Figure 2.2. A Negative-Temperature Coefficient resistor (NTC) has been attached to the battery. This enables the measurement of the battery temperature by contacting it through extra terminals on the battery pack. The term *battery pack* will be explained in more detail in section 2.2.2. The battery voltage can be easily measured, because the battery remains connected to the PM during charging. Depending on the charging algorithm, the current that flows into the battery will sometimes have to be measured. This can be done by measuring the voltage across a low-ohmic resistor in series with the battery.

Based on the measured battery variables, the charging process is controlled inside the PM using a suitable charging algorithm. The addition of a simple identification means (ID in Figure 2.2) to the battery pack is desirable in case batteries of different types and/or chemistries can be used with the portable product. An example is the addition of a resistor to every battery pack, with one connection to the battery minus terminal and another to an extra terminal on the battery pack.

Each pack can then be identified by a certain resistor value, determined by the manufacturer and programmed in, for example, the PM. Changing some parameter values in the charging algorithm will often suffice in case batteries of the same chemistry, but different capacities, can be used with the portable device. Different charging algorithms often have to be used in case batteries of different chemistries can be used. As all charge monitor and control functions are present inside the PM, the DTC will only act as a mechanical and electrical connection between the PM and the battery when it is used. Hence, no intelligence has to be added to it.

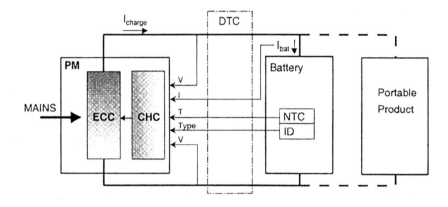

Figure 2.2: Incorporation of the charge monitor and control functions (CHC) inside the power module (PM)

An advantage of this set-up is that no electronics have to be integrated with the battery pack and no extra intelligence has to be added to the portable product. However, the additional complexity of dealing with different types of batteries now has to be added to the PM. A dedicated charge controller IC or a microprocessor can accomplish the CHC function. A microprocessor offers the designer more flexibility with respect to changes in the charging algorithm, for example when newer battery chemistries will be used in future versions of the portable device.

Partitioning of the CHC function: example 2
The charge monitor and control functions have been incorporated inside the portable product in the second example. This situation is depicted in Figure 2.3. As in the first example, only an NTC and a simple identification means (ID) have been added to the battery pack. The latter is only needed in case different types of batteries and/or chemistries may be used with the same portable product. The measurements of the battery voltage, temperature and possibly current are processed in the portable product and a control signal is sent to the PM.

The set-up of Figure 2.3 is especially advantageous when a microcontroller or microprocessor is present inside the portable device. This is for example the case in cellular phones and notebook computers. The charging algorithm can be easily and flexibly programmed in this controller or processor. Moreover, the presence of a micro-controller implies the possibility of programming an SoC algorithm based on the measured battery parameters.

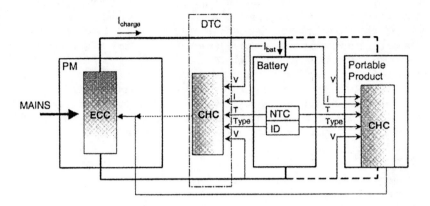

Figure 2.3: Incorporation of the charge monitor and control functions (CHC) inside the portable product

A disadvantage of the set-up of Figure 2.3 is the presence of the CHC inside the DTC. As discussed above, this DTC has to be able to charge a spare battery on its own, without the presence of the portable device. Therefore, the same CHC as programmed inside the portable product has to be added to the DTC, because no charging intelligence is present inside the battery pack nor the PM.

Partitioning of the CHC function: example 3
The charge monitor and control functions have been added to the battery pack in the third example. This situation is depicted in Figure 2.4.

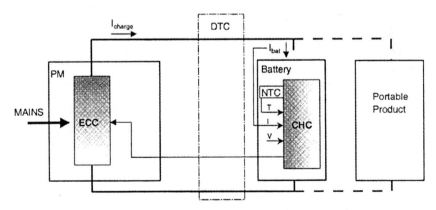

Figure 2.4: Incorporation of the charge monitor and control functions (CHC) inside the battery pack

The measured battery voltage, temperature and/or current are processed inside the battery pack in Figure 2.4. The battery pack controls the PM autonomously through a control wire. This means that it can even be charged directly by the PM, without the need to connect it to the portable device or DTC. An advantage is that each battery pack can now include its own charging algorithm. This obviates the need to identify the battery to determine which charging algorithm should be used.

A disadvantage of the configuration shown in Figure 2.4 is that a considerable amount of hardware has to be added to the battery pack, some of which is already present in the portable product itself. A good example is the micro-controller mentioned above. Also, the user has to pay for the same hardware each time he/she buys such a battery when the CHC function is added to each battery pack.

In practice, the examples shown in Figure 2.2 and Figure 2.3 are encountered most. Combinations of the configurations shown in these figures are also encountered. Some practical examples will be given in section 2.3.

2.2.2   The battery

A battery's basic task is to store energy obtained from the mains or some other external power source and to release it to the load when needed. This enables a portable device to operate without a connection to any power source other than a battery. Different battery systems with different chemistries and different characteristics exist. Examples of some commonly encountered battery systems are nickel-cadmium (NiCd), nickel-metalhydride (NiMH) and lithium-ion (Li-ion) batteries. The characteristics of the various battery systems vary considerably, even for batteries with the same chemistry, but, for example, a different design or different additives. More information will be given in chapter 3.

The term *battery pack* is often used for detachable batteries. Depending on the desired voltage and capacity of the battery, several batteries can be connected in series and/or in parallel inside a battery pack. To avoid confusion, the basic battery building blocks inside the battery pack are often referred to as cells. The series connection of cells yields a higher total battery voltage at the same capacity in [Ah] and the parallel connection yields a higher total battery capacity in [Ah] at the same battery voltage. Apart from one or more cells, a battery pack may also contain other components. Examples are the NTC and ID resistors mentioned in the previous section. Moreover, some electronics might be present inside the pack. All is included inside a plastic container with at least the plus and minus terminals of the battery and possibly some other terminals for temperature measurement and/or identification, as discussed in the previous section. In this book, the term battery will be used for both bare batteries and battery packs.

It is important from a battery-management point of view to realize at this moment that the various battery systems should be approached differently, for example via different charging algorithms. Approaching different battery systems differently is important in determining the electronics that has to be integrated with the battery.

Electronic safety switch for Li-ion batteries

In the case of Li-ion batteries, an electronic safety switch has to be integrated with the battery. The battery voltage, current and temperature have to be monitored and the safety switch has to be controlled to ensure that the battery is never operated in an unsafe region. The reason for this is that battery suppliers are particularly concerned with safety issues due to liability risks. A voltage range, a maximum current and a maximum temperature determine the region within which it is considered safe to use a battery. The battery manufacturer determines these limits. Outside the safe region, destructive processes may start to take place. Generally speaking, in the higher voltage range these processes may eventually lead to a fire or an explosion, whereas in the lower voltage range, they lead to irreversible capacity loss of the battery.

The maximum voltage is dictated by two factors, the maximum battery capacity and its cycle life. The cycle life denotes the number of cycles the battery can be charged and discharged before it is considered to be at the end of its life. A battery's life ends when the capacity drops below a certain level, usually 80% of its nominal capacity. This is illustrated in Figure 2.5, which shows the maximum battery capacity and the cycle life as a function of the voltage applied to the battery during charging. The figure illustrates that the higher this voltage, the higher the maximum battery capacity and the lower the cycle life will be. Around 4.1 to 4.2 V, a 100 mV increase in battery voltage yields a 12% capacity increase, but a sharp decrease in cycle life of 200 cycles.

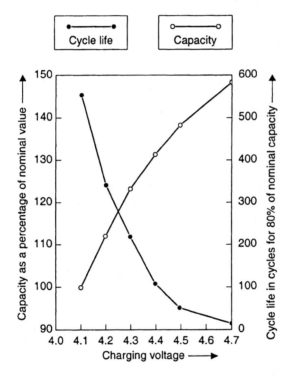

Figure 2.5: Maximum battery capacity and cycle life as a function of battery voltage for a Li-ion battery (source: Panasonic)

The choice of the maximum voltage level for the electronic safety switch is a compromise. The demands on the accuracy of this level are in the order of one percent, because a certain maximum battery capacity and a certain minimum cycle life have to be guaranteed. Obviously, the battery voltage should not reach the maximum voltage level of the electronic safety switch during charging to prevent the risk of current interruption during normal use. The demands on the accuracy of the minimum voltage are less strict, as the voltage drops rather sharply when the battery is almost empty. This means that a deviation in the voltage at which the battery is disconnected from the load by the safety switch does not influence the usable battery capacity significantly. The minimum voltage needed to operate the portable device should be chosen higher than the minimum voltage level of the safety switch. This is to prevent the risk of the safety switch opening during normal operation.

The maximum current specification resembles the specification of a fuse. Below the maximum current level, which value is linked to the maximum capacity of the battery, the safety switch should always conduct. Above this current level, currents are allowed to occur for only a certain amount of time. How long that time will be will depend on the thermal resistance of the battery to its environment and the thermal capacity of the battery. The higher the current, the shorter the time it will be allowed to occur to prevent the risk of the battery temperature becoming too high. This means that the delay time between detection of the current and the opening of the switch is smaller at higher current levels. A general set-up of an electronic safety switch is shown in Figure 2.6.

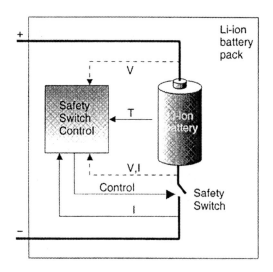

Figure 2.6: General set-up of an electronic safety switch inside a Li-ion battery pack

The battery voltage and the voltage difference across the switch are measured. The latter voltage difference can be used to measure the battery current. This voltage difference also allows start-up of the IC when a charger is connected and the battery voltage is lower than the lowest limit of the safety voltage range, which implies that the switch is open. In that case, there will be a large voltage difference across the opened switch depending on the maximum output voltage of the charger. This voltage difference will allow the operation of some sort of start-up circuit inside the safety-switch control IC. The temperature is usually measured inside the control IC, but could also be measured using an NTC.

The maximum resistance of the electronic switch has to be of the same order of magnitude as other series resistances in the current path. These other series resistances include the internal battery impedance, the resistance of other passive safety devices and the contact resistance between the battery and the load. An example of a passive safety device is a Positive-Temperature Coefficient resistor (PTC). This device gradually inhibits the current flow when the temperature increases. In practice, values in the order of 100 mΩ are found for the electronic switch, for example an MOSFET. The cost aspect determines whether the switch is realized in an IC process other than the control circuitry, or together with the control circuitry on one IC. As the electronic switch is integrated with the battery, it is important that the associated circuitry draws a negligible current from the battery. In practice, values down to several μA are found.

Charge-balancing and pack supervisory electronics inside a battery pack
A different need for electronics arises when several cells are connected in series inside a battery pack. This is for example often the case in notebook computers. Because of differences between the individual cells inside the battery pack, a certain imbalance in SoC may exist. For example, the cells might differ in the maximum amount of charge that can be stored, or in their internal impedance. These differences already occur due to manufacturing variance and will even increase because of aging. Such aging effects may be worse for some cells than for others in a battery pack. A reason could be that some cells become substantially hotter than others. The occurrence of large temperature differences between the cells in a battery pack is largely influenced by the position of the cells relative to the temperature sensor. The weakest cell always determines the characteristics of the battery pack.

Several examples of battery-pack supervisory ICs and charge-balancing ICs can be found in the literature [20]-[22]. The essence of these ICs is to extend a battery pack's useful lifetime by preventing major imbalances between the cells inside the pack. A battery-pack supervisory IC measures each cell voltage individually and interrupts the charge or discharge current when one of the voltages exceeds or drops below a certain level. A charge-balancing circuit distributes the charge applied to or drawn from the battery pack equally between the cells by transferring charge from a cell with a high SoC to a cell with a lower SoC. Alternatively, the cell with the highest SoC can be bypassed by a 'bleeder' resistor, until the other cell(s) reach(es) the same voltage. Although both battery-pack supervisory and charge-balancing circuits prevent major differences between cells, the differences that arise from manufacturing spread can never be eliminated.

Other functions that can be integrated within a battery pack
Especially when electronic circuitry, such as a safety device or a charge-balancing circuit, is already present in a battery pack, it is relatively cheap to add extra functionality. The addition of some intelligence to a battery itself is advantageous, because each type of battery has to be approached in a specific way. This is especially true for portable products that can be used with different battery systems. This eventually leads to a so-called smart battery, which can be defined as a rechargeable battery with a microchip that collects and communicates present, calculated and predicted battery information to the host system under software control [6]. Features that are included in a smart battery are self-monitoring functions, charge-regime control, communication with the system's host, implementation of fault identification and protection and storage of information in a memory. This includes information on the battery characteristics, present status and the history of use. Hence, some of the information stored in the battery pack will be fixed, whereas other information will be continuously updated.

The fact that information is integrated with the battery makes it necessary to standardize the format of the information to allow the use of batteries of manufacturer A in a portable device of manufacturer B. Moreover, standardization facilitates system design as it makes it possible to treat a battery as a standardized system part, regardless of its voltage, chemistry and packaging. An example of this will be presented in section 2.2.5, which deals with the communication channel.

2.2.3   The DC/DC converter

The basic task of a DC/DC converter in a portable product is to connect a battery to the various system parts when the battery voltage does not match the voltage needed. The battery voltage may be either too low or too high. In the first case, DC/DC up-conversion is of course needed. In the latter case, DC/DC down-conversion is needed when the battery voltage is higher than the maximum allowed supply voltage of the load. Apart from this however, from the viewpoint of efficiency it is always a good idea to convert the battery voltage into the minimum supply voltage V_{min} needed by the load for the following reason.

First of all, when a DC/DC converter is used, the design of the load can be optimized for the minimum supply voltage instead of the whole voltage range of the battery discharge curve. This will lead to a higher efficiency of the load in most cases. Secondly, operating a system part with a higher supply voltage than necessary implies a waste of energy. This can be inferred from Figure 2.7, which shows a schematic representation of a typical discharge curve of a battery. Note that the area under this curve represents $V \cdot I \cdot t$, which indeed expresses the energy obtained from the battery in [J]. The minimum voltage V_{min} needed to operate the load is also

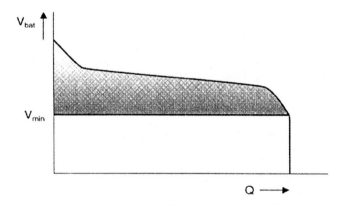

Figure 2.7: Schematic representation of a typical battery discharge curve denoting energy loss (shaded area) when the battery voltage is higher than the minimum supply voltage V_{min} of the load

The shaded area above V_{min} expresses the energy that is lost in heat dissipation when the load is directly connected to the battery due to the fact that the battery voltage V_{bat} is higher than V_{min}. Assuming that the battery voltage can be efficiently converted into V_{min}, the energy loss can be considerably reduced. More information on principles for converting battery voltages into other voltages will be given in chapter 7.

2.2.4 The load

The basic task of the load is to convert the electrical energy supplied by a battery into an energy form that will fulfil the load's function. Many different types of portable products exist, and hence many different types of loads. Some general remarks on the connection of the battery to the load and the implementation of battery management functions will be given below.

Different supply voltages
Different supply voltages are needed in most portable products. These voltages are constantly changing as the technology progresses. The maximum possible supply voltage before an IC breaks down is decreasing, because the feature size of ICs is becoming smaller and smaller. However, the supply voltage needed for some analogue functions will remain higher than the supply voltage needed for digital system parts, for example on account of a need for a specific signal-to-noise ratio of the analogue function. Most IC processes offer the possibility of designing analogue circuits with a higher supply voltage than digital circuits. This can be achieved by using special transistors with a thick gate oxide.

Physical aspects also influence the IC process in which the analogue system part is realized and the supply voltage needed. Examples of such aspects are the need for higher voltages for backlighting LCD screens in notebook computers or the need for negative bias voltages for some types of Power Amplifiers (PAs) in cellular phones. In general, the supply voltages for the analogue system parts encountered in portable devices show a larger variation in values than those of the digital system parts. This means that for a given battery voltage a system designer has to provide this variety of voltages from a single battery voltage using (a combination of) up-conversion, down-conversion or up/down-conversion [23].

Power consumption of system parts
The power consumption of the various system parts of a portable device may vary considerably. Take for example the PA in a cellular Global System for Mobile communication (GSM) phone, which amplifies the Radio-Frequency (RF) signal to the antenna. The PA transmits in bursts in such phones. Currents of up to 2 A can be drawn from the supply pin, depending on the efficiency and supply voltage of the PA. This of course leads to different demands on the design of the power supply circuitry around the PA than in the case of a system part that draws a current in the mA-range.

Use of existing components for BMS functions
The presence of a microcontroller or microprocessor in a portable product will enable the flexible implementation of many of the monitor and control functions needed in a BMS. An example was shown in Figure 2.3.

2.2.5 The communication channel

The basic task of a communication channel in a BMS is to transfer monitor and control signals from one BMS part to another. The complexity of a communication channel in a BMS depends on the complexity of the system and the partitioning of the intelligence needed for battery management. For example, the only communication that takes place between the battery and the charger shown in Figure 2.2 involves the voltage across an NTC and the value of the identification resistor. However, in the case of a smart battery, the batteries can communicate their status to the system host and other system parts, such as the charger. The appearance of smart batteries on the market resulted in a major trend towards standardization of the information stored in batteries and the way it is communicated to other system parts. Battery manufacturers teamed up with semiconductor and computer hardware manufacturers to find a standard architecture for a communication bus between the battery and the system host. The ultimate goal of standardization was to facilitate the development of universal battery management solutions. In these solutions, each battery, regardless of its type or chemistry, communicates information on its status to other system parts in a defined way.

One of several ways of arriving at a standardized interface between a smart battery and the rest of the system was suggested by Duracell in co-operation with Intel in 1994 [5]-[7],[16]-[18]. This co-operation has resulted in the Smart Battery System (SBS) specification. The SBS ownership is in the hands of several companies, including the battery manufacturers Duracell, Energizer, Toshiba and Varta and the semiconductor manufacturers Intel, Benchmarq, Linear Technology, Maxim, Mitsubishi and National Semiconductor. The SBS specification is generic, which means that it is independent of battery chemistry, voltage and packaging. The specification consists of four parts:

- The System Management Bus (SMBus) specification.
- The Smart Battery Data (SBD) specification.
- The Smart Battery Charger (SBC) specification.
- The specification of the interface between the SMBus and the Basic Input Output System (BIOS) in a portable computer.

Although the specification focuses specifically on the portable computer industry, it was designed to be applicable to other portable products too, according to Duracell and Intel. A block diagram illustrating the implementation of the specification is given in Figure 2.8.

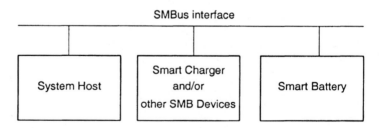

Figure 2.8: Basic block diagram illustrating the implementation of the SBS specification

SMBus specification

The SMBus uses a two-wire communication channel between the components and an electrical layer according to the Philips I^2C protocol with additional software. Each SMBus component can be either a master or a slave, but there can only be one master at a time. For example, the smart battery can signal a warning via the SMBus when the battery is approaching low capacity, or a disk drive can ask the smart battery whether there is enough energy left to save data.

SBD specification

The SBD specification defines the data that flows across the SMBus between the smart battery, the SMBus host, the Smart Battery Charger and other devices. In addition to software definitions, error detection, signalling and smart battery data protocols, it includes a data set of 34 values of battery information. This includes voltage, current, temperature and device type and identification data as well as calculated and stored values, such as:

Run Time To Empty:	the predicted remaining run time at the present rate of discharge
Remaining Capacity:	in units of either charge or energy
Relative State-Of-Charge:	predicted remaining capacity as a percentage of full charge capacity
Full-Charge Capacity:	predicted pack capacity when fully charged
Cycle Count:	number of charge/discharge cycles of the battery

SBC specification

The SBC specification defines three levels of chargers. A level-1 charger can only interpret critical warning messages and cannot adjust its output. For example, it will only interrupt its charge current when a short-circuit occurs. However, it cannot adapt its charge current to the employed battery. This is done by a level-2 charger, which is an SMBus slave device. It is able to dynamically adjust its output according to the charging algorithm stored in the smart battery pack. This means that the battery 'tells' the charger how it has to be charged. An example of such a set-up was shown in Figure 2.4. A level-3 charger is an SMBus master device, which chooses a stored charging algorithm by interrogating the battery. Hence, in the case of a level-3 charger, the charger 'decides' by itself how the battery will be charged.

SMBus/BIOS interface specification

This part of the specification defines how the operating system and applications that run on a portable computer can communicate with SMBus components, using the BIOS layer as an intermediate. Software vendors Phoenix and SystemSoft have developed software that supplies intelligent battery data to a computer user. This enables the user to choose his/her own power management scheme interactively. For example, the SystemSoft Smart Battery Software System and the Phoenix Smart Battery Manager offer the computer users the possibility of choosing between better economy and better performance by using a scroll bar. The software shows the battery performance as a function of the power management scheme chosen by the user. Moreover, the user may install his/her own critical warning messages.

Another example of a specification that defines the connection between software and SMBus devices is the Advanced Configuration and Power Interface (ACPI). It was developed by Intel, Microsoft and Toshiba and enables system manufacturers to supply power management that is directed by the operating system.

Examples are the automatic turning on and off of peripherals, such as hard-disk drives and printers, and the activation of the computer itself by peripherals.

Acceptance of the SBS specification by portable product manufacturers

The SBS specification offers flexibility with respect to the amount of intelligence that is added to a battery. For example, a system designer is free to use only a portion of the battery information specified in the SBD specification. In fact, the system designer can design his/her own tailor-made smart battery system. One can choose from a variety of ICs usable with the SMBus. The original vision of Duracell was to manufacture standard smart batteries with fixed form factors and with all intelligence, fulfilling the SBS specification, inside. This turned out to be too limiting with respect to design freedom for system designers. A system designer prefers to use custom-designed batteries and add intelligence in the form of specific ICs. As a result, Duracell stepped out of the smart battery business for notebook computers in 1997. However, the idea of using the SMBus and a specified interface between the different building blocks has survived.

Although the SBS specification is claimed to be applicable to any portable product, it is only used in notebook computers in practice. A microprocessor is already at hand in this case and the inclusion of a communication protocol in the BMS does not add too much to the cost of the portable product. The added cost of a complicated bus and bus protocol might be too high in the case of other portable devices. Dallas Semiconductor offers a cheaper alternative solution to smart batteries that fulfil the SBS specification [17]. They have developed a range of temperature measurement and identification ICs intended for integration in a battery pack. Information is in this case communicated from the battery to the host processor via a proprietary one-wire interface. This enables the design of a relatively cheap Battery Management System.

2.3   Examples of Battery Management Systems

2.3.1   Introduction

Examples of the BMS implemented in two types of portable products, a shaver and a cellular phone, will be given in this section. The examples have been derived from existing Philips products. In both cases, a distinction will be made between the BMS in two types of the product. A distinction will be made between a low-end and a high-end shaver. This will illustrate the influence of cost and features of the product on the complexity and functionality of the BMS. The impact on the BMS of moving from the NiMH battery technology to a combined use of NiMH and Li-ion batteries will be described for the cellular phone. This will illustrate the impact of the type of battery on the complexity of the BMS.

2.3.2   Comparison of BMS in a low-end and high-end shaver

Figure 2.9 shows the BMS in a low-end shaver with strong emphasis on the cost of the product. The shaver is powered by only one NiCd or NiMH battery, which has been integrated inside the shaver. The PM has also been integrated in the shaver. The battery has been connected to the motor via an ON/OFF switch.

Both the ECC and the CHC control functions have been included in a High-Voltage IC (HVIC). This HVIC regulates the charge current by controlling a high-voltage switch in a Switched-Mode Power Supply (SMPS) in a flyback configuration [24]. The control signals used by the HVIC are the output voltage of the PM, the voltage across resistor R_1 and the voltage across the rectifying diode D. The value of the output current depends on the mode of the HVIC. During charging, the output current is high until 'battery full' detection. The output current is switched to a low value after this detection. A voltage regulation mode exists in addition to these two charge modes. This mode occurs when the motor is switched on during charging. In this case, the PM maintains a constant battery voltage and hence a constant supply voltage for the shaver motor.

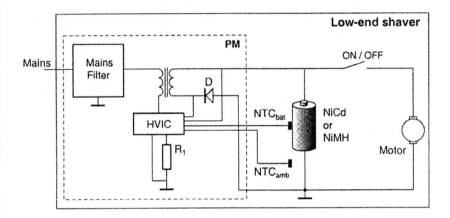

Figure 2.9: Battery Management System inside a low-end shaver

A certain amount of power needs to be transferred to the motor to enable shaving. When the user is shaving with the shaver connected to the mains, the output current of the PM will be relatively high. The reason for this is that the battery voltage is relatively low, as only one NiCd or NiMH battery is used. Current measurement by means of a series resistor is not attractive, because the high current value would lead to a too high temperature in the shaver given the shaver dimensions. Therefore, the value of the output current has to be determined in other measurements. This is achieved by measuring the time for which current flows to the battery through diode D, when the high-voltage switch on the primary side of the transformer is open. This time is obtained by measuring the voltage across diode D [24]. Moreover, the peak current at the start of the current flow through diode D is obtained by measuring the peak voltage across R_1 at the mains side of the transformer at the moment when the high-voltage switch is opened. As the current through diode D will ramp down linearly to zero, this peak current combined with the period of time for which current flows and the total switching period time T can be used to determine the average current.

    One way of determining the 'battery full' status of a NiCd or NiMH battery is to measure the temperature rise of the battery at the end of charging. The difference between the battery temperature and the ambient temperature is measured by two NTCs in Figure 2.9. When the temperature difference between the battery and the environment reaches a certain value that is included in the HVIC, the PM will switch to a low current. This current will maintain the charge level of the battery by compensating self-discharge. Such a low current is commonly referred to as a trickle-charge current.

The BMS for the low-end shaver has proven to be cost-effective. The simple charging algorithm allows for enough charging cycles until the battery capacity drops below an unacceptable level. However, the signalling of battery information to the user is not implemented at all. This is an unacceptable situation for high-end shavers. Therefore, a microcontroller is included in high-end shavers to provide all sorts of signalling of information to the user through an LCD screen. As discussed in section 2.1, the increased functionality will influence the complexity of the BMS. The BMS of a high-end shaver is shown in Figure 2.10. The shaver is powered by two NiCd or NiMH batteries in series. A Timer-Control Module (TCM) has been included. Again, the batteries are connected to the motor via an ON/OFF switch.

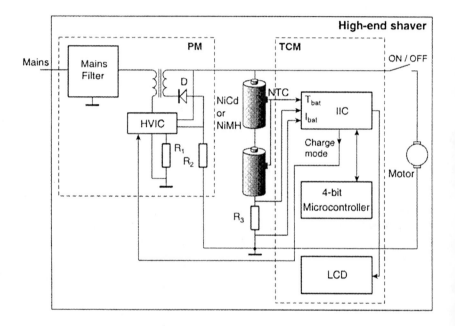

Figure 2.10: Battery Management System inside a high-end shaver

The ECC function has been included in the HVIC in the PM portion as in the previous example. The HVIC controls the output current of the PM to the appropriate value during charging, as communicated by the TCM module. The HVIC switches to output voltage regulation when the motor is switched on during charging. The current supplied by the PM in this mode is lower than in the similar situation in the low-end shaver, which makes current measurement with resistor R_2 attractive.

The CHC function has been included in the TCM. The microcontroller communicates to the rest of the system through the Interface IC (IIC). The IIC measures the battery temperature and current. The temperature is measured through an NTC that is in contact with both batteries. The battery current is measured via the voltage across resistor R_3. The IIC calculates the SoC of the battery by increasing the value of a counter during charging and decreasing it during discharging at a speed proportional to the value of the current. Hence, the charge flowing into and out of the battery is measured, which is commonly referred to as coulomb counting. When no external battery current is flowing, the counter is decreased at a programmed speed to compensate for self-discharge of the battery. This speed is proportional to the measured battery temperature, which means that the speed is higher at higher temperatures. This is in agreement with real battery behaviour, with the self-discharge rate being higher at higher temperatures. The accuracy of the SoC indication method has proven to be sufficient for use in shavers.

The IIC will signal to the HVIC which charge current should be used depending on the value of the counter. The IIC will signal the HVIC to switch from the normal charge current to a lower top-off charge current when the counter reaches its 80% full reference. Because the efficiency of the charging process drops at a high current at the end of charging, lowering the charge current at that moment keeps the charging efficiency high and will prevent a large increase in the battery temperature. The IIC will signal the HVIC to switch to a low trickle-charge current when the counter reaches its 100% full reference. The microcontroller is used to infer the *Minutes Left* or *Shaves Left* value from the SoC value determined by the IIC. The IIC will signal the battery condition to the user through the LCD.

2.3.3   Comparison of BMS in two types of cellular phones

The cost and features are the main drivers for the complexity of the BMS in the shaver example. In the next example, moving from one battery chemistry to a possible combination of two chemistries leads to a complete redesign of the BMS in a cellular phone. The first type of cellular phone, which will be named type A below, can only be powered by a NiMH battery pack. The second type, or type B, can be powered by a NiMH battery pack or a Li-ion battery pack. Moreover, type B is a newer model than type A. Therefore it is smaller and weighs less than type A, which is in line with the trend towards smaller and lighter phones.

Figure 2.11 shows the BMS of a cellular phone of type A, which is powered by a battery pack that includes four NiMH cells in series. This yields a total voltage of the battery pack of 4.8 V. Other NiMH battery packs with higher capacities can be used as accessories. The battery pack is detachable and can be charged separately from the phone on a DTC. The arrangement for controlling the charging process is depicted in Figure 2.11a. The power supply scheme for the various system parts inside the phone is shown in Figure 2.11b. Unlike with the shaver, not all system parts of the cellular phone are powered directly by the battery.

The PM of cellular phone A only includes the ECC. The V-I characteristic of the power module is shown in Figure 2.11a. The output current during charging is 800 mA. The battery pack voltage will not reach the maximum charger voltage of 8.2 V during charging. The charger output will only reach this maximum voltage when an open-circuit occurs. The entire control of the charging process is located inside the cellular phone itself. When the battery pack is connected, its type is read from the value of the identification resistor (ID) inside the battery pack. This value is looked up in a table stored inside the phone, which indicates the corresponding type of battery pack. Depending on the capacity of the pack that has been connected, the charge current of 800 mA that flows into the phone from the PM will be controlled by the PWM switch inside the phone in such a way that the charge time is one hour. So the resulting charge current will be higher in the case of a pack with a higher capacity, the maximum charge current being 800 mA. The NTC inside the battery pack enables temperature measurement. The measured battery pack voltage and temperature are fed to the microcontroller, which calculates the derivatives of the measured values after some internal filtering. As is usually the case with Ni-based rechargeable batteries, use is made of the specific shapes in the battery voltage and temperature curves at the end of charging to determine the full state of the battery. The calculation of the derivatives yields information on the shape of the measured voltage and temperature curves. After a 'battery full' detection, the duty cycle of the PWM switch will be changed to obtain a trickle-charge current that maintains a full battery. As the trickle-charge current has to be small, the duty cycle of the PWM switch will be very low, which means that the switch will be closed for only a small percentage of the time. The battery pack may also be charged separately on the DTC. This is indicated in Figure 2.11a by means of the dashed lines. The PWM switch and the CHC function for controlling this switch are also included in the DTC.

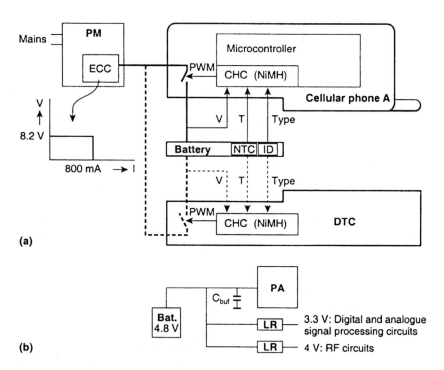

Figure 2.11: BMS of a type A cellular phone with a NiMH battery pack: (a) Control of the charging process (b) Power supply scheme for the various system parts

A 4.8 V PA is used to amplify the RF signals to the antenna in cellular phone A. Hence, this PA can be directly connected to the battery, as can be seen in Figure 2.11b. Because phone A is a GSM phone, this PA will draw current in pulses, as previously mentioned in section 2.2.4. When the battery pack is viewed as a voltage source with series resistance for simplicity, the maximum output power obtainable from this voltage source is proportional to the square of the voltage and inversely proportional to the series resistance. The maximum power that can be supplied by the battery pack will be relatively high, because the voltage is relatively high and the series resistance in the current path from pack to PA is sufficiently low. As a result, a relatively low value of the buffer capacitor C_{buf} of 100 μF is found to be sufficient in practice to guarantee the correct supply voltage at the supply pin of the PA. Hence, most of the peak current drawn by the PA during transmit bursts will be drawn directly from the battery pack. The buffer capacitor serves mainly as a filter that prevents the risk of high-frequency signals interfering with the PA operation. Two Linear Regulators (LR) serve to power the remaining hardware inside the cellular phone. More information on linear regulators will be given in chapter 7.

The BMS of a cellular phone of type B is shown in Figure 2.12. The phone is supplied with a standard Li-ion battery pack that contains a single Li-ion cell of 3.6 V with an electronic safety switch, represented by the dashed box in Figure 2.12. Moreover, a battery pack with three NiMH cells in series with a total voltage of 3.6 V can be bought as an accessory. No safety switch will then be included. The fact that two different battery systems may be used with the cellular phone means that the BMS will have to cover extra issues. The main issue is the difference between charging algorithms for NiMH and Li-ion batteries. Again, the pack can be detached from the phone and can be charged separately on the DTC, indicated by the dashed lines. The arrangement for controlling the charging process is depicted in Figure 2.12a. The power supply scheme for the various system parts inside the phone is illustrated in Figure 2.12b.

In addition to the ECC, the PM also includes part of the charging process control. First of all, the V-I characteristic is a function of the applied battery chemistry. This can be seen in Figure 2.12a, where the maximum voltage limit is lower for a Li-ion battery pack than for a NiMH battery pack. Secondly, the PWM switch is now incorporated inside the PM. This is to avoid heat generation inside the phone, as phone B is smaller than phone A. The partitioning of charge control will depend on whether the battery pack is charged while connected to the phone or separately on the DTC. Note that in addition to power line pair 1, through which the charge current flows, two other wires are present. Wire 2 serves to control the PWM switch inside the PM. Wire 3 is a battery voltage sense line. The use of wires 2 and 3 will be explained below.

When a battery pack is connected to the phone, its type will first be determined by reading the value of the identification resistor (ID). The CHC function is performed in the PM when the battery pack is of type Li-ion. Switch S_1 will be closed first. This will enable the PM to measure the battery voltage currentless via wire 3. The PM will charge the Li-ion battery pack with a constant current of 650 mA until the battery pack voltage reaches the value of 4.1 V. Then, the PM will continue to charge at a constant voltage of 4.1 V. The battery pack temperature, measured by the NTC, will only be used as a safeguard. This means that the charge current will be interrupted by the opening of the PWM switch when the temperature exceeds a certain value. Wire 2 will be used for this.

Switch S_1 will be left open in the case of a NiMH battery pack. A maximum voltage of 6.5 V will be used when the PM detects this on wire 3. The NiMH battery pack voltage will not reach this value during normal charging. The charge current will remain at 650 mA until the microcontroller inside the cellular phone detects a full battery in the same way as in phone A. After this detection, the PWM switch inside the PM will be controlled by the phone to charge the battery with a trickle-charge current. This will take place via wire 2. More information on the charging algorithms for both Li-ion and NiMH batteries will be given in chapter 5.

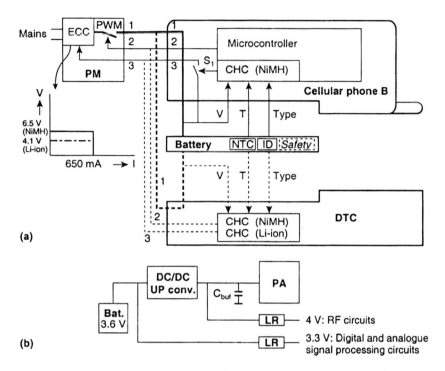

Figure 2.12: BMS of a type B cellular phone with a standard Li-ion battery pack and optional NiMH battery pack (a) Control of the charging process (b) Power supply scheme for the various system parts

When the battery pack is charged on the DTC, its type will first be read. The voltage sense line on wire 3 will be left open-circuit in the case of both types of battery packs, because the DTC is equipped with a dedicated IC capable of controlling the charging process of both NiMH and Li-ion batteries. As a result, the PM will always use the 6.5 V maximum voltage limit, irrespective of the type of battery connected to the DTC. All CHC functions will now be performed inside the DTC, unlike when charging the pack on the phone. The PWM switch inside the PM will only be used to derive the trickle-charge current in the case of NiMH battery packs, to stop charging a Li-ion battery pack or to interrupt the charge current in the case of a fault condition. Again, wire 2 will be used to this end.

Phone B uses a 4.8 V PA. A DC/DC up-converter is used to boost the battery voltage from 3.6 V to 4.8 V. Figure 2.12b shows that the RF circuits have now been connected to the output of the DC/DC converter through a linear regulator. This linear regulator not only controls the battery voltage of 4.8 V to a 4 V supply voltage for RF circuits, but it also serves as a filter to prevent voltage ripple from the output of the DC/DC converter from interfering with the sensitive RF circuitry. More information will be given in chapter 7. The other digital and analogue signal processing circuits are connected to the battery pack through a linear regulator, as was also the case in phone A.

Apart from the inclusion of a DC/DC up-converter between the battery pack and the PA, an important difference with respect to phone A is the value of the buffer capacitor C_{buf}. The battery pack voltage is lower in phone B and the series resistance in the current path is a lot higher. There are several reasons for this. First of all, the safety measures used in Li-ion batteries in the form of electronic safety switches and passive safety devices such as PTCs yield extra series resistance. Secondly, the internal series resistance of a Li-ion battery is higher than the resistance of three NiMH batteries in series. Thirdly, the DC/DC converter introduces some extra series resistance. As a result, the peak current drawn by the PA during the transmit burst now has to be drawn from the buffer capacitor. This results in a value of roughly 2 mF.

2.4 References

[1]   G. Jones, "Battery Management in Modern Portable Systems", *Electronic Engineering*, pp. 43-52, November 1991

[2]   B. Kerridge, "Battery Management ICs", *EDN*, pp. 100-108, May 13, 1993

[3]   F. Goodenough, "Battery-based Systems Demand Unique ICs", *Electronic Design*, pp. 47-61, July 8, 1993

[4]   F. Caruthers, "Battery-management Circuitry Gets Smarter", *Computer Design's OEM Integration*, pp. 15-18, May 1994

[5]   D. Maliniak, "Intelligence Invades the Battery Pack", *Electronic Design*, pp. 153-159, January 9, 1995

[6]   A. Watson Swager, "Smart-Battery Technology: Power Management's Missing Link", *EDN*, pp. 47-64, March 2, 1995

[7]   D. Freeman, "Freeing Portables from Battery Tyranny", *Electronic Design*, pp. 115-121, July 10, 1995

[8]   D. Freeman, "Integrated Pack Management Addresses Smart-Battery Architecture", *Begleittexte zum Design & Elektronik Entwicklerforum*, Batterien, Ladekonzepte & Stromversorgungsdesign, pp. 82-86, München, March 31, 1998

[9]   F. Goodenough, "Microcontroller Grabs Data and Processes Analogue Data", *Electronic Design*, pp. 139-144, April 15, 1996

[10]   R. Cates, R. Richey, "Charge NiCd and NiMH Batteries Properly", *Electronic Design*, pp. 118-122, June 10, 1996

[11]   F. Goodenough, "Battery Management ICs Meet Diverse Needs", *Electronic Design*, pp. 79-96, August 19, 1996

[12]   B. Kerridge, "Mobile Phones Put the Squeeze on Battery Power", *EDN*, pp. 141-150, April 10, 1997

[13]   L.J. Curran, "Charger ICs Reflect Shift to Smart Batteries", *EDN*, pp. S.10-S.11, July 3, 1997

[14]   B. Schweber, "Supervisory ICs Empower Batteries to Take Charge", *EDN*, pp. 61-72, September 1, 1997

[15]   R. Nass, "Gas-Gauge IC Performs Precise Battery Measurements", *Electronic Design*, pp. 39-42, October 13, 1997

[16]   D. Stolitzka, "Smart Battery Standards Simplify Portable System Design", *Electronic Design*, pp. 115-116, May 1, 1997

[17]   C.H. Small, "ICs Put the Smarts in Smart Batteries", *Computer Design*, pp. 35-40, August 1997

[18]   J. Milios, "Harness the Power of the ACPI/Smart-Battery Standard", *Electronic Design*, pp. 113-114, December 1, 1997

[19]   D. Linden, *Handbook of Batteries*, Second Edition, McGraw-Hill, New York, 1995

[20] C. Pascual, P.T. Krein, "Switched-Capacitor System for Automatic Series Battery Equalization", *Proceedings of the Portable Power for Communications '97 Conference*, pp. 1-7, London, 1997

[21] L. Anton, H. Schmidt, "Charge Equalizer-Konzepte zum Optimalen Betrieb Seriell Ver-schalteter Batterien", *Begleittexte zum Design & Elektronik Entwicklerforum*, Batterien, Ladekonzepte & Stromversorgungsdesign, pp. 103-116, München, March 31, 1998

[22] W.F. Bentley, "Cell-Balancing Consideration for Lithium-Ion Battery Systems", *Begleittexte zum Design & Elektronik Entwicklerforum*, Batterien, Ladekonzepte & Stromversorgungsdesign, pp. 168-171, München, March 31, 1998

[23] L. Sherman, "Know the Peculiarities of Portable Power Supply Designs", *Electronic Design*, pp. 94-104, July 7, 1997

[24] K.K. Sum, M.O. Thurston, *Switch Mode Power Conversion; Basic Theory and Design*, Electrical Engineering and Electronics, vol. 22, Marcel Dekker, New York, 1984

Chapter 3
Basic information on batteries

Some basic information on batteries is given in this chapter. A historical overview in section 3.1 describes the developments in battery technology. The characteristics of the most important types of batteries that can be obtained on the market today are summarized in section 3.2. Finally, the basic operational mechanism of batteries is described in section 3.3. The information in this section will be used in chapter 4.

3.1 Historical overview

Batteries have been around for a long time. Earthen containers that served as galvanic cells dating from 250 BC have been found in Baghdad [1]. These containers were filled with iron and copper electrodes, together with an organic acidic solution. This yielded a cell capable of supplying 250 mA at a voltage of 0.25 V for approximately 200 hours. These cells were used to gild silver.

Two names are closely associated with the development of batteries and the related science of electrochemistry. These names are Luigi Galvani and Alessandro Volta [2]. Galvani performed an experiment in 1790, in which he suspended a frog from an iron hook. With a copper probe he measured electric pulses, which he believed originated in contractions of the muscles of the frog's legs. Volta attributed these contractions to the current flowing between the iron and copper metals. To prove that current could flow between two metals with an electrolyte in between, he built a 'pile' consisting of alternating silver and zinc plates interleaved with paper or cloth, which was soaked with an electrolyte. Hence, Volta was the first person in modern times to have built an actual battery. He patented this structure in 1800. In 1834, Michael Faraday derived the laws of electrochemistry based on Volta's work. This established a connection between chemical and electrical energy.

On the basis of Volta's work, other scientists also developed batteries of various designs. A recurring problem with these batteries was gas formation at the electrodes. This limited the available capacity, because the energy used to form the gas could not be recovered. Leclanché found the most successful solution to this problem in 1866 [2],[3]. He used manganese dioxide for the positive electrode, mixed with carbon to improve conductivity. This mixture was fixed around a graphite plate, which served as a current collector. He used zinc for the negative electrode, while ammonium chloride served as the electrolyte. He achieved an open-circuit voltage of 1.5 V with this battery. It could supply a medium current and had a good shelf life. Others later improved the so-called Leclanché cell. The present-day non-rechargeable batteries based on this principle, zinc-carbon or zinc-manganese dioxide ($ZnMnO_2$) batteries, still dominate the household battery market on a world-wide basis [4]. Their performance improved by 700% between 1920 and 1990.

The investigation of the behaviour of various metals in various electrolytes, in particular diluted sulphuric acid, by Gaston Planté was another important event in the development of batteries in 1859 [2],[3]. He constructed batteries as a sandwich of thin layers of lead, separated by sheets of coarse cloth in a cylindrical container filled with diluted sulphuric acid. The thin lead layers were connected as two separate electrodes. He introduced a voltage difference between the two lead electrodes, which charged the battery. By alternating this charge process with a subsequent discharge step, he eventually managed to form a positive electrode that consisted of lead dioxide and a negative electrode consisting of finely distributed lead. Repeated cycling could increase the thickness of these layers. After the formation of the active layers, which eventually took some 24 hours, the charge voltage of the lead/lead-dioxide couple rose to 2.7 V.

In order to prevent breakdown of the lead plates, Planté later used thicker plates, which led to quite heavy and bulky batteries that were used mainly for stationary applications. One of the first applications of a battery of this type was in a house-lighting installation with a dynamo, which served as charger, built by Henri Tudor in 1882 [2]. The time-consuming formation process was later greatly simplified by Camille Faure in France and Charles Brush in the USA in 1881 [2]. An important difference with respect to the Leclanché cell was that the cells constructed by Planté were rechargeable. The so-called lead acid batteries are still widely used nowadays, for example as car batteries for engine starting and vehicle lighting, but they are used very little in consumer electronics.

Waldemar Jungner in Sweden and Thomas Edison in the USA laid the foundation of the nickel-cadmium (NiCd) and nickel-iron alkaline storage battery industry between 1895 and 1905 [2]. Like lead acid batteries, NiCd and nickel-iron batteries are rechargeable. An important advantage of using alkaline solutions for the electrolyte, instead of acidic solutions, is the possibility of using a wide range of materials for the electrodes and containers; metals such as nickel would be affected in an acidic electrolyte solution.

In 1839, William Grove carried out experiments to investigate the decomposition of water into hydrogen and oxygen gas using platinum electrodes [2]. He observed that when the charging was stopped, a current started to flow in the opposite direction due to the recombination of oxygen and hydrogen on the platinum electrodes. These experiments laid the foundation for the development of fuel cells, in which energy stored in fossil fuels is directly converted into electrical energy. The active materials in a fuel cell are continuously supplied from a source external to the cell and the reaction products are continuously removed from the cell. The work on fuel cells also led to a better understanding of the behaviour of gas in porous electrodes. This knowledge was later successfully applied to the development of metal-air batteries, particularly with a zinc negative electrode and a porous air electrode, based on carbon as positive electrode. These batteries are known as zinc-air batteries.

In addition to the types of batteries mentioned so far, a wide variety of batteries have been developed since Volta's 'pile', both rechargeable and non-rechargeable. The advances in the development of portable electronic systems have had a significant effect on the development of new types of batteries. However, only fairly little progress have been made in improving battery characteristics such as energy density, shelf life and reliability in comparison to the advances made in electronic circuits [4]. While existing types of batteries are continuously being improved, new types keep appearing. The introduction of the nickel-metalhydride (NiMH) battery in 1990 and the lithium-ion (Li-ion) battery in 1991 were of a great importance for portable consumer products. Research at Royal Philips Electronics played an important role in the development of NiMH batteries [6],[7]. Apart from the continuous need for higher energy densities, environmental concerns have also boosted the development of these new types of batteries.

3.2 Battery systems

An important distinction that can be made regarding batteries is that between primary and secondary batteries. Primary batteries are non-rechargeable, whereas secondary batteries are rechargeable. Different battery systems are available for both primary and secondary batteries. Each battery system is characterized by its chemistry. Examples of primary batteries are zinc-carbon (Leclanché, also known as zinc-manganese dioxide ($ZnMnO_2$)), zinc-alkaline-MnO_2 (also simply known as alkaline batteries), zinc-air, mercuric-oxide, and lithium batteries [2]-[4]. The focus in this book will be on secondary batteries [1]-[17]. Examples of secondary batteries are sealed lead-acid (SLA) [11], NiCd [8], NiMH [5]-[8],[16], Li-ion [9],[14],[17], Li-ion-polymer, Li-metal, zinc-alkaline-MnO_2 [15], and zinc-air batteries. Note that some battery systems are available in both a non-rechargeable and a rechargeable form, such as zinc-alkaline-MnO_2 batteries. Some characteristics of the most important secondary batteries available on the market today or still under development will be given in the remainder of this section.

3.2.1 Definitions

Various specific terms are often used in the literature and data sheets to specify the characteristics of different battery systems. These terms are defined below [3],[4],[10].

A. *General definitions:*

Cell: The basic electrochemical unit used to generate electrical energy from stored chemical energy or to store electrical energy in the form of chemical energy. A cell consists of two electrodes in a container filled with an electrolyte. Definitions of these terms are given below.

Battery: Two or more cells connected in an appropriate series/parallel arrangement to obtain the required operating voltage and capacity for a certain load. The term battery is also frequently used for single cells. This terminology will also be adopted in this book, except where a distinction between cells and batteries is needed. A good example is a battery pack, which consists of several cells connected in series and/or parallel. The term battery pack was explained in chapter 2.

B. *Capacity-related definitions:*

Energy Density: The volumetric energy storage density of a battery, expressed in Watt-hours per litre (Wh/l).

Power Density: The volumetric power density of a battery, expressed in Watts per litre (W/l).

Rated Capacity: The capacity of a battery, expressed in Ampere-hours (Ah), which is the total charge expressed in [Ah] that can be obtained from a fully charged battery under specified discharge conditions. These conditions are specified by the manufacturer.

Specific Energy: The gravimetric energy storage density of a battery, expressed in Watt-hours per kilogram (Wh/kg).

Specific Power: The gravimetric power density of a battery, expressed in Watts per kilogram (W/kg).

C. Design-related definitions:

Electrode: Basic building block of an electrochemical cell. Each cell consists of a positive and a negative electrode. The cell voltage is determined by the voltage difference between the positive and the negative electrode.

Anode: The electrode at which an oxidation reaction (see section 3.3) occurs. This means that the electrode supplies electrons to an external circuit. The electron flow reverses between charging and discharging. Hence, the positive electrode is the anode during charging and the negative electrode is the anode during discharging. Usually, a cell's anode is specified during discharging and hence the name anode is commonly used for the negative electrode.

Cathode: The electrode at which a reduction reaction (see section 3.3) occurs. This means that the electrode takes up electrons from an external circuit. Hence, the negative electrode is the cathode during charging and the positive electrode is the cathode during discharging. Usually, a cell's cathode is specified during discharging and hence the name cathode is commonly used for the positive electrode. To avoid confusion, the electrodes will be named positive and negative in this book.

Electrolyte: The medium that provides the essential ionic conductivity between the positive and negative electrodes of a cell.

Separator: An ion-permeable, electronically non-conductive material or spacer that prevents short-circuiting of the positive and negative electrodes of a cell.

D. Application-related definitions:

C-rate: A charge or discharge current equal in Amperes to the rated capacity in Ah. Multiples larger or smaller than the C-rate are used to express larger or smaller currents. For example, the C-rate is 600 mA in the case of a 600 mAh battery, whereas the C/2 and 2C-rates are 300 mA and 1.2 A, respectively.

Cycle Life: The number of cycles that a cell or battery can be charged and discharged under specific conditions, before the available capacity in [Ah] fails to meet specific performance criteria. This will usually be 80% of the rated capacity.

Cut-off voltage: The cell or battery voltage at which the discharge is terminated. Also often referred to as End-of-Discharge voltage. The symbol V_{EoD} will be used for this voltage throughout this book.

Self-Discharge: Recoverable loss of capacity of a cell or battery. This is usually expressed in a percentage of the rated capacity lost per month at a certain temperature, because self-discharge rates of batteries are strongly temperature-dependent.

Conditions have to be specified for certain definitions in order to be able to quantify them. For example, the rated capacity is strongly dependent on the discharge current, battery temperature and cut-off voltage. In general, the capacity obtainable from a battery will be lower at higher discharge currents, lower battery temperatures and higher cut-off voltages. Usually, the rated capacity is specified at a current of C/10 and at room temperature for secondary batteries. The specified cut-off voltage is strongly dependent on the battery chemistry.

3.2.2 Battery design

Besides on operational conditions, such as discharge rate and battery temperature, a battery's actual rated capacity will strongly depend on its design. In general, the theoretical energy density, which depends on the battery chemistry, will be much larger than the energy density specified in a battery's data sheet. This is due to the presence of non-reactive components inside a battery, such as the current collectors to which the electrodes are attached, the separator, the electrolyte and the container. These components add to the battery weight and volume, but they do not contribute to its energy density. For example, the energy density of a battery decreases when its volume decreases, because the percentage of 'dead' volume of containers, seals and other structural parts is larger in the case of smaller batteries. The energy density and specific energy of every battery can be found by integrating the voltage-dependent charge $Q(V)$ over the applicable voltage range, similar to what was shown in Figure 2.7, and dividing the result by the battery volume or weight, respectively.

  A distinction can be made between cylindrical and prismatic batteries [3],[4]. The electrodes will be thin strips in most secondary cylindrical batteries. These strips are rolled with the separator in between and placed in a cylindrical can. Rolling the electrodes like this yields a relatively large surface area, enabling the battery to supply larger discharge currents. However, this is achieved at the expense of battery capacity, because the percentage of active material is lower than in a design in which the electrodes are shaped like two concentric cylinders. This is due to the fact that a current collector required to support the thin electrodes takes up volume in a wound structure. A design with concentric cylinders maximizes the amount of active material and is used mainly for primary batteries, e.g. zinc-carbon and Zn-alkaline-MnO_2.

In a prismatic battery, rectangular electrodes are stacked in flat rectangular boxes with the separator in between. Prismatic batteries have a better form factor, because they allow better utilization of the volume available inside a portable device. The shapes and sizes of batteries are subject to standardization described in e.g. the IEC (International Electrotechnical Commission) standard [3].

Most battery systems available on the market, whether cylindrical or prismatic and irrespective of their size, are available in different types [3],[8]. These types have been optimized for use in specific applications. Examples are high-rate batteries, which allow relatively large discharge currents of up to several times the C-rate, high-temperature batteries, which allow operation at higher temperatures, high-capacity batteries and fast-charge batteries, which enable a very quick recharge. These characteristics have been obtained by changing the design of the electrodes or by adding compounds to the electrodes. For example, the use of so-called sintered electrodes in batteries greatly decreases the internal impedance of the cells. This allows the withdrawal of larger discharge currents.

An example of adding compounds to the electrodes of a NiCd battery will be given in chapter 4. This is done to protect the battery during overcharging and overdischarging. Overcharging means the continuation of charging when the battery is full, whereas overdischarging means the continuation of discharging when the battery is empty. The latter situation may occur when cells are connected in series in a battery pack. In that case, the cell with the lowest capacity will be the first to reach its empty state. The empty cell will be the first to be overdischarged when the total battery pack voltage is still high enough to power the load. This is attributable to manufacturing spread in characteristics, e.g. in capacity, of cells of the same type.

3.2.3   Battery characteristics

The main characteristics of the most important secondary battery systems are summarized in Table 3.1 [1],[3],[4],[10],[15]. Ranges are given for the energy density, specific energy, self-discharge rate and cycle life of almost all of the systems in the table, because so many different types of battery systems are available on the market.

NiCd batteries

The NiCd battery is commonly known as relatively cheap and robust. It is universal and can still be found in many portable devices today. Most NiCd batteries can supply large currents. It is possible to charge NiCd batteries in a relatively short period of time because of their robustness. Charge times of only 10 minutes have been reported [1]. The average cell voltage is 1.2 V. The characteristics of high power delivery and short recharge times make NiCd batteries very popular for power tools. Other applications include cordless phones, shavers, camcorders and portable audio products. The positive nickel electrode is a nickel hydroxide/nickel oxyhydroxide ($Ni(OH)_2/NiOOH$) compound, while the negative cadmium electrode consists of metallic cadmium (Cd) and cadmium hydroxide ($Cd(OH)_2$). The electrolyte is an aqueous solution of potassium hydroxide (KOH). Major improvements in energy density and specific energy have been obtained with respect to earlier designs by using high porosity nickel foam instead of the original sintered nickel as substrate for the active materials [4].

Table 3.1: Overview of the main characteristics of the most important secondary battery systems

Battery system:	NiCd	NiMH	Li-ion	Li-ion-polymer	SLA	Rechargeable alkaline
Average operating voltage [V]	1.2	1.2	3.6	3.6	2.0	1.5
Energy density [Wh/l]	90..150	160..310	200..280	200..250	70..90	250
Specific energy [Wh/kg]	30..60	50..90	90..115	100..110	20..40	20..85
Self-discharge rate [%/month] at $20^{\circ}C$	10..20	20..30	5..10	1	4..8	0.2
Cycle life	300..700	300..600	500..1000	200	200..500	15..25
Temperature range [$^{\circ}C$]	-20..50	-20..50	-20..50	?	-30..60	-30..50

Although very suitable for, for example, power tools, NiCd batteries have some drawbacks. First of all, their energy density and specific energy are relatively low. Secondly, NiCd batteries suffer from the so-called memory effect [1],[3],[4],[10]. This effect can be defined as a decline in effective capacity with repeated partial charge/discharge cycles. As partial cycling continues, the battery will eventually only be able to supply the capacity retrieved from the partial cycling. The battery voltage drops significantly after this capacity has been removed and most portable devices will stop functioning at that moment. However, putting the battery through one or more complete charge/discharge cycles can restore full capacity. Most battery experts attribute the memory effect to the Cd electrode. Although the effect is hard to reproduce in laboratory experiments, it seems to result from the growth of large crystals on the Cd electrode, which reduces the effective area [10]. As a final drawback, the use of cadmium in NiCd batteries involves serious environmental problems.

NiMH batteries
In response to the relatively low energy density and specific energy as well as the environmental concerns associated with NiCd batteries, Sanyo Electric in Japan introduced the first NiMH battery in 1990 [16]. A lot of useful research was performed at Philips Research in the early seventies [6]. Table 3.1 illustrates that NiMH batteries offer the same average operating voltage as NiCd batteries, with the great advantage of a higher energy density. Applications include notebook computers, cellular phones and shavers. In NiMH batteries a metalhydride (MH) alloy has replaced the cadmium electrode. The positive electrode and the electrolyte are more or less the same as in NiCd batteries.

The MH alloy is capable of storing hydrogen in a solid state. Two classes of metal alloys are generally used in NiMH batteries, AB_2 and AB_5 alloys [3],[5],[6]. The AB_2 class of alloys consists of titanium and zirconium, while the alloys of class AB_5 include rare-earth alloys based on lanthanum nickel. Almost 100% of the commercially available NiMH batteries are based on the AB_5 class of alloys [5],[6]. These alloys possess better rate capability and better stability characteristics, at the cost of a lower specific energy than AB_2 alloys [3]. Although the chemistry of a NiMH battery is similar to that of a NiCd battery, there are differences between the two:

- NiMH batteries have a better energy density than NiCd batteries. This is due to the fact that the MH electrode has a higher energy density than the Cd electrode in NiCd batteries [3].
- The self-discharge rate of NiMH batteries is somewhat higher than that of NiCd batteries. One of the factors that influence the self-discharge rate of a NiMH battery is the ability of the MH electrode to retain the stored hydrogen under storage conditions. The more hydrogen will be released under storage conditions, the higher the self-discharge rate will be.
- Most manufacturers claim that NiMH batteries do not suffer from the memory effect known from NiCd batteries, although examples can be found in literature of the occurrence of the memory effect in NiMH batteries [3],[10],[12].
- In general, NiMH batteries are less robust with respect to overcharging than NiCd batteries. This means that the charging algorithm of NiMH batteries should be more accurate to prevent overcharging, especially when the charge current is higher.
- A difference between the charging process of NiMH batteries and that of NiCd batteries is that the net charging reaction in a NiMH battery is exothermic. This means that heat is generated continuously during the charging. On the other hand, the net charging reaction of a NiCd battery is endothermic. This means that heat is consumed during the first phase of charging [3]. In the case of a NiMH battery, this means that the battery temperature rises continuously during charging, with a steep rise at the end of the charging. However, in a NiCd battery the temperature remains relatively constant for most of the charging process, with a steep rise at the end of the charging.
- Another difference between the charging process of NiMH and NiCd batteries concerns the battery voltage profile, which is less pronounced in the case of NiMH batteries than in the case of NiCd batteries at the start of overcharging. This makes it more difficult to detect the 'battery full' condition. More information on differences between the charging algorithms of NiMH and NiCd batteries will be given in chapter 5.

Li-ion batteries
The first Li-ion battery was introduced by Sony in Japan in 1991. The chemistry of Li-ion batteries differs significantly from that of Ni-based batteries [3],[9],[14]. Li-ion cells offer the advantage of a high average operating cell voltage of 3.6 V, because of the very negative standard potential of lithium with respect to the standard hydrogen reference electrode (SHE). Moreover, Li-ion batteries have a relatively high specific energy, which results in batteries that are lighter than Ni-based batteries at the same capacity. Applications include palmtop and notebook computers, cellular phones and camcorders. The electrodes in a Li-ion battery are intercalation electrodes. Intercalation electrodes have a lattice structure in which guest species may be inserted and extracted without any great structural modifications in the host material. The operation of Li-ion batteries is based on the transfer of lithium ions from the positive electrode to the negative electrode during

charging and *vice versa* during discharging. This is generally referred to in the literature as the 'rocking chair' principle [3],[10],[14].

The positive electrode of a Li-ion battery consists of one of a number of lithium metal oxides, which can store lithium ions. The oxides encountered most frequently in commercially available batteries are lithium cobalt oxide ($LiCoO_2$), lithium nickel oxide ($LiNiO_2$) and lithium manganese oxide ($LiMn_2O_4$). $LiCoO_2$ and $LiNiO_2$ offer the advantage of a slightly higher capacity. $LiMn_2O_4$ is less toxic and less expensive than the other materials [3]. The negative electrode of a Li-ion battery is a carbon electrode, with the maximum ratio of the number of lithium ions and the number of carbon atoms in the lattice being 1:6. The carbon electrode can be made from graphite or petroleum coke. The use of graphite results in a higher capacity and a flatter discharge curve than the use of petroleum coke [3].

The electrolytes used in Li-ion batteries are non-aqueous, as opposed to the aqueous electrolytes used in Ni-based batteries. A salt dissolved in an organic solvent serves as the electrolyte in Li-ion batteries. The choice of the organic solvent is limited to those based on ethylene carbonate when graphite is used for the negative electrode. Other solvents such as diethylene carbonate and propylene carbonate can also be used when petroleum coke is used [10]. A popular choice for the salt is lithium hexafluorophosphate ($LiPF_6$). Important concerns with respect to the electrolyte are compatibility with the chosen electrode materials, good ionic conductivity and thermal and electrochemical stability. In general, the conductivity of the mixtures used as electrolytes in Li-ion batteries is some two orders of magnitude lower than the conductivity of the aqueous electrolytes used in Ni-based batteries [3],[10].

Table 3.1 shows that, apart from a higher specific energy, Li-ion batteries also have considerably lower self-discharge rates than Ni-based batteries. Moreover, Li-ion batteries do not suffer from the memory effect. In applications, Li-ion batteries have to be approached differently than Ni-based batteries, as far as both charging and discharging are concerned. First of all, Li-ion batteries need a different charging algorithm than Ni-based batteries. More information will be given in chapter 5. Furthermore, Li-ion batteries are less capable of delivering large currents, expressed in C-rate, than Ni-based batteries. Overdischarging Li-ion batteries leads to a decrease in cycle life. Without further precautions, overcharging Li-ion batteries leads to dangerous situations and may even cause a fire or an explosion of the battery. Hence, it can be generally stated that overcharging and overdischarging of Li-ion batteries is not allowed. This is an important difference from NiCd and NiMH batteries. Therefore, strict control of the charging and discharging of the battery is essential for safety reasons and for retaining a long cycle life. As described in chapter 2, this leads to an increase in electronics for implementing the necessary monitor and control functions in the form of an electronic safety switch. Combined with a Positive-Temperature Coefficient (PTC) resistor used as a passive safety device and a lower conductivity of the electrolyte, the total battery series resistance of a Li-ion battery, including safety measures, can add up to around 300 mΩ in practical consumer batteries. When this series resistance is compared with the relatively low series resistance of around 20 mΩ in practical consumer Ni-based batteries, the fact that the discharge capability of Li-ion batteries is lower than those of Ni-based batteries is partly explained.

Li-ion-polymer batteries

As a successor of the Li-ion battery with a liquid organic electrolyte, the 'solid-state' Li-ion battery started to appear on the market at the end of 1997. The basic difference with respect to Li-ion batteries is that the electrolyte consists of a solid ion-conducting polymer material [3],[4],[10]. The polymer electrolyte also serves as a separator. Batteries of this type were referred to as Li-ion-polymer batteries in Table 3.1.

The conductivity of polymer electrolytes is even lower than that of liquid organic electrolytes [3]. Most liquid organic electrolytes have a conductivity in the order of 10^{-3} $(\Omega cm)^{-1}$ at 20°C. A polymer electrolyte, such as the commonly used polyethylene oxide (PEO) electrolyte, has a conductivity in the order of 10^{-8} $(\Omega cm)^{-1}$ at 20°C. As a result, the polymer electrolytes have to be kept very thin to achieve a reasonable conductivity. Operation of the battery at higher temperatures also leads to an improved conductivity of the polymer electrolyte. This is impractical for use in many portable products. Improvements in conductivity at room temperature have been obtained by adding liquid plasticizers, such as polypropylene carbonate. This increases the conductivity to 10^{-4} $(\Omega cm)^{-1}$ at 20°C. Another approach involves the use of so-called 'gelled' electrolytes. These electrolytes are obtained by trapping a liquid solution of a lithium salt in an organic solvent in a polymer matrix. Using such 'gelled' electrolytes, conductivities of up to 10^{-3} $(\Omega cm)^{-1}$ at 20°C have been obtained.

Polymer electrolytes are less reactive with respect to lithium than liquid electrolytes. This is beneficial for the safety of the battery. However, adding liquid plasticizers or using 'gelled' electrolytes to increase the conductivity of the polymer electrolyte leads to an electrolyte that is more reactive with respect to lithium and part of the safety benefit is consequently lost. The fact that the polymers have to be made very thin to have a reasonable conductivity moreover makes them vulnerable to mechanical stress and pin-hole defects.

The use of a polymer electrolyte offers the possibility of fast production of the cells using web equipment and the fabrication of thin cells [4],[10],[17]. The electrodes, electrolyte and current collector are simply stacked in a sandwich structure. Because the polymer keeps the structure together, no outside pressure of a can is needed to form a battery. Instead, the housing can be made from a thin laminated foil material. The Li-ion-polymer cells can be configured in many possible dimensions, because they can be made with a flexible width, length and thickness. This increases the energy density for a given battery cavity in a portable product. Some battery manufacturers, such as Ultralife, have recently introduced Li-ion-polymer batteries on the consumer market. The Ultralife batteries are based on the use of a 'gelled' type of electrolyte. The technology of such batteries must however still be substantially improved for them to acquire a significant market share. The numbers cited in Table 3.1 are consequently subject to change.

Li-metal batteries

A further development of lithium-based batteries that is still in the research phase is the replacement of the carbon negative electrode by a metallic lithium electrode. Such batteries are commonly referred to as Li-metal batteries. The literature gives an example of Li-metal batteries under development by the battery company Tadiran [12],[13]. The biggest advantage of storing lithium in its metallic form, instead of its ionic form surrounded by carbon atoms in the maximum ratio of 1:6, is a gain in energy density and specific energy. However, the use of metallic lithium introduces the severe problem of its very high reactivity. For example, metallic lithium will react with any liquid electrolyte and forms a passivation film on the electrode after every charge/discharge cycle. This process consumes a lot of lithium. Therefore, lithium has to be present in excess amounts, which lowers the energy density and specific energy.

Research was initiated into the possibility of using a polymer electrolyte in Li-metal batteries, because lithium reacts less with polymer electrolytes than with liquid electrolytes. However, a problem that still arises in this case is the build-up of surface irregularities on the lithium electrode during charging of the battery. These irregularities are known as dendrites. There is a risk of short-circuiting by dendrites that puncture the solid polymer film, because the solid-polymer electrolyte has to be rather thin to ensure sufficient conductivity. At best, such short-circuiting shortens the battery's cycle life. The substantial increase in temperature as a result of the heat generated by the short-circuit current however also implies safety problems such as flaming. Extensive research will have to be carried out to solve such safety problems and increase the cycle life of Li-metal batteries.

SLA batteries
A relatively old battery technology, which can still be found on the secondary battery market today, is that of the SLA battery [3],[4]. The chemistry of an SLA battery is basically the same as that of an ordinary car battery. SLA batteries are however maintenance-free. This means that the electrolyte does not have to be replaced. SLA batteries contain only a limited amount of electrolyte, which is absorbed in the separator or a gel. The positive electrode of an SLA battery is formed by lead dioxide (PbO_2), while metallic lead (Pb) in a high-surface-area porous structure is used for the negative electrode. A sulphuric acid (H_2SO_4) solution is used for the electrolyte. The average operating voltage of an SLA cell is 2 V.

Advantages of the SLA battery are its good rate capability and relatively low self-discharge rate; see Table 3.1. Moreover, SLA batteries do not suffer from the memory effect. They can be used under a continuous float charge. The reason for this is that a starved electrolyte and an excess amount of material in the negative electrode are used in the design. This facilitates the recombination of oxygen that is formed during overcharging. As a result, large pressure build-up inside the battery is prevented [3]. Moreover, the batteries are manufactured with a pressure valve. This valve is resealable and opens only under extreme pressure build-up. The valve prevents the introduction of ambient oxygen. The possibility of withstanding long periods of float charge is advantageous for use in a back-up system. A final advantage is that SLA batteries are low in costs.

A major disadvantage of SLA batteries is their low energy density and specific energy. Especially when compared with Li-ion batteries, SLA batteries are heavy and take up a large amount of space. An additional disadvantage is the problem of irreversible capacity loss under deep discharge conditions. The deep discharge condition may even occur due to self-discharge of the battery. This means that the cycle life of an SLA battery drastically decreases when it is not frequently recharged [1].

The term SLA is generally reserved for cylindrical batteries. The term Valve-regulated lead-acid (VRLA) is used for prismatic types. VRLA batteries generally vent at lower pressures. An example of an ultra-thin VRLA battery was introduced by the company Bolder [11]. This battery is manufactured using a thin-metal film technology. The rate capability is extremely high and these batteries can be charged in short times of only several minutes, because the technology results in a very low battery impedance. However, as the technology is based on lead acid, the energy density and specific energy of such batteries are relatively low. This limits widespread use in portable devices.

Rechargeable alkaline batteries

Secondary zinc-alkaline-MnO_2 batteries based on the same chemistry as their primary counterpart have become available on the market [3],[15]. Batteries of this type were introduced by the Renewal company in the USA in 1993 [4]. As with the primary type, the positive electrode consists of manganese dioxide (MnO_2), the negative electrode of zinc (Zn), and KOH is used for the electrolyte. An increase in the capacity of the Zn electrode, a change in the separator design and additives to both electrodes allow recharging [3]. Hence, although some charger manufacturers claim to be able to recharge ordinary alkaline batteries, only the specially adapted secondary types are really rechargeable.

The average operating voltage of a rechargeable alkaline cell during practical use is 1.3 V. The rechargeable alkaline battery offers the advantage of a low self-discharge rate and low cost. Disadvantages are the poor cycle life and the fact that the initial capacity is lower than that of primary alkaline batteries. For example, the initial capacity at 20°C is about 70% of the capacity of the primary battery [3]. Twenty discharge cycles can be realized before the capacity drops to 50% of the initial capacity when the battery is discharged and subsequently charged from a completely discharged state. This means that the available capacity decreases rapidly with cycling. The achievable recharged capacity moreover decreases when the battery has previously been deeply discharged.

Zinc-air batteries

In addition to the secondary battery systems described so far, various other systems can be found in the literature. An interesting new development is the secondary zinc-air battery [3],[10]. This battery system is still in the pioneering phase. Its chemical structure is the same as that of its primary counterpart. The positive electrode is made of carbon, the negative electrode consists of Zn, while KOH is used as the electrolyte. The carbon electrode is exposed to the air and is only used as a reactive surface, which is why it is known as an 'air' electrode. The voltage of a zinc-air cell ranges from 1 V to 1.2 V.

The simplest way of 'recharging' a zinc-air battery is by replacing the Zn negative electrode when all the Zn has been consumed. In a practical consumer situation the battery should however be recharged electrically. The carbon 'air' electrode then produces oxygen during charging and consumes oxygen during discharging. An air-management system is needed for proper operation of a secondary zinc-air battery. This system ensures the flow of air into the battery required to supply current. On the other hand, the flow of air from outside is blocked when no current is drawn. This prevents crystallization of carbonate in the porous air electrode caused by carbon dioxide absorbed from the air. Such crystallization could impede the introduction of air. Also, blocking the airflow prevents dehydration of the cell or the opposite situation of flooding of the air electrode pores. It also reduces the self-discharge rate of the battery.

Obviously, the air-management system adds to the volume and weight of the battery. This decreases the resulting energy density and specific energy. The present zinc-air batteries are moreover very sensitive to electrical abuse. To overcome this problem, zinc-air batteries have built-in monitoring and charging circuitry. Some manufacturers aim at the portable computer market, but no commercial battery has so far been introduced.

3.3 General operational mechanism of batteries

3.3.1 Introduction

The general operational mechanism of a battery will be discussed in this section. In its simplest definition, a battery is a device capable of converting chemical energy into electrical energy and *vice versa*. The chemical energy is stored in the electro-active species of the two electrodes inside the battery. The conversions occur through electrochemical reduction-oxidation (redox) or charge-transfer reactions [2],[3],[18],[19]. These reactions involve the exchange of electrons between electro-active species in the two electrodes through an electrical circuit external to the battery. The reactions take place at the electrode/electrolyte interfaces. When current flows through the battery, an oxidation reaction will take place at the anode and a reduction reaction at the cathode. The oxidation reaction yields electrons to the external circuit, while a reduction reaction takes up these electrons from the external circuit. The electrolyte serves as an intermediate between the electrodes. It offers a medium for the transfer of ions. Hence, current flow is supported by electrons inside the electrodes and by ions inside the electrolyte. Externally, the current flows through the charger or load.

The charging and discharging of a battery are schematically illustrated in Figure 3.1a and b, respectively. In both cases, the negative electrode (-) is shown on the left and the positive electrode (+) on the right. Figure 3.1 shows that oxidation occurs at the positive electrode during charging, whereas reduction occurs at the negative electrode. The reverse takes place during discharging.

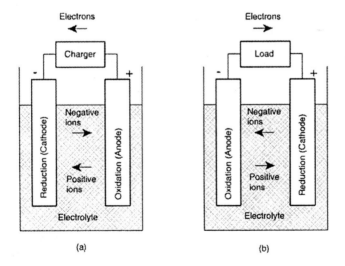

Figure 3.1: Electrochemical operation of a battery during charging (a) and discharging (b)

The charge-transfer reactions taking place at the positive and negative electrode can be represented in a simplified form by:

Positive electrode (+):

$$A \ Red_+ \ \underset{k_{c,+}}{\overset{k_{a,+}}{\rightleftarrows}} \ B \ Ox_+ + n \ e \tag{3.1}$$

Negative electrode (-):

$$C \ Ox_- + m \ e \ \underset{k_{a,-}}{\overset{k_{c,-}}{\rightleftarrows}} \ D \ Red_- \tag{3.2}$$

where Ox and Red denote the oxidized and reduced species, respectively, and $k_{a,+/-}$ and $k_{c,+/-}$ denote the reaction rate constants for the anodic and cathodic reactions, respectively, whoses dimensions depend on the reaction equation. In each electrode, oxidized and reduced species are present, which form a redox couple. Note that (3.1) and (3.2) have been simplified for convenience. Other species will often be involved, resulting in more complex reactions. During the charging of the battery, A molecules of the reduced species Red_+ are converted into B molecules of the oxidized species Ox_+ at the positive electrode, yielding n electrons (e). At the same time, C molecules of the oxidized species Ox_- are converted into D molecules of the reduced species Red_- at the negative electrode, taking up m electrons. The reactions at both electrodes are reversed during discharging. The net reaction can be found by adding (3.1) and (3.2). When $n = m$ is assumed, this yields:

$$A \ Red_+ + C \ Ox_- \ \underset{d}{\overset{ch}{\rightleftarrows}} \ B \ Ox_+ + D \ Red_- \tag{3.3}$$

3.3.2 Basic thermodynamics

The battery is in a state of equilibrium when no external current flows and the reaction rates of the electrode reactions given by (3.1) and (3.2) in the right and left directions are the same. The equilibrium potential (E^{eq}) of each electrode can be obtained with Nernst's equation [2],[3],[18],[19]:

$$E_+^{eq} = E_+^o + \frac{RT}{nF} \ln \frac{\left(a_{Ox_+}\right)^B}{\left(a_{Red_+}\right)^A} \tag{3.4}$$

for the positive electrode and

$$E_-^{eq} = E_-^o + \frac{RT}{mF} \ln \frac{\left(a_{Ox_-}\right)^C}{\left(a_{Red_-}\right)^D} \tag{3.5}$$

for the negative electrode, where E_i^o is the standard redox potential of electrode i in [V], R is the gas constant (8.314 J/(mol.K)), T is the temperature in [K], n and m denote the number of electrons involved in the charge-transfer reactions, F is Faraday's constant (96485 C/mol), and a_i is the activity of species i in [mol/m^3].

The activity a_i is linearly proportional to its concentration (c_i) and molar amount (m_i) according to:

$$a_i = \gamma.c_i = \frac{\gamma.m_i}{Volume} \tag{3.6}$$

where γ is the dimensionless activity coefficient, which will be assumed to be unity throughout this book, c_i denotes the concentration of species i in [mol/m^3], m_i is the molar amount of species i in [mol] and the *Volume* in which species i resides is given in [m^3]. From (3.4) and (3.5) the following expression is obtained for the equilibrium potential of the complete battery:

$$E_{bat}^{eq} = E_+^{eq} - E_-^{eq} = E_+^o - E_-^o + \frac{RT}{nF} \ln \frac{(a_{Ox_+})^\beta (a_{Red_-})^\rho}{(a_{Red_+})^\alpha (a_{Ox_-})^\varsigma} \tag{3.7}$$

where again $n = m$ has been assumed for simplicity. (3.7) shows that the value of the equilibrium potential of a battery depends on the ratio of the activities of the oxidized and reduced species in the two electrodes. Therefore, the value of the equilibrium potential depends on the State-of-Charge (SoC) of the two electrodes and hence on the battery's SoC.

The Gibbs free energy change (ΔG^o) of the battery under standard conditions is the driving force enabling a battery to supply electrical energy to an external circuit. ΔG^o in [J/mol] is given by [3],[19]:

$$\Delta G^o = -nF(E_+^o - E_-^o) = -nFE_{bat}^o \tag{3.8}$$

The expression $\Delta G^o = +nFE^o{}_{bat}$ is also encountered in the literature, depending on the convention that is used. This sign confusion is due to the fact that the sign of ΔG^o depends on whether the reaction is an oxidation or a reduction reaction and the sign of $E^o{}_{bat}$ is fixed [19]. More information will be given in chapter 4.

3.3.3 Kinetic and diffusion overpotentials

Electrical energy is supplied to or taken up from an external circuit in a non-equilibrium situation. The rates in the left and right directions of the electrode reactions of (3.1) and (3.2) are then not the same. For example, the reactions from left to right prevail at both electrodes during the charging. The actual battery voltage differs from the equilibrium value dictated by (3.7) when current flows through the battery. This is due to polarization at the electrode/electrolyte interfaces, resulting from the charge-transfer reactions. The difference between the actual electrode potential E during current flow and the equilibrium electrode potential E^{eq} of each electrode is denoted as the overpotential η^{ct} of the charge-transfer reaction [2],[3],[6],[18],[19]. Hence,

$$\eta^{ct} = E - E^{eq} \tag{3.9}$$

where the sign of the overpotential η^{ct} depends on the direction of the current that flows into or out of the electrode. An oxidation reaction results in a current I_a, which means that electrons flow out of the electrode. A reduction reaction results in a current I_c, which means that electrons flow into the electrode. These currents are equal and opposite in direction in a state of equilibrium, resulting in a zero net current; see Figure 3.2. In a state of non-equilibrium however, one current is larger than the other. For example, a net oxidation reaction will occur when I_a is larger than I_c. A net current I_a-I_c will then flow into the electrode and the overpotential will be positive. A net reduction reaction will occur when I_a is smaller than I_c and a current I_a-I_c will then flow out of the electrode. The overpotential will then be negative. For example, the overpotential will be positive at the positive electrode and negative at the negative electrode during charging.

The overpotential can be seen as the driving force of the electrode reaction. Both kinetic aspects and mass transport phenomena contribute to its value. Kinetic aspects result in a kinetic overpotential (η^k) and mass transport phenomena result in a diffusion overpotential (η^d). This means that

$$\eta^{ct} = \eta^k + \eta^d \tag{3.10}$$

The Butler-Volmer equation is valid for the relation between η^k and the net reaction current I for each charge-transfer reaction [2],[3],[18],[19], i.e.

$$I = I_a - I_c = I^o \left\{ \exp\left(\alpha \frac{nF}{RT} \eta^k \right) - \exp\left(-(1-\alpha) \frac{nF}{RT} \eta^k \right) \right\} \tag{3.11}$$

where I^o expresses the exchange current for the charge-transfer reaction in [A] and α denotes the dimensionless transfer coefficient, $0 < \alpha < 1$. In a state of equilibrium, I and η^k are both zero and $I_a = -I_c = I^o$. The exchange current I^o_j of a charge-transfer reaction j is given by [6],[19],[20]:

$$I^o_j = nFA \left(k_{a,j} \right)^{1-\alpha} \left(k_{c,j} \right)^{\alpha} (a^b_{Ox})^{\alpha x} (a^b_{Red})^{(1-\alpha)y} \tag{3.12}$$

where A expresses the surface area of the electrode at which the reaction occurs in [m^2], a^b_i stands for the bulk activity of species i in [mol/m^3] and x and y express the dimensionless reaction orders of the Ox and Red species. For example, assuming that the reaction given in (3.1) is the rate-determining reaction occurring at the positive electrode, the values of $x=B$ and $y=A$ have to be used for obtaining I^o_+ from (3.12). The bulk activity is equal to the surface activity a^s at the electrode/electrolyte interface in the case of all species when the reaction rate is not limited by mass transport processes. Hence, the concentration of each species that takes part in the reaction is uniformly distributed throughout the electrode or the electrolyte. (3.12) shows that the value of the exchange current depends on the values of the bulk activities of the oxidized and reduced species, and hence on the electrode's SoC.

The current-voltage relations for both electrodes of the battery of Figure 3.1 are shown in Figure 3.2. It is assumed that no mass transport limitation occurs, so η^d is zero. The arrows for I, V_{bat} and $\eta^k_{+/-}$ denote the situation during charging. Figure 3.2 shows that the net reaction current, represented by a solid line, results from currents I_a and I_c, represented by dashed curves, at each electrode. This is in agreement with (3.11). The overpotential is indeed positive at the positive electrode and negative at the negative electrode when I flows in the directions indicated for the charging. The battery voltage V_{bat} equals the difference between the positive electrode potential E_+

and the negative electrode potential E_-. This means that, during the charging, V_{bat} is larger than the battery equilibrium potential E^{eq}_{bat}. The direction of I at both electrodes is reversed during discharging. In that case, V_{bat} will be smaller than E^{eq}_{bat}. The net reaction current and the overpotentials are zero at both electrodes in a state of equilibrium. V_{bat} then equals E^{eq}_{bat}.

The exchange current I_+^o was chosen to be smaller than I_-^o in the example shown in Figure 3.2. As a result, the overpotential η_+^k is larger than η_-^k. This also follows from (3.11), where η^k has to be larger at smaller I^o values to allow a certain value of I at a constant temperature T. This means that in this example, the kinetics of the reaction at the positive electrode are worse than at the negative electrode. The battery voltage V_{bat} depends on the battery's SoC, because E^{eq}_{bat} depends on the battery's SoC through (3.7), and η_+^k and η_-^k depend on the SoC of the respective electrodes through (3.11) and (3.12).

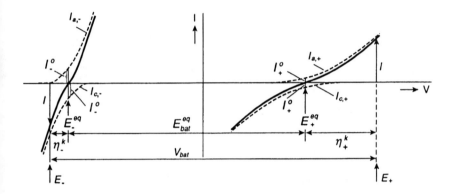

Figure 3.2: Current-voltage relation for the positive (+) and negative (-) electrode without mass transport limitations, $\eta^d=0$ [6]. The battery voltage V_{bat}, current I and kinetic overpotentials $\eta^k_{+/-}$ during charging are shown

If the value of the overpotential η^k in (3.11) is large enough, either I_a or I_c can be neglected. In that case, the equation can be rewritten by taking the ln-form. This way the Tafel relation is found, which is often encountered in the literature [3],[19]. When $I_a \gg I_c$ the Tafel relation reads:

$$\eta^k = \frac{RT}{\alpha n F}\ln(I) - \frac{RT}{\alpha n F}\ln(I^o) \tag{3.13}$$

The value of α can be inferred from the slope and the value of I^o is found by extrapolating the plot to $\eta^k=0$ when η^k and I are plotted according to (3.13) and n is a known quantity. In principle, this makes it easy to derive these values from measured overpotentials and electrode currents.

So far it has been assumed that the bulk and surface concentrations are the same. In practice, mass transport limitations may lead to a difference between the surface and average concentration of reacting species. The reaction rate is influenced when the electrode surface becomes depleted of or accumulated with species consumed in or generated by the reaction, respectively. Mass transport of species to and from electrode surfaces can occur in essentially three processes [3],[18],[19]:

(1) diffusion as a result of a concentration gradient;
(2) migration as a result of an electrical field;
(3) convection.

Both charged and uncharged species can diffuse. Typically, species will diffuse from a region with a high concentration to a region with a low concentration. Only charged particles can migrate. The potential change across a medium and so the electrical field will be low when the medium through which the charged species is transported is highly conductive. So migration then contributes little to mass transport of the species. Convection occurs in, for example, zinc-air batteries, where the air-management system ensures the circulation of air through the system.

One-dimensional diffusion in the x direction can be described using Fick's first and second law [3],[18],[19]:

$$J_i(x,t) = -D_i \frac{\delta c_i(x,t)}{\delta x} \tag{3.14}$$

and

$$\frac{\delta c_i(x,t)}{\delta t} = D_i \frac{\delta^2 c_i(x,t)}{\delta x^2} \tag{3.15}$$

where $J_i(x,t)$ denotes the flux of the diffusing species i in [mol/(m^2.s)] at location x [m] and time t [s], D_i is the diffusion coefficient of the diffusing species i in [m^2/s] and $c_i(x,t)$ is the concentration of the diffusing species i in [mol/m^3] at location x and time t.

Initial and boundary conditions have to be defined for each diffusion problem. The initial condition specifies the concentration of the diffusing species at $t = 0$ at all locations x. For example, a species can be uniformly distributed throughout the medium through which diffusion occurs at $t = 0$. The boundary conditions specify the concentrations or concentration gradients at certain positions x for all times t. For example, boundary conditions can be valid at an electrode/electrolyte interface or for x approaching infinity. The solution of the diffusion equations will depend on the initial and boundary conditions. Consider the situation depicted in Figure 3.3, where the diffusing species i is extracted at $x = 0$, for example at the electrode surface, in a charge-transfer reaction. After some time t, c^s may become close to zero, as indicated in the figure, and the concentration profile becomes linear.

The species is supplied from the bulk through diffusion. The concentration of the diffusing species is constant and equal to the bulk concentration c^b at a distance x larger than the diffusion layer thickness δ. An expression for the diffusion flux J can be easily derived from Fick's first law (3.14) when the concentration profile is stationary. This occurs when the concentrations at $x = 0$ and $x = \delta$, as well as δ itself remain constant. The maximum diffusion flux J_i occurs when c^s approaches zero:

$$J_i = \frac{-D_i(c^b - c^s)}{\delta} = \frac{-D_i c^b}{\delta} = \frac{-D_i a^b}{\delta \gamma} \qquad (3.16)$$

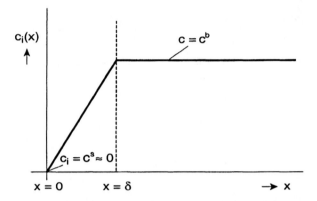

Figure 3.3: Concentration $c_i(x)$ of a diffusing species i during stationary diffusion in the x direction

Because the flux is negative, it will flow from right to left in Figure 3.3. So, the species indeed diffuses from a high to a low concentration.

The concentration profile of a species that takes part in a charge-transfer reaction will give rise to an overpotential η^d in addition to the overpotential η^k. This additional overpotential is given by [19]:

$$\eta^d = \frac{RT}{nF} \ln\left(\frac{c^s}{c^b}\right) \qquad (3.17)$$

When the concentration of a reacting species approaches zero at the electrode surface, the reaction rate becomes limited by diffusion and hence by the speed at which the species is supplied at the electrode surface. Therefore, the current levels off to a maximum value in the current-voltage characteristic of a reaction for which diffusion is important, for example proportional to the maximum flux in (3.16) in the case of stationary diffusion. The current-voltage relation described by (3.11) will then no longer be valid. This is schematically shown in Figure 3.4, where the anodic reaction is assumed to be limited by diffusion. The dashed curves denote currents I_a and I_c according to (3.11). As can be clearly seen in the figure, the anodic current levels off to a diffusion-limited current $I_{d,max}$.

Taking a closer look at Figure 3.4, we see that the overpotential η may approach infinity when the anodic reaction becomes diffusion-limited. In practice, this will not be the case, because different charge-transfer reactions may take place at each electrode. The competition between these charge-transfer reactions is determined by the thermodynamics and kinetics of these reactions. For example, when the rate of an oxidation reaction decreases because of kinetic or diffusion limitations, a different oxidation reaction that involves different species will start to compete with this oxidation reaction. Such competition between charge-transfer reactions is very important in determining the characteristics of aqueous battery systems. This will be described in chapter 4 for NiCd batteries.

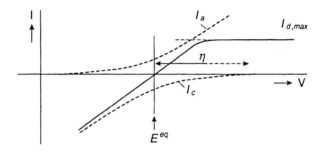

Figure 3.4: Current-voltage relationship for an electrode reaction in which the anodic reaction is diffusion-limited at higher anodic reaction currents

The diffusion problem shown in Figure 3.3 can be easily solved because of its stationary nature. However, it will often be impossible to solve diffusion equations (3.14) and (3.15) analytically, for example because the diffusion layer thickness cannot be considered constant. In such cases, the diffusion equations can be solved numerically by taking the appropriate initial and boundary conditions into account. Examples of ways of solving the diffusion problem for different diffusing species can be found in the next chapter.

3.3.4   Double-layer capacitance

Charge separation will occur at the electrode/electrolyte interface when an electrode is immersed in an electrolyte [3],[19]. The electrical charge present on the electrode surface will attract ions of opposite charge in the electrolyte and will orient the solvent dipoles. This will result in an electrical double-layer, described by the Helmholtz theory in its simplest form. In this theory, the potential changes linearly from the electrode potential φ^s to the electrolyte potential φ^l in a thin layer with thickness δ_H. In its simplest form, δ_H equals the radius of the adsorbed hydrated ions in the electrolyte. This layer is commonly referred to as the Helmholtz layer and can be described by a constant capacitance C_H.

In a more detailed approach, a layer consisting of ions of both positive and negative charges exists in addition to the Helmholtz layer. This layer is denoted as the Gouy-Chapman diffuse layer and forms a space-charge layer with a charge density that decreases gradually into the electrolyte. This layer is wider for more dilute solutions. The Helmholtz and Gouy-Chapman theories are schematically illustrated in Figure 3.5.

The electrostatic potential in the Gouy-Chapman layer changes exponentially with the distance from the electrode/electrolyte interface. As a result, the corresponding capacitance is voltage-dependent, and hence non-linear. This non-linear capacitance C_{G-C} is connected in series with C_H. In addition to the Helmholtz and Gouy-Chapman theories, other more complicated theories concerning the electrical double-layer can be found in the literature [18]. These theories are however beyond the scope of this book.

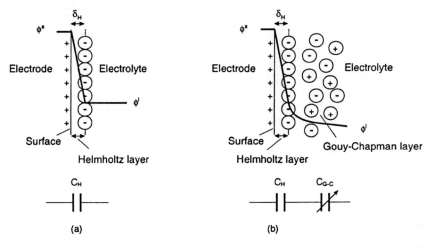

Figure 3.5: Helmholtz model (a) and Gouy-Chapman model (b) of the electrical double-layer at an electrode/electrolyte interface. The electrode is assumed to be positively charged and hence to attract negative ions from the electrolyte in this example. The electrode potential is ϕ^s (s=solid) and the bulk electrolyte potential is ϕ^l (l=liquid)

The charging of the electrical double-layer with a current I_{dl} can be described by:

$$I_{dl} = \frac{d}{dt}(AQ) = Q(E)\frac{dA}{dt} + A\frac{d}{dt}Q(E) \qquad (3.18)$$

where Q denotes the electrical charge stored in the double-layer per unit area in [C/m²], A represents the surface area of the electrode/electrolyte interface in [m²] and E is the electrode potential in [V], which equals ϕ^s - ϕ^l. When, as in the simple Helmholtz theory, the double-layer capacitance is considered to be constant, and hence independent of the electrode potential, and the surface area of the electrode in contact with the electrolyte does not change, (3.18) reduces to the simple equation of charging a capacitor, i.e.:

$$I_{dl} = AC^{dl}\frac{dE}{dt} \qquad (3.19)$$

where C^{dl}, the double-layer capacitance per unit area, is equal to dQ/dE. The double-layer capacitance is almost entirely determined by C_H in the case of concentrated electrolyte solutions commonly encountered in battery systems [2]. The value of the double-layer capacitance is found to be in the order of 0.2 F/m² in the case of a smooth electrode/electrolyte surface [2],[18].

3.3.5   Battery voltage

To summarize the theory presented in this section, the total battery voltage V_{bat} is given by the difference between the electrode potentials E_+ and E_-, i.e.:

$$V_{bat} = E_+ - E_- = \left(E_+^{eq} \pm \eta_+^k \pm \eta_+^d\right) - \left(E_-^{eq} \mp \eta_-^k \mp \eta_-^d\right) \pm I \sum R_\Omega \qquad (3.20)$$

The overpotentials η^k and η^d are positive at the positive electrode and negative at the negative electrode when the battery is charged. This situation is reversed during discharging. As a result of ohmic resistances in the electrodes and the electrolyte (ΣR_Ω), the battery voltage increases with $I\Sigma R_\Omega$ during charging. This ohmic voltage drop is subtracted during discharging. The potential development from the positive to the negative electrode during charging is shown in Figure 3.6.

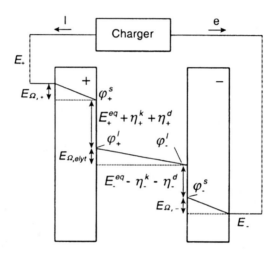

Figure 3.6: Potential development inside a battery during charging; see (3.20)

Figure 3.6 shows an ohmic voltage drop $E_{\Omega+}$ in the positive electrode, $E_{\Omega elyt}$ across the electrolyte and $E_{\Omega-}$ in the negative electrode. The equilibrium and overpotentials exist across the electrode/electrolyte interfaces.

3.4   References

[1]   M Prochaska, C. Fabich, "Stromkonserve, Akkutechnologien: Von Blei bis Lithium-Ion", *ELRAD*, Heft 12, pp. 54-75, 1994
[2]   M. Barak (Ed.), T. Dickinson, U. Falk, J.L. Sudworth, H.R. Thirsk, F.L. Tye, *Electrochemical Power Sources: Primary & Secondary Batteries*, IEE Energy Series 1, A. Wheaton &Co, Exeter, 1980
[3]   D. Linden, *Handbook of Batteries*, Second Edition, McGraw-Hill, New York, 1995
[4]   R.A. Powers, "Batteries for Low Power Electronics", *Proceedings of the IEEE*, vol. 83, no. 4, pp. 687-693, April 1995
[5]   J.R. van Beek, H.C. Donkersloot, J.J.G. Willems, "Rechargeable Hydride Electrodes for Ni-H$_2$ Batteries Based Upon Stable Hydrogen-Storing Materials", *Proc. of the 14th Int.*

Power Sources Symposium (Brighton, United Kingdom, September 1984), chapter 22 in Power Sources, vol. 10, L. Pearce, Ed., pp. 317-338, London, 1985

[6]   P.H.L. Notten, "Rechargeable Nickel-MetalHydride Batteries: a Successful New Concept", *Interstitial Intermetallic Alloys*, chapter 7 in NATO ASI Series E 281, vol. 151, F. Grand-jean, G.J. Long and K.H.J. Buschow, Ed., 1995

[7]   P.H.L. Notten, J.R.G. van Beek, "Nickel-MetalHydride Batteries: from Concept to Characteristics", *Chemical Industry*, vol. 54, pp. 102-115, 2000

[8]   N. Furukawa, "Developers Spur Efforts to Improve NiCd, NiMH Batteries", *JEE*, pp. 46-50, October 1993

[9]   I. Kuribayashi, "Needs for Small Batteries Spur Progress in Lithium-Ion Models", *JEE*, pp. 51-54, October 1993

[10] M.J. Riezenman, "The Search for Better Batteries", *IEEE Spectrum*, pp. 51-56, May 1995

[11] B. Nelson, "High-Power Rechargeable Cells will be Available for Wireless Applications", *Wireless System Design*, pp. 58-60, January 1997

[12] P. Dan, "Recent Advances in Rechargeable Batteries", *Electronic Design*, PIPS Rechargeable Battery Technologies, pp. 112-116, February 3, 1997

[13] V. Biancomano, "Various Chemistries Protect PC Charges", *Electronic Engineering Times*, pp. 107-119, July 14, 1997

[14] G.H. Wrodnigg, B. Evers, I. Schneider, J.O. Besenhard, M. Winter, "Entwicklungstendenzen von Lithium-Batterien", *Begleittexte zum Design & Elektronik Entwicklerforum*, Batterien, Ladekonzepte & Stromversorgungsdesign, pp. 17-26, München, March 31, 1998

[15] R. Müller, "Grundlagen der Neuen Wiederaufladbaren AccuCell-Batterien", *Begleittexte zum Design & Elektronik Entwicklerforum*, Batterien, Ladekonzepte & Stromversorgungsdesign, pp. 59-66, München, March 31, 1998

[16] N. Furukawa, T. Ueda, "Sanyo NiMH batteries Outperform Li-ion", *Nikkei Electronics Asia*, pp. 80-83, May 1996

[17] G. Smith, "Li-ion Units Lead Portable Charge", *Electronic Engineering Times*, pp. 114-120, July 14, 1997

[18] J. O'M. Bockris, D.M. Drazic, *Electro-Chemical Science*, Taylor&Francis Ltd., London, 1972

[19] A.J. Bard, L.R. Faulkner, *Electrochemical Methods: Fundamentals and Applications*, John Wiley&Sons, New York, 1980

[20] P.H.L. Notten, P. Hokkeling, "Double-Phase Hydride Forming Compounds: A New Class of Highly Electrocatalytic Materials", *J. Electrochem. Soc.*, vol. 138, no. 7, pp. 1877-1885, July 1991

Chapter 4
Battery modelling

This chapter describes the development of simulation models of rechargeable batteries. Battery behaviour is a result of a complex interaction between various electrochemical and physical processes. Therefore, simulation models based on the mathematical description of these processes are a useful tool in optimizing the Battery Management System (BMS) used in portable devices. By applying all sorts of external electronic and thermal stimuli to the model during simulations, the designer of a BMS can not only investigate the development of the battery voltage and temperature, but also the course of each of the various reactions that take place inside the battery. This chapter discusses the development of battery models of this type, which have not yet been reported in the literature in this form.

The general approach to modelling rechargeable batteries adopted in this book is based on the theory of physical system dynamics. This theory is based on the analogy between the definition of energy and power in several physical domains, such as the electrical domain, the chemical domain and the thermal domain [1]. The general approach is described in section 4.1. On the basis of this approach, the development of a simulation model of a rechargeable NiCd battery and a rechargeable Li-ion battery is described in sections 4.2 and 4.3, respectively. It is shown that these models are based on a wide variety of parameters. Finding the correct values for all the parameters that occur in the battery models is an important task, because a close quantitative agreement between simulated and measured battery characteristics is essential. The approach adopted for this task and its application to the parameters of the NiCd model is described in section 4.4. Some simulation examples based on the NiCd and Li-ion models are described in section 4.5. Finally, conclusions are drawn in section 4.6.

4.1 General approach to modelling batteries

Modelling rechargeable batteries based on a mathematical description of the processes that occur inside the battery has often been reported in the literature. Apart from electrode models, such as Ni-electrode models [2],[3], complete battery models that are based on the mathematical description of electrochemical and physical processes can be found for SLA [4], NiCd [5],[6], Li-ion [7] and Li-polymer batteries [8]. No use is made of an equivalent network and these models only describe the batteries' voltage characteristics. None of these models considers the battery temperature and internal gas pressure development during operation of the battery and the interaction between these characteristics. In fact, the battery voltage, pressure and temperature are closely linked in a sealed rechargeable battery and their mutual influence is considerable. For example, both the internal partial oxygen pressure and the battery temperature influence the rate of oxygen production and recombination in NiCd batteries. These reactions do not contribute to effective energy storage inside the battery. Therefore, the internal partial oxygen pressure and the battery temperature influence the charging and discharging efficiency.

Besides purely mathematical models, battery models in the form of equivalent electronic networks can also be found in the literature [9]-[12]. Three approaches can be distinguished:

(i)   The equivalent electronic network may consist of linear passive elements, such as resistances, capacitances and inductances, to account for the impedance of a battery [9]. However, these models generally fail to describe the most important battery characteristics due to the complex non-linear behaviour of batteries.

(ii)   The equivalent electronic network may consist of linear passive elements in combination with voltage sources, which describe the battery behaviour by means of look-up tables [10],[11]. In this approach, the simulated battery characteristics can only be realized through interpolation of the tabulated data. Moreover, the accuracy of this type of model depends strongly on the amount and reliability of the tabulated data.

(iii)   An equivalent electronic-circuit model for a NiCd battery based on non-linear passive and active elements configured around a PIN-diode model has been reported [12]. Again, this model does not take into account the influence of the internal partial oxygen pressure and battery temperature on the battery's behaviour. Also, many parameters in this model do not have any electrochemical meaning.

An advantage of these models is that they can be simulated together with surrounding electronic circuits, such as a charger, using a conventional electronic-circuit simulator. This enables a system designer to simulate a battery, together with its surrounding BMS. However, the electronic-network models above have some drawbacks. Therefore, a new approach will be presented in this chapter.

An electrochemical/physical model for rechargeable batteries will be derived first in the approach of battery modelling adopted in this book. This model describes the thermodynamics, charge-transfer kinetics and mass transport limitations of the various reactions. It also describes the physical processes that occur during (over)charging, resting and (over)discharging. All reactions and processes will be described in the form of mathematical equations. Then, the mathematical equations will be clustered for each process and each cluster will be represented by linear and/or non-linear equivalent network elements. These network elements are combined in a single network.

On the basis of the principles of physical system dynamics [1], three domains are distinguished, notably the electrical, chemical and thermal domains. These domains are coupled, which enables the transfer of energy from one domain to another. The electronic-network elements, electrical resistance R_{el} in [Ω] and electrical capacitance C_{el} in [F], form the electrical domain, where voltage in [V] and current in [A] determine the battery behaviour. The (electro)chemical-network elements, chemical resistance R_{ch} in [Js/mol^2] and chemical capacitance C_{ch} in [mol^2/J], form the chemical domain, where battery behaviour is based on chemical potential in [J/mol] and chemical flow in [mol/s]. Key factors in this domain are, for example, diffusion profiles inside an electrode and the amount of oxygen present inside an aqueous battery. This amount can evidently be translated into a partial oxygen gas pressure. Finally, the thermal-network elements, thermal resistance R_{th} in [K/W] and thermal capacitance C_{th} in [J/K], form the thermal domain, where battery behaviour is dictated by temperature in [K] and heat flow in [W]. The battery temperature development is determined in this domain in conjunction with the electrical and chemical domains. A similar approach has also been successfully applied to model transformers, taking into account the electrical and magnetic domains [13].

The advantage of the modelling approach adopted in this book is the possibility of simulating a battery in its electronic and thermal environment. The battery voltage, temperature, internal gas pressure, the state of all the charge-transfer reactions and the mutual interaction between all these variables are taken into account. This as opposed to the electronic-network models found in the literature mentioned above, which only describe electrical battery behaviour. The origin of specific battery behaviour can be examined for example by studying the rates of the various battery reactions, because the models developed in this book rely on electrochemical and physical battery theory. In simple terms, the models act as 'transparent batteries', enabling us to view the course of all relevant processes that take place under various user conditions. An advantage is that all the parameters in the models have an electrochemical meaning and can therefore, in principle, be determined in experiments, obtained from the literature or in some cases estimated within reasonable limits. This section deals with the general principles applied in modelling batteries on the basis of the principles of physical system dynamics. The basic building blocks that can be used to build the equivalent network of a complete battery are described [14]. Examples in which these building blocks will be used to model specific battery systems will be given in later sections for NiCd and Li-ion batteries.

The basic principle of physical system dynamics is that several domains can be distinguished in physics, in which an analogy exists between the definitions of energy and power. An effort and a flow variable can be defined in each domain, the product of these variables expressing power in [W]. Energy can be stored in each domain or exchanged with other domains. The analogy between energy and power definitions in the three relevant domains in batteries, the electrical, chemical and thermal domains, is clarified in Table 4.1. The effort variables of these domains are given in the first row, the flow variables in the second row. The power variables are shown in the third row. The displacement variables in the fourth row of Table 4.1 are defined as the integral of the flow variables over time. The energy, capacitance and resistance variables are listed in the fifth, sixth and seventh row, respectively.

Table 4.1: Analogy between the electrical, chemical and thermal domains

Electrical domain			Chemical domain		
Quantity	*Symbol*	*Unit*	*Quantity*	*Symbol*	*Unit*
Voltage	V	V	Chemical potential	μ	J/ mol
Current	I	A	Chemical flow	J_{ch}	mol/s
Electrical power	$P_{el}=VI$	W	Chemical power	$P_{ch}=\mu J_{ch}$	W
Charge	$Q=It$	C	Molar amount	$m=J_{ch}t$	mol
Electrical energy	$E_{el}=P_{el}t$	J	Chemical energy	$E_{ch}=P_{ch}t$	J
Electrical capacitance	$C_{el}=Q/V$	F	Chemical capacitance	$C_{ch}=m/\mu$	mol^2/J
Electrical resistance	$R_{el}=V/I$	Ω	Chemical resistance	$R_{ch}=\mu/J_{ch}$	Js/ mol^2

Thermal domain		
Quantity	*Symbol*	*Unit*
Temperature	T	K
Heat flow	J_{th}	W
"Thermal power"	$P_{th}=TJ_{th}$	WK
Heat	$Q_{th}=J_{th}t$	J
"Thermal energy"	$E_{th}=P_{th}t$	JK
Thermal capacitance	$C_{th}=Q_{th}/T$	J/K
Heat resistance	$R_{th}=T/J_{th}$	K/W

Note that with the choice of heat as the displacement variable in the thermal domain, the terms "thermal power" and "thermal energy" do not have any physical meanings. The product of effort and flow indeed has the unit [W], as in the other domains, when the entropy S is used as the displacement variable and hence the flow variable becomes the entropy flow rate dS/dt. However, it is common practice to use heat as the displacement variable, as is shown in Table 4.1, because it is generally not practical to use entropy as the displacement variable. The analogy with other domains is in this case more mathematical than physical [1].

4.1.1   Chemical and electrochemical potential

The chemical potential μ_i of an uncharged species i in [J/mol] is defined as [15],[16]:

$$\mu_i = \mu_i^o + RT \ln\left(\frac{a_i}{a_i^{ref}}\right) \tag{4.1}$$

where μ_i^o denotes the standard chemical potential of species i in [J/mol], R is the gas constant (8.314 J/(mol.K)), T stands for temperature in [K], a_i denotes the activity of species i in [mol/m^3] and a_i^{ref} is the activity of species i in the reference state in [mol/m^3].

A similar definition holds for the chemical potential of a charged ionic species i. Then the term electrochemical potential is used, with symbol $\overline{\mu}_i$. An extra term that involves the electrostatic or Galvani potential ϕ [16], valid in the place where the ionic species i is present, is added to the normal chemical potential definition in this case [15],[16].

$$\overline{\mu}_i = \mu_i^o + RT \ln\left(\frac{a_i}{a_i^{ref}}\right) + z_i F\phi = \mu_i + z_i F\phi \tag{4.2}$$

where z_i denotes the valence of ionic species i. For example, $z = +n$ for a positive ion A^{n+} and $z = -m$ for a negative ion B^{m-}. F is Faraday's constant (96485 C/mol) and ϕ is the electrostatic potential in [V].

The activity of a species is linearly proportional to its concentration, and hence to its molar amount, in a certain volume; see chapter 3. On the basis of the analogy of Table 4.1, both the chemical potential of an uncharged species i and the electrochemical potential of a charged ionic species i can be modelled by means of a chemical capacitance, where the 'charge' on the capacitance corresponds to the molar amount m_i of the species and the 'voltage' across the capacitance amounts to the species' chemical potential μ_i. The potential of the bottom plate of the capacitance with respect to ground equals the term $z_i F\phi$ in (4.2) in the case of a charged species. The chemical capacitance models the storage of energy in the chemical domain for both uncharged and charged species. The chemical capacitances that store uncharged and charged species are shown in Figures 4.1(a) and (b), respectively.

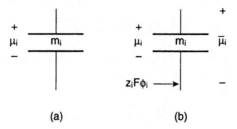

Figure 4.1: Chemical capacitance representing (a) chemical potential and storage of uncharged species i and (b) electrochemical potential and storage for charged ionic species i

An expression for the inverse of both capacitances in Figure 4.1 can be inferred from the derivative of the chemical potential with respect to the molar amount, i.e.

$$
\frac{1}{C_{ch}(m)} = \frac{d\mu}{dm} = \frac{d}{dm}\left(\mu^o + RT \ln\left(\frac{a}{a^{ref}}\right)\right)
$$

$$
= \frac{d}{dm}\left(\mu^o + RT \ln\left(\frac{m}{m^{ref}}\right)\right) = \frac{RT}{m} \tag{4.3}
$$

As can be inferred from (4.3), the capacitances in Figure 4.1 are non-linear and have the unit [mol^2/J], which is in accordance with Table 4.1.

4.1.2 Modelling chemical and electrochemical reactions

In addition to the electrochemical reactions involving the exchange of electrons between electroactive species described in chapter 3, chemical reactions may also take place inside a battery. These reactions involve uncharged species only. Consider, for example, the following chemical reaction:

$$
A + B \underset{k_b}{\overset{k_f}{\rightleftarrows}} C + D \tag{4.4}
$$

where k_f denotes the forward and k_b the backward reaction rate constant, respectively. The unit of both constants depends on the details of the reaction equation.

The reactions in both directions take place at the same rate in a state of equilibrium and the net molar amounts of species A, B, C and D remain unchanged. The sum of the chemical potentials of species A and B equals the sum of the chemical potentials of species C and D in this case. However, one reaction takes place at a higher rate than the other in a non-equilibrium situation and the net result is a chemical flow from the two species with the highest total chemical potential to the species with the lowest total chemical potential. The chemical flow for the chemical reaction in (4.4) in non-equilibrium situations is given by [15]:

$$J_{ch} = k_f a_A a_B - k_b a_C a_D \qquad (4.5)$$

The chemical reaction of (4.4) can be represented by a network model by combining four chemical capacitances, described by (4.3), with a chemical resistance. The 'charges' on the chemical capacitances represent the molar amounts m_i of the four species taking part in the reaction and the chemical resistance represents the chemical flow J_{ch} according to (4.5). This is shown in Figure 4.2.

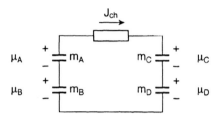

Figure 4.2: Simulation model in the chemical domain of a chemical reaction according to (4.4)

The chemical resistance is defined by μ_R/J_{ch}, where μ_R is the chemical potential across the resistance and J_{ch} is the chemical flow through the resistance, see Table 4.1. Therefore, the chemical resistance in Figure 4.2 depends on k_f, k_b and the molar amounts of all the reacting species, which makes it non-linear. The chemical flow J_{ch} is directed from left to right in Figure 4.2 when the sum of the chemical potentials of species A and B is higher than the sum of the chemical potentials of species C and D. The molar amounts of species A and B will decrease in this case, whereas the molar amounts of species C and D will increase. The chemical flow J_{ch} is directed from right to left in the opposite situation. The relation between k_f and k_b can be inferred from the state of equilibrium, when the activities a_i of all the species are equal to their equilibrium or bulk activities a^b and the chemical flow J_{ch} is zero. Hence,

$$k_f a_A^b a_B^b = k_b a_C^b a_D^b \Rightarrow \frac{k_f}{k_b} = \frac{a_C^b a_D^b}{a_A^b a_B^b} \qquad (4.6)$$

Moreover, the following equation holds for the chemical potentials in equilibrium:

$$\mu_A + \mu_B = \mu_C + \mu_D \Rightarrow \mu_A^o + \mu_B^o - \mu_C^o - \mu_D^o$$

$$= RT \ln\left(\frac{a_C^b a_D^b}{a_A^b a_B^b}\right) - RT \ln\left(\frac{a_C^{ref} a_D^{ref}}{a_A^{ref} a_B^{ref}}\right) \qquad (4.7)$$

and hence:

$$\frac{k_f}{k_b} = \frac{a_C^{ref} a_D^{ref}}{a_A^{ref} a_B^{ref}} \exp\left(\frac{\mu_A^o + \mu_B^o - \mu_C^o - \mu_D^o}{RT}\right) \tag{4.8}$$

The ratio k_f/k_b is constant at a constant temperature. This means that the ratio of the activities of C and D and A and B, as shown in (4.6), remains constant when the reaction is in equilibrium and the temperature remains constant.

A chemical reaction will take place in the chemical domain only. As described in chapter 3, electrochemical reactions enable the conversion of electrical energy into chemical energy and *vice versa*. Hence, interaction between the electrical and chemical domains is realized through an electrochemical reaction. Therefore, the electrical and chemical domains must be coupled in a network model of an electrochemical reaction.

Consider the following simple electrochemical reaction that takes place at the surface of a well-conducting electrode, immersed in an electrolyte solution:

$$Ox^{n+} + ne^- \underset{k_a}{\overset{k_c}{\rightleftarrows}} Red \tag{4.9}$$

Consider, for example, that the Ox species is present in the electrolyte, while the Red species is present in the electrode. An example is an iron electrode (Fe) that is immersed in an electrolyte containing Fe^{2+} ions. Consider a net reduction reaction (R), with the reaction directed to the right prevailing. Electrons are consumed by the reaction in this case, which results in a cathodic electrode current. Ions Ox^{n+} from the electrolyte transform into Red atoms in the electrode by taking up n electrons from the electrode. Adopting the sign convention proposed by Vetter [16], the change in free energy ΔG_R for the reduction reaction R equals $-nFE^{eq}$ in this case, with E^{eq} being the equilibrium potential of the electrode. By definition, the value of ΔG can be inferred from $\Sigma v_i \mu_i$ for both the reduction reaction R and the oxidation reaction O, with v_i being the stoichiometric factors for species i in the reaction equation [16]. By definition, these factors v_i are positive for reaction products and negative for reactants. The Red species is the reaction product and Ox^{n+} and electrons are the reactants in the case of a reduction reaction. ΔG depends only on chemical potentials and not on electrochemical potentials.

It also holds that $\Sigma v_i \overline{\mu}_i = 0$ for any electrochemical reaction in a state of equilibrium [16]. In the literature, the summation $\Sigma v_i \overline{\mu}_i$ is defined as $\Delta \overline{G}$, which is the change in free electrochemical energy [15],[16]. The electrochemical potentials and hence also the electrostatic potentials are taken into account in this term. In a state of equilibrium, $k_a = k_c$ and an expression for the equilibrium potential can be found by considering either the reduction reaction or the oxidation reaction. In the case of the change in free electrochemical energy $\Delta \overline{G}_R$ (in [J/mol]) for the reduction reaction of (4.9) this yields:

$$\Delta \overline{G}_R = \overline{\mu}_{Red} - \left(\overline{\mu}_{Ox} + n\overline{\mu}_e \right)$$

$$= \mu_{Red} - \left(\mu_{Ox} + z_{Ox} F \phi_{Ox} + n\mu_e + n z_e F \phi_e \right) \qquad (4.10a)$$

$$= \mu_{Red} - \left(\mu_{Ox} + nF \phi^l + n\mu_e - nF \phi^s \right) = 0$$

where $\phi^s = \phi_e$ is the electrostatic potential of the electrode in [V], i.e. the electrostatic potential valid for the electrons, $\phi^l = \phi_{Ox}$ is the electrostatic potential of the electrolyte in [V], i.e. the electrostatic potential valid for the Ox^{n+} ions, $z_{Ox} = +n$ is the valence of the Ox^{n+} ions and $z_e = -1$ is the valence of the electrons. The change in free energy ΔG_R (in [J/mol]) can now be inferred from:

$$\Delta G_R = \mu_{Red} - \mu_{Ox} - n\mu_e$$

$$= \mu_{Red}^o + RT \ln \left(\frac{a_{Red}}{a_{Red}^{ref}} \right)$$

$$- \left(\mu_{Ox}^o + RT \ln \left(\frac{a_{Ox}}{a_{Ox}^{ref}} \right) + n\mu_e^o + nRT \ln \left(\frac{a_e}{a_e^{ref}} \right) \right) \qquad (4.10b)$$

$$= \left(\mu_{Red}^o - \mu_{Ox}^o - n\mu_e^o \right) + RT \ln \left(\frac{a_{Red} a_{Ox}^{ref}}{a_{Red}^{ref} a_{Ox}} \right)$$

$$= nF \left(\phi^l - \phi^s \right)$$

where (4.10a) was used in the last step. The activity of electrons has been assumed equal to a_e^{ref} in (4.10b), because the activity of electrons does not change noticeably in an electrode with a very good electronic conductivity [15]. An expression E^{eq} can now be found from (4.10b):

$$E^{eq} = \frac{-\Delta G_R}{nF} = \frac{\mu_{Ox}^o + n\mu_e^o - \mu_{Red}^o}{nF} + \frac{RT}{nF} \ln \left(\frac{a_{Red}^{ref} a_{Ox}}{a_{Red} a_{Ox}^{ref}} \right)$$

$$= E^o + \frac{RT}{nF} \ln \left(\frac{a_{Red}^{ref} a_{Ox}}{a_{Red} a_{Ox}^{ref}} \right) = \phi^s - \phi^l = E_{electrode} \qquad (4.10c)$$

(4.10c) shows that the electrode is indeed in equilibrium, because the electrode potential $E_{electrode}$ equals the equilibrium potential E^{eq}. The expression for E^{eq} is in agreement with Nernst's equation given in chapter 3. In addition, it can be inferred from (4.10c) that:

$$\Delta G_R^o = \mu_{Red}^o - \mu_{Ox}^o - n\mu_e^o \qquad (4.11)$$

The signs of ΔG and $\Delta \overline{G}$ reverse for the oxidation reaction of (4.9), in which case the reaction directed to the left prevails, which means that the changes in free energy and free electrochemical energy become $\Delta G_O = \mu_{Ox} + n\mu_e - \mu_{Red} = -\Delta G_R$ and $\Delta \overline{G}_O = \overline{\mu}_{Ox} + n\overline{\mu}_e - \overline{\mu}_{Red} = -\Delta \overline{G}_R$. In that case, ΔG_O equals nFE^{eq}, because the sign of E^{eq} is fixed. Hence, the same expression for E^{eq} in (4.10c) is found when the oxidation reaction is considered. The sign convention of ΔG may lead to confusion when reading the literature. The sign convention proposed by Vetter and already used above will be adopted throughout this book [16].

The chemical flow associated with the electrochemical reaction in (4.9) is given by [15]:

$$
J_{ch} = A(k_a)^{1-\alpha}(k_c)^{\alpha}(a_{Ox}^b)^{\alpha}(a_{Red}^b)^{(1-\alpha)y} \cdot \left\{ \exp\left(\frac{-\alpha\Delta\overline{G}_O}{RT}\right) - \exp\left(\frac{-(1-\alpha)\Delta\overline{G}_R}{RT}\right) \right\}
$$

(4.12)

where A denotes the electrode surface area in [m^2] and k_a and k_c denote the reaction rate constants for the anodic (O) and cathodic (R) reactions, respectively. Their unit will depend on the reaction equation. α denotes the transfer coefficient, which is dimensionless, and $0<\alpha<1$; x and y reflect the reaction orders. It will be shown later in this section that (4.12) is in fact the Butler-Volmer equation, expressed in the chemical domain. Note that when $k_c>k_a$, the absolute value of $\Delta\overline{G}_R$ is larger than that of $\Delta\overline{G}_O$. When $k_a>k_c$, the reverse situation occurs and the direction of J_{ch} reverses. Neither the change in free electrochemical energy $\Delta\overline{G}$ nor the chemical flow J_{ch} can be measured directly. However, they can be derived from the electrode potential and current, respectively.

The coupling between the electrical and chemical domains can be modelled by means of an ideal transformer that couples the electrical effort variable *voltage V* to the chemical effort variable *chemical potential μ* and the electrical flow variable *current I* to the chemical flow variable *chemical flow J$_{ch}$*; see Table 4.1. The transformer coefficient is $1/nF$, where n is the number of electrons involved in the electrochemical reaction that couples the two domains.

$$
V = \frac{\mu}{nF}
$$

(4.13)

$$
I = nFJ_{ch}
$$

The transformer is assumed to be ideal, which means that no energy is stored in the transformer. This means that $VI=\mu J_{ch}$. The ideal transformer is a purely mathematical component used to link different physical domains [1]. The network model for an electrochemical reaction can be found using the ideal transformer. This is shown in Figure 4.3. The direction of the current $I_{electrode}$ and chemical flow J_{ch}, as well as the sign of η^k and the relation between ΔG_O and E^{eq} are shown for $k_a>k_c$ in (4.9). This means that the oxidation reaction is assumed to prevail.

The electrical and chemical domains separated by an ideal transformer can be clearly recognized in Figure 4.3. The coupling between the equilibrium potential E^{eq}

in the electrical domain and the change in free energy ΔG_O in the chemical domain through the transformer coefficient $1/nF$ is indicated in the figure. Moreover, the coupling between the electrode current I and the chemical flow J_{ch} is indicated. The direction of $I_{electrode}$ and chemical flow J_{ch} and the signs of ΔG and η^k are reversed for the reduction reaction.

The electrode and equilibrium potentials are denoted by the symbol E in the electrical domain, the kinetic overpotential by the symbol η^k and the electrostatic potentials of the electrode and electrolyte by ϕ^s and ϕ^l, respectively. The electrode/electrolyte interface is represented by a linear capacitance C^{dl}, which denotes the double-layer capacitance. The electrode potential $E_{electrode} = \phi^s - \phi^l$ occurs across this double-layer capacitance, as described in chapter 3. The anti-parallel diodes O and R model the exponential relation between the electrode current I and the kinetic overpotential η^k according to the Butler-Volmer equation, while the electrical port of the ideal transformer models the equilibrium electrode potential E^{eq} as defined in (4.10c).

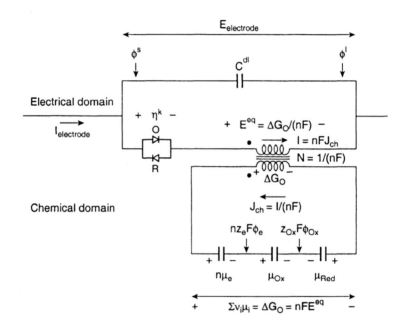

Figure 4.3: Simulation model in the electrical and chemical domains for an electrochemical reaction according to (4.9). The definition of E^{eq}, the signs of ΔG and η^k and the direction of the current and chemical flow hold for an oxidation reaction (O)

Each capacitance in the series connection in the chemical domain corresponds to a species that takes part in the charge-transfer reaction; see (4.9). The polarity of the capacitances depends on whether the species appear on the left side or the right side of (4.9) and the chemical potential of the electrons is multiplied by n, in accordance with (4.10a). The chemical flow J_{ch} is given by (4.12). In accordance with Figure 4.1, the potential at the bottom plate of the capacitance that stores the molar amount of electrons is $nz_eF\phi_e$ and the potential at the bottom plate of the capacitance that stores the molar amount of oxidized species is $z_{Ox}F\phi_{Ox}$. From Figure 4.3, the change in free energy ΔG_O at the chemical port of the ideal transformer is inferred to be:

$$\Delta G_O = -\Delta G_R = n\mu_e + \mu_{Ox} - \mu_{Red}$$

$$\Rightarrow E^{eq} = \frac{\Delta G_O}{nF} = \frac{-\Delta G_R}{nF} \tag{4.14}$$

This means that the network shown in Figure 4.3 is in accordance with the sign methodology discussed earlier for both the oxidation and the reduction reaction. An expression for η^k in terms of $\Delta\overline{G}_O$ can be inferred from Figure 4.3:

$$\eta^k = E_{electrode} - E^{eq} = \phi^s - \phi^l - \frac{\Delta G_O}{nF}$$

$$= \phi^s - \phi^l - \frac{1}{nF}\left(\Delta\overline{G}_O - nF\left(\phi^l - \phi^s\right)\right) \tag{4.15}$$

$$= -\frac{\Delta\overline{G}_O}{nF}$$

In a similar way, it can be found for the reduction reaction that $\eta^k = \dfrac{\Delta\overline{G}_R}{nF}$.

(4.12) can be easily transferred to the electrical domain by realizing that $I = nFJ_{ch}$ and using (4.15). The resulting equation links I and η^k and is in accordance with the Butler-Volmer equation defined in chapter 3. The analogy between the various parameters in the electrical and chemical domains is summarized in Table 4.2.

Instead of modelling an electrochemical reaction in the chemical and electrical domains, it is also possible to represent all the elements of the model in the electrical domain only. This can be accomplished by transferring the model equations from the chemical domain to the electrical domain. An example was given above for (4.12). Chemical capacitances can be represented by electrical capacitances with a voltage μ/nF across them and an electrical charge $q = nFm$ on them. The following equation is then obtained for the voltage across each capacitance:

$$V = \frac{\mu}{nF} = \frac{\mu^o}{nF} + \frac{RT}{nF}\ln\left(\frac{a}{a^{ref}}\right)$$

$$= V^o + \frac{RT}{nF}\ln\left(\frac{m}{m^{ref}}\right) \tag{4.16}$$

$$= V^o + \frac{RT}{nF}\ln\left(\frac{q}{q^{ref}}\right)$$

where γ has again been assumed to be unity.

Table 4.2: Analogy between variables in the electrical and chemical domains shown in Figure 4.3

Electrical domain			Chemical domain			Relation
Description	*Symbol*	*Unit*	*Description*	*Symbol[1]*	*Unit*	
Current	I	A	Chemical flow	J_{ch}	mol/s	$I = nFJ_{ch}$
Equilibrium potential	E^{eq}	V	Change in free energy	$\Delta G_j = \sum_i v_i \mu_i$	J/mol	$E^{eq} = \dfrac{\Delta G_O}{nF}$ $= \dfrac{-\Delta G_R}{nF}$
Standard redox Potential	E^o	V	Change in standard free energy	$\Delta G_j^o = \sum_i v_i \mu_i^o$	J/mol	$E^o = \dfrac{\Delta G_O^o}{nF}$ $= \dfrac{-\Delta G_R^o}{nF}$
Kinetic over-potential	η^k	V	Change in free electro-chemical energy	$\Delta \overline{G}_j = \sum_i v_i \overline{\mu}_i$	J/mol	$\eta^k = \dfrac{-\Delta \overline{G}_O}{nF}$ $= \dfrac{\Delta \overline{G}_R}{nF}$

Note 1: $j = O$ or R, i is species in reaction equation

The same result as that obtained in transferring equations from the chemical to the electrical domain is obtained when using the simulation model presented in Figure 4.3 when all circuit elements in the chemical domain are "shifted" through the ideal transformer to the electrical domain. Series and parallel connections of elements in the chemical domain remain series and parallel connections in the electrical domain, respectively. The impedances of circuit elements in the chemical domain have to be multiplied by N^2, with $N=1/nF$, as can be understood by dividing the top by the bottom equation in (4.13). Hence, it can be found for the electrical capacitances, using (4.3) and $q=nFm$:

$$\frac{1}{C_{el}(q)} = N^2 \cdot \frac{1}{C_{ch}(m)} = \frac{RT}{(nF)^2 m} = \frac{RT}{nFq} \tag{4.17}$$

where $C_{ch}(m)$ is inferred from (4.3). (4.17) can also be found by calculating dV/dq from (4.16). The representation of the simulation model of Figure 4.3 in the electrical domain only is shown in Figure 4.4.

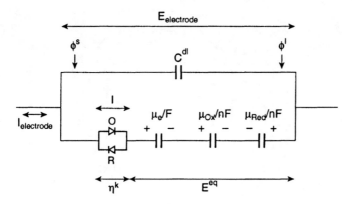

Figure 4.4: Simulation model of Figure 4.3 in the electrical domain only

Figure 4.4 illustrates that the transformer has now vanished from the model.

4.1.3 Modelling mass transport

Mass transport to and from the electrode surfaces inside a battery can influence the rate at which the electrochemical reactions take place, as was described in chapter 3. In electrochemistry, it is quite common to model mass transport of species by means of electronic-network models. For example, the low-frequency behaviour of the battery impedance is described by the Warburg impedance in most textbooks that deal with basic electrochemistry [17],[18]. As early as 1900, Warburg derived an equation for the impedance of diffusion-controlled processes at a planar electrode. The Warburg impedance is basically an RC-ladder network or transmission line, which describes the diffusion of species to and from the electrode surface. An expression for the Warburg impedance can be found under certain limiting conditions, using expressions for the non-linear capacitances and resistances in the RC-ladder network [19].

An example of an approach to modelling ionic mass transport under the influence of joint diffusion and migration using equivalent network models can be found in the literature [20]. However, this approach is not based on the principles of physical system dynamics. Therefore, a different approach is presented in this section, which is based on the theory of physical system dynamics. It will be shown that the expressions derived for the resistances and capacitances in the RC-ladder network, which describes the one-dimensional diffusion of species, yield the same RC time as the RC time calculated in [20]. The approach adopted in the derivation is the same as that employed in arriving at a numerical solution to the problem. Diffusion in one dimension is considered and the system is divided into spatial elements of the thickness Δx. Unlike in efforts aimed at arriving at a mathematical numerical solution to the problem, it is not necessary to use discrete time steps because time discretization has already been performed by a circuit simulator during the simulation of the network model. The division into spatial elements of a chemical system in which mass transport takes place in the x-direction only is clarified in Figure 4.5.

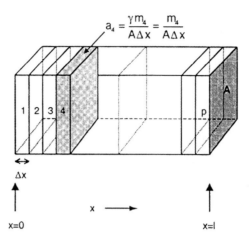

Figure 4.5: Division of a chemical system into p spatial elements with a thickness Δx and a volume $A\Delta x$ between $x=0$ and $x=l$

The activity a_4 of the species in element 4 is shown as an example in Figure 4.5, where the activity coefficient γ has again been assumed to be unity for simplicity.

Mass transport law
The driving force behind the movement of a species is a gradient in the electrochemical potential. Mass transport by convection will not be considered in this book. The direction of the resulting chemical flow is from a region in which the electrochemical potential is high to a region in which the electrochemical potential is low. For an uncharged species, an expression for the chemical flow $J_{ch}(x)$ at position x can be inferred from [15]:

$$
\begin{aligned}
J_{ch}(x) &= -\frac{ADc(x)}{RT}\frac{\partial \bar{\mu}}{\partial x} = -\frac{ADc(x)}{RT}\frac{\partial \mu}{\partial x} \\
&= -\frac{ADc(x)}{RT}\frac{\partial}{\partial x}\left(\mu^o + RT \ln\left(\frac{c(x)}{c^{ref}}\right)\right) \\
&= -\frac{ADc(x)}{RT}\left(\frac{RT}{c(x)}\frac{\partial c(x)}{\partial x}\right) = -AD\frac{\partial c(x)}{\partial x} \\
&= \frac{-\Delta\mu}{R_{ch}(x)}
\end{aligned}
\tag{4.18}
$$

where $J_{ch}(x)$ denotes the chemical flow of the transported species at position x in [mol/s], as defined in Table 4.1, $R_{ch}(x)$ is the associated chemical resistance at position x in [Js/mol^2], see Table 4.1, $c(x)$ is the concentration of the uncharged species at location x in [mol/m^3], c^{ref} is the reference concentration in [mol/m^3] and D denotes the diffusion coefficient of the transported species in [m^2/s]. (4.18) shows that a chemical potential difference $\Delta\mu$ across a chemical resistance R_{ch} yields a chemical flow J_{ch}. This is in fact Fick's first law, which was introduced in chapter 3. For a charged species, (4.18) becomes:

$$J_{ch}(x) = -\frac{ADc(x)}{RT}\frac{\partial\bar{\mu}}{\partial x} = -\frac{ADc(x)}{RT}\left(\frac{RT}{c(x)}\frac{\partial c(x)}{\partial x} + zF\frac{\partial\phi(x)}{\partial x}\right)$$

$$= -AD\frac{\partial c(x)}{\partial x} - \frac{zF}{RT}ADc(x)\frac{\partial\phi(x)}{\partial x} = \frac{-\Delta\bar{\mu}}{R_{ch}(x)} \qquad (4.19)$$

(4.19) is known as the Nernst-Planck equation [15]. It expresses the mass transport of a charged species by combined diffusion and migration. Using (4.2), the term $\Delta\bar{\mu}$ can be rewritten as:

$$\Delta\bar{\mu} = \frac{d\bar{\mu}}{dx}\cdot\Delta x = \left(RT\frac{1}{a(x)}\frac{da(x)}{dx} + zF\frac{d\phi(x)}{dx}\right)\Delta x \qquad (4.20)$$

A similar expression can be found for $\Delta\mu$ using (4.1). In that case, the term that contains ϕ is zero. Using (4.20) for $\Delta\bar{\mu}$ or a similar equation for $\Delta\mu$, both (4.18) and (4.19) yield the same expression for $R_{ch}(x)$ at position x. It should be realized that chemical resistance is defined as a positive variable, which means that a positive chemical flow is directed from the terminal with the highest electrochemical potential to the terminal with the lowest electrochemical potential. Suppose a species flows from x_1 to x_2, where $x_2 > x_1$ and $\bar{\mu}_1 > \bar{\mu}_2$. The gradient in electrochemical potential from x_1 to x_2 is negative, whereas the associated chemical flow is positive. By definition, $\Delta\mu$ and $\Delta\bar{\mu}$ are negative in this case.

$$-\Delta\bar{\mu} = R_{ch}(x)J_{ch}(x)$$

$$= -R_{ch}(x)\left(AD\frac{\partial c(x)}{\partial x} + \frac{zF}{RT}ADc(x)\frac{\partial\phi(x)}{\partial x}\right)$$

$$= -\left(RT\frac{1}{a(x)}\frac{da(x)}{dx} + zF\frac{d\phi(x)}{dx}\right)\Delta x \qquad (4.21a)$$

in which (4.19) was used in the second line with consideration of the sign and (4.20) was used in the third line. The activity a will now be introduced in the second line, with again $\gamma=1$, while the third line will be re-arranged such that a similar term appears as in the second line. Realizing that the second line equals the third line in (4.21a), the following expression is then obtained:

$$R_{ch}(x)\left(AD\frac{\partial a(x)}{\partial x} + \frac{zF}{RT}ADa(x)\frac{\partial\phi(x)}{\partial x} \right)$$
$$= \left(AD\frac{\partial a(x)}{\partial x} + \frac{zF}{RT}ADa(x)\frac{\partial\phi(x)}{\partial x} \right)\frac{RT\Delta x}{ADa(x)}$$

(4.21b)

Now, an expression for $R_{ch}(x)$ can be inferred from (4.21b):

$$R_{ch}(x) = \frac{RT\Delta x}{ADa(x)} = \frac{RT\Delta x}{AD\left(\dfrac{m(x)}{A\Delta x}\right)} = \frac{RT\Delta x^2}{m(x)D}$$

(4.21c)

Mass transport at position x through a chemical resistance $R_{ch}(x)$ described by (4.21c) is indicated in Figure 4.6.

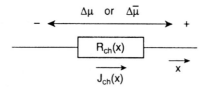

Figure 4.6: Representation of mass transport at position x by a chemical resistance $R_{ch}(x)$, using the analogy of Table 4.1

Conservation of mass
The conservation of mass is expressed by the following equation:

$$\frac{dm}{dt} = \Delta J_{ch}$$

(4.22)

In Figure 4.5, conservation of mass means that the change in time of the molar amount m_i in a spatial element i, between locations $(i-1)\Delta x$ and $i\Delta x$, is determined by the net chemical flow ΔJ_{ch}^{i}. This flow is the difference between the chemical flow at location $x=(i-1)\Delta x$ and that at location $x=i\Delta x$. Using the analogy between the electrical and chemical domain of Table 4.1, the conservation of mass can be described by a chemical capacitance C_{ch}^{i} for each spatial element i, as shown in Figure 4.1, where the amount m_i denotes the molar amount in spatial element i. This is visualized in Figure 4.7. An expression for the chemical capacitance C_{ch}^{i} was derived in (4.3).

Initial and boundary conditions
Initial and boundary conditions have to be defined for each mass transport problem to enable its solution. For example, an initial condition could be that the activity, and hence the molar amount under the assumption that $\gamma=1$, of the transported species is uniformly distributed throughout the system and is the same for all x at $t=0$.

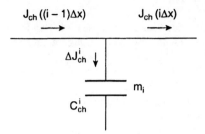

Figure 4.7: Representation of mass conservation law in spatial element *i* (see Figure 4.5) by means of a chemical capacitance

Two surface boundaries occur for the chemical flow in a chemical system in which one-dimensional mass transport is considered, for example at $x=0$ and $x=l$ in Figure 4.5. The transported species can either be able to cross the boundary to a neighbouring system or be restricted to the chemical system through which it is transported. In the first case, the chemical flow across the boundary usually will depend on the concentration gradient across the boundary, whereas in the second case the flow across the boundary will be zero.

Mass transport in the absence of an electrical field

Transport by migration will not occur when no electrical field is present. Fick's law can then be used for mass transport; see (4.18). Combining Fick's law, the condition of mass conservation of (4.22) and initial and boundary conditions, one-dimensional mass transport of a species in the x-direction in the absence of an electrical field can now be described with the aid of an RC-ladder network. This is shown in Figure 4.8 for the considered chemical system of Figure 4.5.

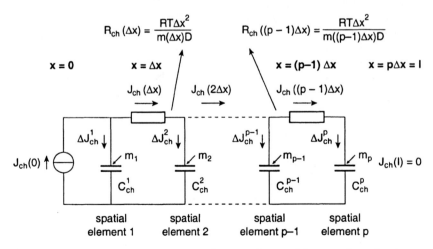

Figure 4.8: Simulation model in the chemical domain describing one-dimensional mass transport of a species in the x-direction in the absence of an electrical field

The molar amounts m_i in spatial elements i are present in chemical capacitances $C_{ch}{}^i$ in Figure 4.8. A chemical potential, described by (4.1), is present across each chemical capacitance. A difference in chemical potentials between two neighbouring spatial elements will give rise to the transport of the species by pure diffusion from one chemical capacitance to its neighbour through a chemical resistance R_{ch}. With R_{ch}, defined in (4.21c), and C_{ch}, defined in (4.3), an RC time of $\Delta x^2/D$ is found for each spatial element. This is indeed in agreement with the derivations of Horno et al., where $R=\Delta x/D$ and $C=\Delta x$ [20].

As an example, the boundary conditions have been taken so that the species is injected from a neighbouring system in the form of a chemical flow $J_{ch}(0)$ at the surface at $x=0$. The species is not allowed to cross the boundary at $x=l$. The system's initial condition can be enforced by the initial values of m_i at $t=0$.

Consider again the electrochemical reaction given by (4.9). When both the oxidized and the reduced species can diffuse through the system and no electrical field is present, the network model of Figure 4.3 changes into the network model of Figure 4.9. As in Figure 4.3, the directions of currents and flows as well as the polarities of ΔG and η^k are shown for an oxidation reaction. The directions and polarities reverse for a reduction reaction. An RC-ladder network, as shown in Figure 4.8, has been added in the chemical domain in Figure 4.9. The boundary at $x=0$ is the electrode/electrolyte interface at which the electrochemical reaction takes place, while the boundary at $x=l$ is the electrode/current-collector interface. The molar amounts on the capacitances in the RC-ladder network represent the molar amounts of the oxidized and reduced species throughout the electrode, respectively. Spatial element 1 starts at $x=0$, spatial element p ends at $x=l$. A difference between the molar amounts in spatial element 1 and those inside the electrode will give rise to mass transport by diffusion through the electrode. The chemical resistances $R_{ch}(x)$ are given by (4.21c), in which D_{Ox} is used for the oxidized species and D_{Red} for the reduced species. Neither the oxidized nor reduced species can cross the boundary at $x=l$. At the boundary at $x=0$, the species are allowed to cross the boundary by means of the electrochemical reaction. The bulk activity of electrons is assumed to be equal to the surface activity. Therefore, no ladder network is used for electrons.

Figure 4.9 shows that the chemical potentials across the capacitances that represent spatial element 1 at $x=0$ are denoted as surface chemical potentials μ^s. As a result, the potential across the transformer winding in the electrical domain is now obtained from the following equation. Again, the activity of electrons is assumed to be equal to the reference activity $a_e{}^{ref}$ as in (4.10b). In addition, the chemical potential of electrons is assumed uniform throughout the electrode, which means that $\mu_e^s = \mu_e$.

$$
\begin{aligned}
E^{eq*} &= \frac{\Delta G_O}{nF} = \frac{n\mu_e + \mu_{Ox}^s - \mu_{Red}^s}{nF} \\
&= \frac{\Delta G_O^o}{nF} + \frac{RT}{nF}\ln\frac{a_{Ox}^s a_{Red}^{ref}}{a_{Red}^s a_{Ox}^{ref}}
\end{aligned}
\tag{4.23}
$$

The equilibrium potential now depends on the surface activities a^s of the oxidized and reduced species, according to (4.23). This potential can be denoted as the apparent equilibrium potential E^{eq*}. The difference between E^{eq*} and the true equilibrium potential E^{eq}, which was defined in (4.10c), is the fact that E^{eq} is related to the bulk activities a^b of the reacting species. Bulk activities are the same as surface activities in a state of true equilibrium. The difference between E^{eq*} and E^{eq} is the diffusion overpotential η^d, which can be found by rewriting (4.23).

$$
\begin{aligned}
E^{eq*} &= \frac{\Delta G_O^o}{nF} + \frac{RT}{nF} \ln \frac{a_{Ox}^s a_{Red}^{ref} a_{Ox}^b a_{Red}^b}{a_{Red}^s a_{Ox}^{ref} a_{Ox}^b a_{Red}^b} \\
&= \frac{\Delta G_O^o}{nF} + \frac{RT}{nF} \ln \frac{a_{Ox}^b a_{Red}^{ref}}{a_{Red}^b a_{Ox}^{ref}} + \frac{RT}{nF} \ln \frac{a_{Ox}^s a_{Red}^b}{a_{Ox}^b a_{Red}^s} \\
&= E^{eq} + \frac{RT}{nF} \ln \frac{a_{Ox}^s a_{Red}^b}{a_{Ox}^b a_{Red}^s} = E^{eq} + \eta^d
\end{aligned}
\tag{4.24}
$$

The expression for the diffusion potential is in agreement with the general expression for a diffusion overpotential given in chapter 3; see (3.17). E^{eq} equals E^{eq*} and η^d is zero, when a^b equals a^s for the Ox and Red species.

It is not possible to discern between E^{eq} and η^d when the simulation model is shown in the form of Figure 4.9, because both variables are present in E^{eq*}. It is sometimes convenient to make this distinction. It becomes possible to distinguish between E^{eq} and η^d when the definition of the chemical capacitances is altered. This is shown in Figure 4.10, in which the model has also been represented in the electrical domain as additional illustration. As was discussed in section 4.1.2, the representation of the complete network in the electrical domain can be achieved by "shifting" the elements in the chemical domain through the ideal transformer to the electrical domain.

The electrical capacitance C_{bulk} represents the series connection of three capacitances which store the bulk molar amounts of all species that take part in the reaction. Only one capacitance has been drawn for simplicity. The voltage across this capacitance equals E^{eq}, as defined in (4.10c). By definition, the voltage across each capacitance in the RC-ladder network is proportional to the difference between the chemical potential in the applicable spatial element and the bulk chemical potential μ^b. This bulk chemical potential is the same for all spatial elements. The chemical potential will be μ^b in all the spatial elements at true equilibrium. As a result of this definition, the capacitances describe the amount of material in spatial element i relative to the bulk amount of material. The following equation illustrates this for the voltage across the surface capacitances C_j, where j can be Ox or Red.

$$
V_j = \frac{1}{nF}(\mu_j^s - \mu_j^b) = \frac{RT}{nF} \ln \frac{a_j^s}{a_j^b} = \frac{RT}{nF} \ln \frac{m_j^s}{m_j^b}
\tag{4.25}
$$

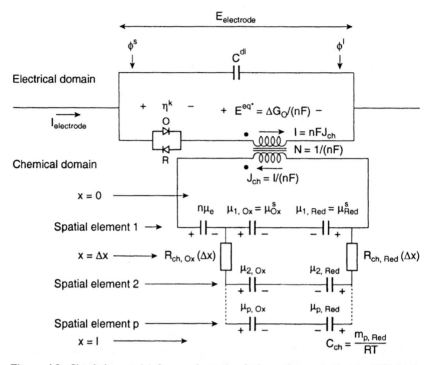

Figure 4.9: Simulation model for an electrochemical reaction according to (4.9) in the electrical and chemical domains, where mass transport of the Ox and Red species takes place in the electrode in the absence of an electrical field. The definition of E^{eq}, the signs of ΔG and η^k and the directions of the current and chemical flow hold for an oxidation reaction

It can be easily verified that the diffusion overpotential defined by (4.24) is present across these two surface capacitances:

$$\eta^d = \frac{\mu_{Ox}^s - \mu_{Ox}^b}{nF} - \frac{\mu_{Red}^s - \mu_{Red}^b}{nF} = \frac{RT}{nF} \ln \frac{a_{Ox}^s a_{Red}^b}{a_{Ox}^b a_{Red}^s} \tag{4.26}$$

For the Ox and Red species, $\mu = \mu^b$ in all the spatial elements in a state of equilibrium. Therefore, the voltage across all capacitances in the RC-ladder network is zero and η^d is zero in a state of equilibrium. The value of each resistance R_{ch} has been multiplied by $N^2 = (1/nF)^2$ because of the transition from the chemical to the electrical domain. Likewise, the value of each capacitance C_{ch} has been divided by N^2 in the electrical domain.

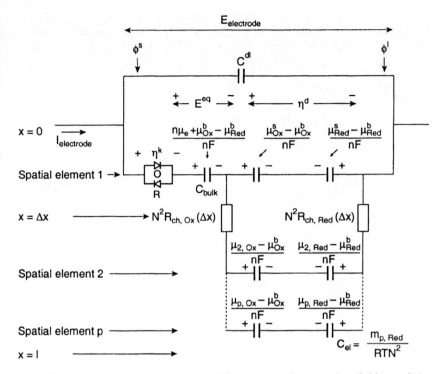

Figure 4.10: Adapted simulation model of Figure 4.9 with altered definitions of the capacitances to allow distinction between the true equilibrium potential E^{eq} and the diffusion overpotential η^d. The model is represented entirely in the electrical domain

Mass transport in the presence of an electrical field
Mass transport of a charged species will take place by combined diffusion and migration when an electrical field is present. An expression for the electrical field is needed to describe this mass transport, which can be found by solving the Poisson equation. Assume that the transported species are positive and negative ions with equal and opposite charges, for example ions of a fully dissociated binary salt $A^{n+}B^{n-}$. In a spatial element with volume $A\Delta x$, the Poisson equation reads [15]:

$$\frac{d^2\phi}{dx^2} = -\frac{F}{\varepsilon_r\varepsilon_o}\sum_i z_i a_i = -\frac{nF}{\varepsilon_r\varepsilon_o A\Delta x}(m_+ - m_-) \qquad (4.27)$$

where ε_o is the electrical permittivity in free space ($8.85.10^{-12}$ C^2/(N.m^2)), ε_r is the dielectric constant of the medium through which the ions are transported, z_i denotes the valence of the transported species, $z=+n$ for the A^{n+} ions and $z=-n$ for the B^{n-} ions. m_+ and m_- denote the molar amounts of A^{n+} and B^{n-} ions in volume $A\Delta x$, respectively.

The basic building block for modelling the Poisson equation for a fully dissociated binary salt $A^{n+}B^{n-}$ is shown in Figure 4.11 for both the chemical and the electrical domain. The model describes a spatial element with a volume $A\Delta x$, as introduced in Figure 4.5, and its connection to the neighbouring spatial elements. It consists of four capacitances connected together in a central node point. The capacitances C^ε to the left and right of the central-node point denote geometric parallel-plate capacitances with a surface area A and a thickness Δx. The electrostatic potential ϕ_c is valid at the centre of the spatial element, whereas ϕ_l is valid at the centre of the neighbouring spatial element to the left and ϕ_r is valid at the centre of the neighbouring spatial element to the right. The geometric capacitances denote the capacitances between the centre points of the spatial elements, because the distance between the centres of the spatial elements is Δx.

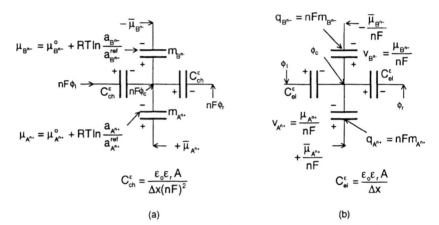

(a)   (b)

Figure 4.11: Simulation model of the Poisson equation in (a) the chemical domain and (b) the electrical domain

The ratio of the capacitances in the chemical and electrical domains is N^2, as discussed before. The 'charge' on the capacitance above the central-node point in Figure 4.11a denotes the molar amount of negative ions B^{n-}, while the 'charge' on the capacitance below the central-node point represents the molar amount of positive ions A^{n+}. The unit of these capacitances is [mol^2/J], whereas it is [F] in Figure 4.11b. It can be seen that the networks illustrated in Figure 4.11 indeed describe the Poisson equation of (4.27). The mass conservation law of (4.22) holds for the central-node point for the network in the chemical domain. Hence, it can be found from Figure 4.11a that:

$$(nF\phi_l - nF\phi_c) \cdot C_{ch}^\varepsilon + m_{A^{n+}} = (nF\phi_c - nF\phi_r) \cdot C_{ch}^\varepsilon + m_{B^{n-}} \qquad (4.28)$$

This is in fact the equivalent of Kirchhoff's current law in the electrical domain. (4.28) can be re-arranged, where the equation for C_{ch}^ε in Figure 4.11a is used in the second line of the following equation:

$$\left(-nF\phi_l + 2nF\phi_c - nF\phi_r\right)C_{ch}^\varepsilon = m_{A^{n+}} - m_{B^{n-}}$$

$$\Rightarrow \left(-\phi_l + 2\phi_c - \phi_r\right)nF\frac{\varepsilon_0\varepsilon_r A}{\Delta x(nF)^2} = m_{A^{n+}} - m_{B^{n-}} \qquad (4.29)$$

$$\Rightarrow \left(-\phi_l + 2\phi_c - \phi_r\right) = \frac{nF\Delta x}{\varepsilon_0\varepsilon_r A}(m_{A^{n+}} - m_{B^{n-}})$$

The Poisson equation according to (4.27) can be found from (4.29), with the aid of a basic mathematical equation for the second derivative of the electrostatic potential ϕ in the centre of the spatial element:

$$\frac{\Delta^2\phi}{\Delta x^2} = \frac{\dfrac{\phi_l - \phi_c}{\Delta x} - \dfrac{\phi_c - \phi_r}{\Delta x}}{\Delta x} \Rightarrow \Delta^2\phi = \phi_l - 2\phi_c + \phi_r$$

$$\Rightarrow -\Delta^2\phi = -\phi_l + 2\phi_c - \phi_r = \frac{nF\Delta x}{\varepsilon_0\varepsilon_r A}\left(m_{A^{n+}} - m_{B^{n-}}\right) \qquad (4.30)$$

$$\Rightarrow \frac{\Delta^2\phi}{\Delta x^2} = -\frac{nF}{\varepsilon_0\varepsilon_r A\Delta x}\left(m_{A^{n+}} - m_{B^{n-}}\right)$$

The Poisson equation in the electrical domain can be found in an analogous way.

Now that the basic building block for modelling the Poisson equation has been defined, the mass transport of charged species by combined diffusion and migration under the influence of an electrical field can be modelled. As an example, consider an electrochemical cell with two metal electrodes and an ideal solution of a binary salt $M^{n+}X^{n-}$ that acts as electrolyte. Assume that only the metal ions M^{n+} can react with metal M at the electrodes and that the X^{n-} ions do not react with any species. M^{n+} ions are injected into the ideal solution by an oxidation reaction at the positive electrode, whereas M^{n+} ions are withdrawn from the ideal solution by a reduction reaction at the negative electrode when an external voltage source is applied to the cell. This is illustrated in Figure 4.12. Both M^{n+} and X^{n-} ions are transported by combined diffusion and migration in the electrolyte.

The electrical field is directed from the positive electrode to the negative electrode. The positive ions M^{n+} will migrate in the direction of the electrical field, whereas the negative ions X^{n-} will migrate in the opposite direction. The concentration gradients of positive and negative ions are identical throughout the electrolyte, because charge neutrality is sustained. The concentration gradient for both ions is directed from the positive electrode to the negative electrode, because positive ions are injected at the positive electrode and removed at the negative electrode. Therefore, mass transport by diffusion is directed from the positive electrode to the negative electrode for both ions. The model for mass transport in the absence of an electrical field illustrated in Figure 4.8 can be combined with the model shown in Figure 4.11 for modelling the Poisson equation. For this purpose, the electrolyte is divided into p spatial elements, as shown in Figure 4.5. This yields the model for one-dimensional mass transport in the x-direction in the electrolyte by combined diffusion and migration for a system such as that depicted in Figure 4.12. This is illustrated in Figure 4.13. The network model of Figure 4.13 can also be modelled in the electrical domain, as was discussed in section 4.1.2.

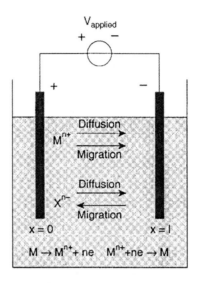

Figure 4.12: Schematic representation of an electrochemical cell with two metal (M) electrodes placed in an ideal solution using a $M^{n+}X^{n-}$ binary salt as the electrolyte, with the M^{n+} and X^{n-} ions being transported by combined diffusion and migration

The upper part of the network in Figure 4.13 describes the mass transport of the positive ions M^{n+}, while the lower part describes the mass transport of the negative ions X^{n-}. The molar amounts of the positive ions M^{n+} in spatial elements i are present on the capacitances $C_{ch}^{i,M^{n+}}$, while the capacitances $C_{ch}^{i,X^{n-}}$ store the molar amounts of the negative ions X^{n-}. A chemical potential μ is present across each capacitance. The interface between the positive electrode $(+)$ and the electrolyte is located at $x=0$. Here, positive ions M^{n+} are injected into the electrolyte by an electrochemical reaction. This electrochemical reaction takes place at a rate of $J_{ch}^{M^{n+}}(0)$. The electrostatic positive electrode potential is $\phi_{e,+}$. The interface between the negative electrode and the electrolyte is located at $x=l$. Here, ions M^{n+} are removed from the electrolyte by an electrochemical reaction. The rate at which this electrochemical reaction takes place is $J_{ch}^{M^{n+}}(l)$. The electrostatic negative electrode potential is $\phi_{e,-}$. The negative ions X^{n-} are not allowed to react at the electrode/ electrolyte interfaces.

The electrostatic potentials, multiplied by nF, at the centres of all the spatial elements are present at the central node points, as indicated in Figure 4.13. The electrostatic electrode potentials $\phi_{e,+}$ and $\phi_{e,-}$ are shown on the left and right, respectively. The sum of the electrostatic potential at a certain location multiplied by zF and the chemical potential μ yields the electrochemical potential $\overline{\mu}$ at that location, as described by (4.2). Note that for the positive ions, $z=+n$, whereas for the negative ions, $z=-n$. Therefore, a positive $\overline{\mu}_+$ is present at the top node points, whereas a negative $\overline{\mu}_-$ is present at the bottom node points. Differences between the electrochemical potentials $\overline{\mu}_+$ of neighbouring spatial elements yield a chemical flow through the chemical resistances in the upper part of Figure 4.13. In an analogous manner, differences in $\overline{\mu}_-$ lead to a chemical flow through the chemical resistances in the lower part of Figure 4.13. The chemical resistances were defined in (4.21c), in which the appropriate diffusion coefficients have to be used for positive and negative ions, respectively. The expression for the geometric capacitances C_{ch}^{ε} is given in Figure 4.11a. The distance from the centre to the electrode surface in spatial elements 1 and p is only $\Delta x/2$. Therefore, $\Delta x/2$ instead of Δx should be used in the expressions of $C_{ch}^{\varepsilon,edge}$.

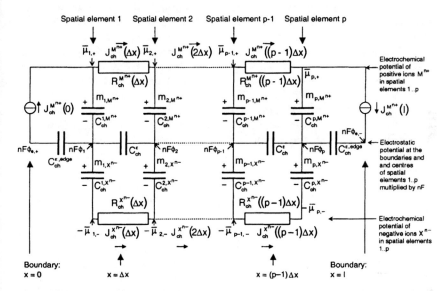

Figure 4.13: Simulation model in the chemical domain for one-dimensional mass transport of a charged species in the x-direction in the presence of an electrical field in, for example, the system depicted in Figure 4.12

Figure 4.13 shows that the potential across the capacitances $C_{ch}^{\varepsilon,edge}$ equals $nF(\phi^s-\phi^l)$ for both electrodes. The potential of the positive electrode ϕ^s equals $\phi_{e,+}$ and the electrolyte potential ϕ^l in spatial element 1, which is directly adjacent to the electrode surface, equals ϕ_l. For the negative electrode, $\phi^s=\phi_{e,-}$ and $\phi^l=\phi_p$. This means that the capacitances $C_{ch}^{\varepsilon,edge}$ are actually the double-layer capacitances in the chemical domain! As a result, the distance $\Delta x/2$ is actually the thickness of the Helmholtz layer; see chapter 3. It will be shown in section 4.3 how the network of Figure 4.13 can be used to model an electrolyte and how it has to be combined with the positive and negative electrode models.

The direction of migration and diffusion flows in the network of Figure 4.13 will now be considered. A complicating factor is that species with positive and negative signs have to be considered in one network. Chemical flows are commonly defined as a flow of positively charged particles. Therefore, all the species will be considered to be positively charged for simplicity in the explanations below, which means that chemical flows in the lower part of the network of Figure 4.13 express the flow of the ions X as if they were positive. In reality, the flow of the negative ions X^{n-} will be in the opposite direction. The electrostatic potentials ϕ_i will decrease from $x=0$ to $x=l$, because the positive electrode potential is higher than the negative electrode potential. As a result, a chemical migration flow from the left to the right occurs in the upper part of the network, which means that positive ions migrate from left to right under the influence of the electrical field. The chemical migration flow in the lower part of the network is also directed from the left to the right, which means that negative ions will migrate from right to left in reality. At the same time, the injection of positive ions at $x=0$ and their removal at $x=l$ will lead to a concentration gradient of positive ions from the left to the right. This will lead to a diffusion flow of positive ions from the left to the right.

The electrostatic potentials in the spatial elements will be constant when the electrical field throughout the electrolyte is stationary, which means that it does not change in time. Hence, the flow through the geometric capacitances C_{ch}^{ϵ} will be zero. As a result, all chemical flow through the chemical capacitances in the upper part of Figure 4.13 will flow through the capacitances in the same spatial element in the lower part. Assume for the moment that this has been the case since $t=0$, which means that the molar amount of the M^{n+} ions is the same as that of the X^{n-} ions in a spatial element. As an example, assume that $nF\phi_l$ equals 10 J/mol in spatial element 1 and that the chemical potential across capacitances C_{ch}^{1} for both the M^{n+} and the X^{n-} ions is 1 J/mol. The latter statement implies that μ^0 has been assumed equal for both ions for simplicity. Furthermore, assume that $nF\phi_2$ equals 9 J/mol in spatial element 2, because the electrostatic potential decreases from the left to the right, and that the chemical potential across both the capacitances C_{ch}^{2} in spatial element 2 is 0.5 J/mol. This means that the concentration gradient will indeed be directed from the left to the right. As a result, $\overline{\mu}_{1,+}$ equals 11 J/mol and $\overline{\mu}_{2,+}$ equals 9.5 J/mol, which can be inferred from applying Kirchhoff's voltage law in the chemical domain. This leads to a chemical flow from the left to the right of 1.5 mol/s when R_{ch} is assumed to be 1 Js/mol^2 for simplicity. This flow can be separated in a migration flow of 1 mol/s and a diffusion flow of 0.5 mol/s, which both flow from the left to the right, which is in agreement with Figure 4.12. In the bottom part of the network, an electrochemical potential of 9 J/mol is present on the left-hand side of R_{ch} in spatial element 1, while an electrochemical potential of 8.5 J/mol is present at the right-hand side in spatial element 2. Assuming the same value of 1 Js/mol^2 for R_{ch}, this leads to a chemical flow of 0.5 mol/s from the left to the right. This flow can be separated in a migration flow of 1 mol/s from the left to the right and a diffusion flow of 0.5 mol/s from the right to the left. This leads to a net chemical flow by combined diffusion and migration of 0.5 mol/s from the left to the right. The actual migration and diffusion flows of the negative ions will be in the opposite direction, as stated above. Hence, the migration is directed from the right to the left and the diffusion is directed from the left to the right. This is in accordance with Figure 4.12. In summary, the migration part of the chemical flow will be caused by a difference in electrostatic potentials between two spatial elements for each chemical resistance in Figure 4.13, while the diffusion part will be caused by a difference in chemical potential.

In the case described above, in which the concentration profiles of positive and negative ions are the same, charge neutrality holds. This means that the molar amounts of positive ions equal the molar amounts of negative ions. The Poisson equation of (4.27) shows that the second derivative of the electrostatic potential is zero in this case. This means that the electrical field inside the electrolyte is not a function of the location x. Hence, the difference between ϕ_i and ϕ_j across each geometrical capacitance C_{ch}^{ε} is constant, so a linear electrostatic potential gradient is present across the electrolyte.

The second derivative of the electrostatic potential is not zero in the Gouy-Chapman diffuse layer, as described in chapter 3. This means that, according to the Poisson equation of (4.27), there will be no charge neutrality in the spatial elements present in this Gouy-Chapman layer. In order to avoid charge neutrality inside a spatial element, the current through the geometrical capacitance C_{ch}^{ε} on the left-hand side must have been unequal to the current through the capacitance on the right-hand side for at least some time. This leads to a difference in molar amounts of positive and negative ions and hence to the absence of charge neutrality. It has been assumed above that a stationary situation occurred since $t=0$. In reality, the electrical field has to be built up since $t=0$. Let us therefore consider the build-up of an electrical field inside an electrolyte in some more detail. Starting at $t=0$, the current that flows through the capacitance C_{ch}^{ε} on the left-hand side of a spatial element in one of the two Gouy-Chapman layers at the electrode/electrolyte interfaces will differ from the current that flows through the capacitance C_{ch}^{ε} on the right-hand side. The current flowing through the capacitances C_{ch}^{ε} on the left- and right-hand sides will be equal for spatial elements outside the Gouy-Chapman layers. As soon as a stationary situation occurs at $t=t_1$, and the electrical field no longer changes in time, the current through all capacitances C_{ch}^{ε} will be zero. Because the current through all capacitances C_{ch}^{ε} outside the Gouy-Chapman layers has been equal from $t=0$ onwards, the electrostatic potential across each capacitance will have the same value. Therefore, the electrostatic potential gradient will indeed be linear. However, this will not be the case for capacitances C_{ch}^{ε} inside the Gouy-Chapman layers. Therefore, the second derivative of the electrostatic potential in these two regions will not be zero. Moreover, although the current flowing through the upper chemical capacitance will equal the current flowing through the lower chemical capacitance from $t=t_1$ onwards, charge neutrality will never be attained in those spatial elements inside the Gouy-Chapman layer. The reason for this is that the current through the upper capacitance has been unequal to the current through the lower capacitance from $t=0$ to $t=t_1$.

The theory of mass transport presented in this section can be summarized as follows:

* The chemical resistance, defined in (4.21c), models the general mass transport laws of (4.18) and (4.19).
* The combination of geometric capacitances and chemical capacitances placed around a central node point models the Poisson equation; see (4.27) for the chemical domain.
* The principle of mass conservation is modelled by means of an RC-ladder network, in which the difference between the chemical flows entering and leaving a spatial element is integrated in a chemical capacitance.
* Boundary conditions at the interface of the considered system through which species are transported, such as an electrolyte solution, and neighbouring systems, such as electrodes, can be modelled by means of chemical flow sources at these interfaces. The chemical flow sources have a zero value and can be omitted when the species cannot be transported across the interface.

- Initial conditions of the mass transport problem can be introduced by placing initial molar amounts on all the chemical capacitances at all locations in the system.

4.1.4   Modelling thermal behaviour

Heat production or consumption of an electrochemical cell has not been considered in the previous sections. A well-known thermodynamic law links the change in free energy ΔG of a cell to the change in enthalpy ΔH, the change in entropy ΔS and the temperature T at which the reactions takes place [15],[16],[21],[22]:

$$\Delta G = \Delta H - T\Delta S \qquad (4.31)$$

A simple example will be given to explain the meaning of this law. Consider a simple experiment in which a battery is charged from an equilibrium situation. The charge current is chosen infinitesimally small. As a result, the overpotentials can be considered zero, and hence no losses occur in the system when the battery is successively charged and discharged. This means that the amount of energy that has been supplied during charging can be fully retrieved during discharging. The equilibrium potentials of the electrodes can be considered constant, because the current is so small that the equilibrium hardly changes. The battery can be regarded as a system on which electrical work is performed by a current source that charges the battery. The battery is to be considered to be in contact with a thermal reservoir [21]. Heat transfer can occur between the battery and the thermal reservoir. As an arbitrary choice, the charging reaction is considered to be exothermic, which means that heat is transferred from the battery to the thermal reservoir. A similar derivation as given below can be made when the charging reaction is assumed to be endothermic. The situation is sketched in Figure 4.14a in which subscripts c mean charging. The battery is shown in contact with a thermal reservoir and a current source that performs work on the battery. Besides signs indicating energy and entropy changes, potentials and currents, the figure also shows the effect of the thermodynamic law of (4.31) on the values of ΔG, ΔH and $(-T\Delta S)$ of the battery.

Figure 4.14a shows that an oxidation reaction (O) occurs at the positive electrode and a reduction reaction (R) at the negative electrode. Note that $E^{eq,+} > 0$ and $E^{eq,-} < 0$. This has been chosen for simplicity, but of course the explanation below will also hold for other polarities. The polarities simply depend on which potential is chosen as a reference in the system. Positive electrical energy ΔG_O is supplied to the positive electrode, because both the electrode potential and the current directed into the electrode are positive. Positive electrical energy ΔG_R is also supplied to the negative electrode, because both the electrode potential and the current directed out of the electrode are negative. The changes in free energy ΔG_O and ΔG_R are indeed positive when the sign convention described in section 4.1.2 is adopted. This is shown directly below the battery in Figure 4.14a, with a plus sign being used for an oxidation reaction and a minus sign for a reduction reaction. The total electrical energy supplied to the battery is positive and equals $\Delta G_c = E_{bat}I_c t$, which is simply the product of the supplied electrical power $E_{bat}I_c$ and time. It can also be inferred from the summation of the energy ΔG_O supplied to the positive electrode and the energy ΔG_R supplied to the negative electrode, because zero losses have been assumed. This leads to the expression $\Delta G_c = nFE_{bat}$ illustrated below the battery in Figure 4.14a. This is in accordance with the sign convention adopted in this book, because $\Delta G=+nFE$ holds in the case of a non-spontaneous reaction in all the electrochemical literature. This will indeed be the case in a charging reaction, as work has to be performed on the battery to 'force' the charging reactions to occur.

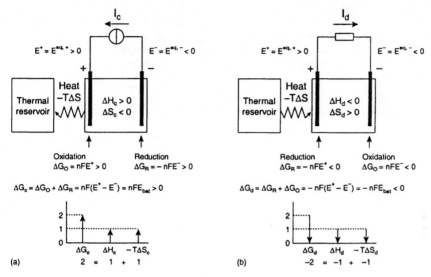

Figure 4.14: (a) Charging a battery (subscripts c mean charging) with an infinitesimally small current, so that overpotentials are zero, so no losses occur and the equilibrium potential does not change. The energy ΔG_c supplied during charging can be fully retrieved (ΔG_d) (subscripts d mean discharging) during discharging. Changes in free energy (ΔG), enthalpy (ΔH) and entropy (ΔS) of the battery are shown. The charge reaction has arbitrarily been chosen to be exothermic ($\Delta S_c<0$) (b) Same as (a), but for discharging the battery with an infinitesimally small current, the reaction is now endothermic ($\Delta S_d>0$) by definition

The change in entropy in the battery during charging will be negative ($\Delta S_c<0$), which means that the entropy decreases, because of the arbitrary choice of an exothermic charging reaction. This decrease in entropy in the battery must be balanced by transfer of heat to the thermal reservoir [21]. This is shown in Figure 4.14a. The consequence of (4.31) is depicted at the bottom of Figure 4.14a. The change in free energy of the battery ΔG_c, supplied by the current source, is divided between a change in enthalpy ΔH_c and transfer of heat $-T\Delta S$ to the thermal reservoir. Note that $-T\Delta S$ will be positive, because $\Delta S_c<0$ and a positive temperature T have been assumed. The numbers given in the figure are arbitrary and merely serve to illustrate the principle of (4.31). The battery's free energy changes from 0 to 2 J/mol ($\Delta G_c=2$ J/mol>0), which leads to a change in enthalpy from 0 to 1 J/mol ($\Delta H_c=1$ J/mol>0) and a change in heat from 0 to 1 J/mol ($-T\Delta S=1$ J/mol>0), which is delivered to the thermal reservoir.

The battery will be discharged when the current source is replaced by a resistor. Again, an infinitesimally small current is assumed. Now, a reduction reaction will occur at the positive electrode and an oxidation reaction at the negative electrode. Both ΔG_O and ΔG_R will now be negative, as indicated below the battery in Figure 4.14b in which subscripts d mean discharging. The change in the battery's free energy during discharging ΔG_d, which equals the work performed on the environment, is again inferred from the summation of ΔG_O and ΔG_R, leading to $\Delta G_d = -nFE_{bat}$. This means that the minus sign convention ($\Delta G = -nFE$) that is often encountered in the literature actually holds for the spontaneous reaction direction. Note that $\Delta G_d = -\Delta G_c$, which means that the energy supplied during charging will be retrieved completely. This is in agreement with the assumption of zero losses. As (4.31) is of course still valid, the discharge reaction has to be endothermic, or $\Delta S_d > 0$. This means that the battery's entropy will increase. This can be seen at the bottom of Figure 4.14b, in which the battery free energy changes from 2 J/mol to 0 ($\Delta G_d = -2$ J/mol<0) resulting from a change in enthalpy from 1 J/mol to 0 ($\Delta H_d = -1$ J/mol<0) and a change in heat from 1 J/mol to 0 ($-T\Delta S_d = -1$ J/mol<0). This heat ($-T\Delta S$) is obtained from the thermal reservoir.

The above experiment merely served to give an idea of what happens with respect to heat in an ideal case. In reality, losses will occur in a battery when it is charged or discharged. The heat flow generated inside a battery by electrochemical reactions i is given by [22],[27]:

$$J_{th}^{in,elch} = \sum_i \frac{-T\Delta S_i}{n_i F} |I_i| + \sum_i I_i \eta_i + I^2 R_{total}^{ohmic} \tag{4.32}$$

where $J_{th}^{in,elch}$ is the heat flow in [W], generated by the battery, I the battery current in [A], I_i is the partial current of electrochemical reaction i in [A], T the battery temperature in [K], ΔS_i denotes the entropy change associated with electrochemical reaction i in [J/(mol.K)], n_i the number of electrons exchanged in electrochemical reaction i, η_i the total overpotential due to kinetic aspects and mass transport associated with electrochemical reaction i in [V] and R_{total}^{ohmic} denotes the total ohmic series resistance inside the battery in [Ω].

The heat flows defined in the second and third terms of (4.32) will always be positive, directed from the battery to the thermal reservoir in Figure 4.14. Note that in the second term, I_i and η_i will always have the same sign by definition. This means that, in practice, the battery will always heat up because of these two terms, irrespective of the direction of the current. This makes sense, as these two terms represent the losses that occur inside the battery as a result of overpotentials and ohmic series resistance, respectively. These heat flows can be seen as 'one-way' flows of heat from the battery to its thermal environment. These 'one-way' flows occur during both charging and discharging, and hence the amount of energy supplied during charging will never be fully retrieved during discharging.

The first term can be either positive or negative. It is positive when $\Delta S_i < 0$, which is valid for an exothermic reaction. This means that the battery will heat up as a result of all the exothermic reactions that take place. The battery will cool down as a result of the endothermic reactions, for which $\Delta S_i > 0$. This was explained in Figure 4.14. A reaction that is exothermic during charging is endothermic during discharging and vice versa. The heat flow in the first term is defined as the amount of heat, $-T\Delta S$ in [J/mol], multiplied by a chemical flow I/nF in [mol/s]. This is in agreement with the theory presented earlier in this section. In a purely chemical reaction, the second and third terms in (4.32) will be zero, because no electrical current I is involved. The first term will be a heat flow defined by $-T\Delta S J_{ch}$.

The temperature of a battery is determined by the net amount of heat inside the battery and the thermal or heat capacitance of the battery, according to the analogy between the electrical and thermal domains in Table 4.1. The net amount of heat is built up by a net heat flow J_{th}^{net} inside the battery, which is determined by the total heat flow $\Sigma J_{th}^{in,i}$, generated internally by all the chemical and electrochemical reactions, and the heat flow J_{th}^{out} from the battery to the environment by means of conduction, convection and radiation. Hence:

$$J_{th}^{net} = \sum_i J_{th}^{in,i} - J_{th}^{out} \tag{4.33}$$

The heat flow from the battery to the environment is given by [1]:

$$J_{th}^{out} = \alpha_{th} A_{bat} \left(T - T_{amb} \right) = \frac{1}{R_{th}^{bat}} \cdot (T - T_{amb})$$

$$\Rightarrow R_{th}^{bat} = \frac{1}{\alpha_{th} A_{bat}} \tag{4.34}$$

where J_{th}^{out} denotes the total heat flow from the battery to the environment in [W], α_{th} the battery heat transfer coefficient in [W/(K.m^2)], A_{bat} the battery surface area in [m^2], T_{amb} stands for the ambient temperature in [K], and R_{th}^{bat} denotes the thermal resistance from the battery to the environment in [K/W]. This thermal resistance is the sum of the conductive, convective and radiative thermal resistances that describe the heat transfer from the battery to the environment.

The net heat flow from inside and outside the battery yields the accumulated amount of heat inside the battery. This net amount of heat can be calculated using:

$$Q_{th}^{net} = \int J_{th}^{net}(t)dt = C_{th}^{bat} \cdot T \tag{4.35}$$

where Q_{th}^{net} is the net amount of heat inside the battery in [J] and C_{th}^{bat} the thermal or heat capacitance of the battery in [J/K].

A simple first-order network model can be constructed that models the thermal behaviour of a battery using the analogy between the electrical and thermal domains in Table 4.1 and the equations described in this section. This is illustrated in Figure 4.15.

As can be seen in Figure 4.15, the net heat flow J_{th}^{net} inside the battery is integrated in the thermal capacitance C_{th}^{bat}. The battery temperature T_{bat} is present across the battery thermal capacitance. In a similar way, the ambient temperature is present across the ambient thermal capacitance C_{th}^{amb}. A temperature difference between the battery and the environment will give rise to a heat flow J_{th}^{out} between the battery and the environment, with the sign depending on the sign of the temperature difference. Changes in the battery temperature will have virtually no effect on the ambient temperature when the ambient thermal capacitance is much larger than the battery thermal capacitance. Then the battery will not be able to heat up the environment. Only a single battery temperature is calculated in this simple thermal-network model. This means that the battery temperature is considered to be uniform throughout the battery. More sophisticated thermal-network models could take into account the changes in temperature occurring from one location inside the battery to the other.

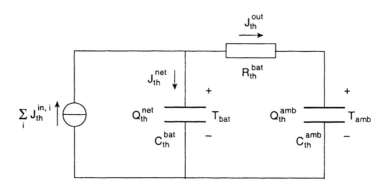

Figure 4.15: Simulation model in the thermal domain that describes the temperature development in a battery

4.2 A simulation model of a rechargeable NiCd battery

4.2.1 Introduction

The general principles described in the previous section will be used in this section to construct a network model for a rechargeable NiCd battery [23]-[28]. As many different types of NiCd batteries exist, there is no single model that is valid for all types.

A schematic representation of a mature design of a sealed rechargeable NiCd battery is given in Figure 4.16. This design was chosen as the basis for deriving the battery model in this section. An example of a NiCd battery based on this design is the P60AA manufactured by Panasonic. The three basic elements in Figure 4.16 are: the positive nickel electrode, the negative cadmium electrode and a concentrated KOH electrolyte solution that provides ionic conductivity between the electrodes. To prevent electrical contact between the positive and negative electrodes, the two electrodes are separated by a thin porous layer of insulating material, the separator.

Main storage reactions that take place during charging and discharging
The electrochemical storage reactions are given in the region denoted as *storage capacity* in Figure 4.16. The arrows pointing upwards denote the charging reactions, while the arrows pointing downwards denote the discharging reactions. All reaction equations have been arranged with the same amount of electrons for comparison. In their simplest form, the main storage reactions are given by:

$$Ni(OH)_2 + OH^- \xrightleftharpoons[k_{c,Ni}]{k_{a,Ni}} NiOOH + H_2O + e^- \qquad (4.36)$$

for the nickel reaction and

$$Cd(OH)_2 + 2e^- \xrightleftharpoons[k_{a,Cd}]{k_{c,Cd}} Cd + 2OH^- \qquad (4.37)$$

for the cadmium reaction. During charging, $Ni(OH)_2$ is oxidized to NiOOH at the nickel electrode. At the same time, $Cd(OH)_2$ is reduced to metallic Cd at the cadmium electrode. The reverse reactions take place during discharging. Current flow is supported by electrons inside the electrodes and by OH⁻ ions inside the electrolyte. Externally, the current flows through the charger or load. K⁺ ions do not react at the electrode interfaces. Their behaviour can be compared to that of the X^{n-} ions in Figure 4.12, i.e. the net current supported by K⁺ ions can become zero.

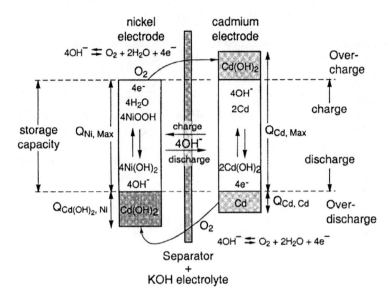

Figure 4.16: Schematic representation of a sealed rechargeable NiCd battery

Side-reactions taking place during overcharging and overdischarging
Besides the main storage reactions, electrochemical side-reactions also take place, especially during overcharging and overdischarging. These side-reactions include oxygen production and reduction reactions, schematically represented by the curved arrows in Figure 4.16, and a cadmium side-reaction at the nickel electrode, represented by the grey $Cd(OH)_2$ area in the nickel electrode. All grey areas in Figure 4.16 symbolize the excess material that is added to the NiCd battery to protect it from excessive pressure build-up during overcharging and overdischarging. How this is achieved will be discussed below.

Overcharging
An overcharge reserve of excess $Cd(OH)_2$ is added to the cadmium electrode. Hence, the nickel electrode is the capacity-determining electrode, which is reflected by the fact that $Q_{Ni,Max}$ is smaller than $Q_{Cd,Max}$ in Figure 4.16. As a result, the amount of $Ni(OH)_2$ will approach zero before all the $Cd(OH)_2$ has been consumed at the cadmium electrode during the charging process. An alternative oxidation reaction has to occur at the nickel electrode. This reaction is the production of oxygen, according to:

$$4OH^- \xrightarrow{k_{a,O_2}} O_2 + 2H_2O + 4e^- \qquad (4.38)$$

Oxygen is produced at the nickel electrode through the oxidation of OH⁻ ions into oxygen and water molecules. Metallic Cd is still formed at the cadmium electrode, because of the excess $Cd(OH)_2$. Oxygen molecules can diffuse to the cadmium electrode when the amount of oxygen inside the battery increases. There, the oxygen-reduction reaction starts to compete with the Cd-forming reaction. The oxygen molecules can be reduced to OH^- ions, according to:

$$O_2 + 2H_2O + 4e^- \xrightarrow{k_{c,O_2}} 4OH^- \tag{4.39}$$

The production of oxygen at the nickel electrode and removal of oxygen at the cadmium electrode are schematically represented by the top curved arrow in Figure 4.16.

Overdischarging
An overdischarge reserve of excess $Cd(OH)_2$, generally denoted as a depolarizer, is added to the nickel electrode in the amount $Q_{Cd(OH)_2,Ni}$, as indicated in Figure 4.16. Moreover, some metallic cadmium is added to the cadmium electrode in an amount $Q_{Cd,Cd}$ smaller than $Q_{Cd(OH)_2,Ni}$. This is done to enforce similar oxygen production and reduction processes as those occurring during overcharging. In this way, the production of hydrogen gas at the nickel electrode is prevented and a hydrogen recombination cycle is avoided [29]. This is important because of the poor electrocatalytic activity of the cadmium electrode with respect to the H_2 oxidation reaction, which would lead to an undesirable very high internal H_2 pressure. Overdischarging takes place in two steps.

Because of the excess of metallic Cd in the cadmium electrode, the amount of NiOOH in the nickel electrode will diminish before all the metallic Cd has been consumed in the cadmium electrode. An alternative reduction reaction now has to occur at the nickel electrode. The excess $Cd(OH)_2$ in the nickel electrode will be reduced to metallic Cd during the first overdischarging step and the reverse oxidation reaction will still occur at the cadmium electrode. Hence, the following reactions occur during the first overdischarging step.

$$Cd(OH)_2 + 2e^- \xrightarrow{k_{c,Cd}} Cd + 2OH^- \tag{4.40}$$

at the nickel electrode, and

$$Cd + 2OH^- \xrightarrow{k_{a,Cd}} Cd(OH)_2 + 2e^- \tag{4.41}$$

at the cadmium electrode.

The second overdischarging step starts when discharging proceeds to the stage at which all the metallic Cd has been consumed at the cadmium electrode. Oxygen gas will be produced at the cadmium electrode in an alternative oxidation reaction. This leads to an increase in the amount of oxygen inside the battery. As a result, the oxygen reduction reaction starts to compete with the $Cd(OH)_2$ reduction reaction at the nickel electrode and a similar situation as during overcharging occurs. The reactions of (4.39) and (4.40) compete at the nickel electrode, while the reaction of (4.38) takes place at the cadmium electrode. The production of oxygen at the cadmium electrode and the removal of oxygen at the nickel electrode during the second overdischarging step are schematically represented by the bottom curved arrow in Figure 4.16.

Figure 4.16 shows that three types of reactions can take place inside a NiCd battery. More details on these reactions will be given in the next three sections, in which use is made of the theory described in chapter 3 and section 4.1. The activity effects of electrons will be neglected, or a_e and a_e^{ref} will be assumed to be the same, as discussed in section 4.1.2.

4.2.2   The nickel reaction

The main storage reaction for the nickel electrode was given above in (4.36). It is assumed that the volume of the active material does not change during charging and discharging and that NiOOH, which is the Ox species, and $Ni(OH)_2$, which is the Red species, are the only species present within the region of the nominal storage capacity. In other words, it is assumed that the sum of the mol fractions x_i of the species i is unity, or:

$$x_{NiOOH} + x_{Ni(OH)_2} = 1 \qquad (4.42)$$

where the mol fractions x_i are defined by ($\gamma = 1$ assumed):

$$x_i = \frac{a_i}{a_{NiOOH} + a_{Ni(OH)_2}} \qquad (4.43)$$

where i is either NiOOH or $Ni(OH)_2$. This means that the electrode is considered to be a solid-solution electrode, with the activities of the species being variable and their sum constant. Mol fraction x_{NiOOH} corresponds to the State-of-Charge (SoC) of the NiCd battery, because NiOOH is the species that is formed during charging.

Thermodynamics of the nickel reaction
The change in free energy $\Delta G_{Ni,O}$ of the oxidation reaction (O) that occurs at the nickel electrode during charging can be found in the same way as shown above in (4.10a); see also (4.36):

$$\Delta G_{Ni,O} = \mu_{NiOOH} + \mu_{H_2O} + \mu_{e^-} - \mu_{Ni(OH)_2} - \mu_{OH^-}$$

$$= \mu_{NiOOH}^o + \mu_{H_2O}^o + \mu_{e^-}^o - \mu_{Ni(OH)_2}^o - \mu_{OH^-}^o$$

$$+ RT \ln\left(\frac{a_{NiOOH}\, a_{H_2O}\, a_{Ni(OH)_2}^{ref}\, a_{OH^-}^{ref}}{a_{Ni(OH)_2}\, a_{OH^-}\, a_{NiOOH}^{ref}\, a_{H_2O}^{ref}} \right) \qquad (4.44)$$

$$= \Delta G_{Ni,O}^o + RT \ln\left(\frac{a_{NiOOH}\, a_{H_2O}\, a_{Ni(OH)_2}^{ref}\, a_{OH^-}^{ref}}{a_{Ni(OH)_2}\, a_{OH^-}\, a_{NiOOH}^{ref}\, a_{H_2O}^{ref}} \right)$$

The a_{OH^-} activity can be considered to be constant when, as a simplification, the electrolyte concentration is considered to be constant and uniformly distributed across the battery. Activity a_{H_2O} is also constant, because of this simplification. The simplification is reasonable when the currents applied to the battery are not too high. The ln term containing these constant activities can then be added to $\Delta G^o_{Ni,O}$. The contribution of the a^{ref} terms is constant and will also be added to $\Delta G^o_{Ni,O}$. The following expression is now found for the equilibrium potential of the nickel reaction:

$$
\begin{aligned}
E_{Ni}^{eq} &= \frac{\Delta G_{Ni,O}}{n_{Ni}F} \\[2ex]
&= \frac{\Delta G_{Ni,O}^{o} + RT \ln\left(\dfrac{a_{H_2O}}{a_{OH^-}}\right)}{n_{Ni}F} + \frac{RT}{n_{Ni}F} \ln\left(\frac{a_{NiOOH}}{a_{Ni(OH)_2}}\right) \\[2ex]
&= \frac{\Delta G_{Ni,O}^{o'}}{n_{Ni}F} + \frac{RT}{n_{Ni}F} \ln\left(\frac{a_{NiOOH}}{a_{Ni(OH)_2}}\right) \\[2ex]
&= E_{Ni}^{o} + \frac{RT}{n_{Ni}F} \ln\left(\frac{a_{NiOOH}}{a_{Ni(OH)_2}}\right)
\end{aligned}
\tag{4.45}
$$

where $n_{Ni}=1$, according to (4.36). The same expression can be found for the equilibrium potential when the reduction or discharge reaction of (4.36) is considered. This was shown in section 4.1.2. All activities are bulk activities, which are equal to the surface activities due to true equilibrium, because (4.45) expresses the true equilibrium potential. As will be described below, H^+ ions are removed from the nickel electrode during charging and inserted during discharging. It has been found in experiments that electrical and elastic interaction between these H^+ ions and between the H^+ ions and the nickel and oxygen atoms in the electrode does not lead to a deviation from the Nernst-like behaviour of the equilibrium potential in (4.45) [30].

The value of $E^o{}_{Ni}$ in a strong alkaline solution has to be used, which is 0.52 V vs the Standard Hydrogen Electrode (SHE) at 298 K, because the electrolyte in the NiCd battery is a concentrated 8 M KOH solution and the constant concentration of the electrolyte is present in $E^o{}_{Ni}$ through a_{OH^-} [31].

Kinetics of the nickel reaction

The kinetics of an electrochemical reaction are determined by the value of the exchange current through the Butler-Volmer relationship, as was discussed in chapter 3. In agreement with this theory and (4.12), the exchange current value for the nickel reaction is determined by:

$$
\begin{aligned}
I_{Ni}^{o} &= n_{Ni}FA_{Ni}\left(k_{a,Ni}\right)^{1-\alpha_{Ni}}\left(k_{c,Ni}\right)^{\alpha_{Ni}} \cdot \\[1ex]
&\quad \left(a_{NiOOH}^{b}\right)^{\alpha_{Ni}w_{Ni}}\left(a_{Ni(OH)_2}^{b}\right)^{(1-\alpha_{Ni})x_{Ni}}\left(a_{H_2O}^{b}\right)^{\alpha_{Ni}y_{Ni}}\left(a_{OH^-}^{b}\right)^{(1-\alpha_{Ni})z_{Ni}}
\end{aligned}
\tag{4.46}
$$

where A_{Ni} refers to the surface area of the nickel electrode, $k_{a,Ni}$ and $k_{c,Ni}$ are the anodic and cathodic rate constants of the nickel reaction, respectively, and w_{Ni}, x_{Ni}, y_{Ni} and z_{Ni} denote the reaction orders of the corresponding species in the nickel reaction. Again, the bulk activities of the reacting species are used in (4.46), because the exchange current is derived from the equilibrium situation.

Diffusion limitation of the nickel reaction

The bulk activities of the reacting species in (4.36) may differ from the activities at the electrode surface. The nickel reaction is generally assumed to proceed through diffusion of H^+ ions through the active electrode material. H^+ ions are formed during charging, which diffuse to the electrode/electrolyte interface, where they recombine with OH^- ions to form H_2O. The opposite occurs during discharging. This is described by the following reaction equations, which yield (4.36) by addition.

$$Ni(OH)_2 \underset{k_{c,Ni}}{\overset{k_{a,Ni}}{\rightleftharpoons}} NiOOH + H^+ + e^- \qquad (4.47)$$

$$H^+ + OH^- \underset{k_b}{\overset{k_f}{\rightleftharpoons}} H_2O \qquad (4.48)$$

A schematic one-dimensional representation of the nickel electrode during charging is shown in Figure 4.17. It is assumed that the positions of the nickel atoms are fixed and that charge transport occurs by means of electrons and H^+ ions only. The movement of electrons is considered to be much faster than that of H^+ ions, so mass transport limitation is attributed to H^+ ions only. It is further assumed that the nickel electrode is a perfect conductor. As a result, the electrical field inside the electrode will be negligible and mass transport through migration can be neglected. Mass transport through convection can also be neglected and hence only diffusion of H^+ ions is considered.

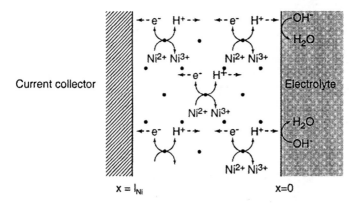

Figure 4.17: Schematic one-dimensional representation of the nickel electrode during charging

Electrons are removed from the lattice at the nickel atom sites during charging and H^+ ions are formed in the oxidation of $Ni(OH)_2$ to $NiOOH$. The H^+ ions move through the lattice from one nickel atom site to another, which leads to an apparent flux of $Ni(OH)_2$ and $NiOOH$ species. Concentration profiles of $Ni(OH)_2$ and $NiOOH$ arise in the electrode when the charge-transfer kinetics is fast in comparison with the mass transport rate of H^+ ions. The H^+ ions eventually arrive at the electrode/electrolyte interface, where they form H_2O with OH^- ions. The diffusion of H^+ ions through the electrode in the absence of an electrical field, as shown in Figure 4.17, can be described as explained in section 4.1.3. An apparent equilibrium potential E_{Ni}^{eq*} can be defined, which is the summation of the true equilibrium potential E_{Ni}^{eq}, defined in (4.45), and a diffusion overpotential η_{Ni}^d according to (4.26), when the surface activities of the NiOOH and $Ni(OH)_2$ species differ from the bulk activities. The apparent equilibrium potential can be inferred from (4.45) by using the surface activities for the NiOOH and $Ni(OH)_2$ species. Hence,

$$
\begin{aligned}
E_{Ni}^{eq*} &= E_{Ni}^o + \frac{RT}{n_{Ni}F}\ln\left(\frac{a_{NiOOH}^s}{a_{Ni(OH)_2}^s}\right) \\
&= E_{Ni}^o + \frac{RT}{n_{Ni}F}\ln\left(\frac{a_{NiOOH}^b}{a_{Ni(OH)_2}^b}\right) + \frac{RT}{n_{Ni}F}\ln\left(\frac{a_{NiOOH}^s a_{Ni(OH)_2}^b}{a_{NiOOH}^b a_{Ni(OH)_2}^s}\right) \\
&= E_{Ni}^{eq} + \eta_{Ni}^d
\end{aligned}
\tag{4.49}
$$

4.2.3 The cadmium reactions

The cadmium reaction equation was given in (4.37). The reaction can occur at both electrodes, as described in section 4.2.1. Only the cadmium electrode will be considered in the remainder of this section, for simplicity. The same equations also hold for the cadmium reaction that may occur in the nickel electrode.

The cadmium electrode is considered to be a heterogeneous system composed of two separate phases, which are metallic Cd, the *Red* species, and an oxide layer consisting of $Cd(OH)_2$, the *Ox* species. This as opposed to the nickel electrode, where two species NiOOH and $Ni(OH)_2$ co-exist in a single phase. It is assumed that electron transfer takes place at the interface of the Cd species and the electrolyte only and not at the $Cd(OH)_2$/electrolyte interface. This assumption is based on the fact that $Cd(OH)_2$ is a poor electronic conductor [32]. The reaction of (4.37) takes place via a dissolution-precipitation mechanism [33]. This mechanism is a two-step reaction according to:

$$
Cd(OH)_2 + 2OH^- \underset{k_b}{\overset{k_f}{\rightleftharpoons}} Cd(OH)_4^{2-}
$$

$$
Cd(OH)_4^{2-} + 2e^- \underset{k_{a,Cd}}{\overset{k_{c,Cd}}{\rightleftharpoons}} Cd + 4OH^-
\tag{4.50}
$$

Soluble intermediate $Cd(OH)_4^{2-}$ ions are formed by a chemical reaction during charging. Subsequently, these ions can be reduced at the solid-solution interface, resulting in the formation of a Cd layer. The actual transfer of electrons takes place at the Cd surface only, because electrons are only exchanged here. During discharging, the reverse reaction sequence occurs.

Thermodynamics of the cadmium reactions

The change in free energy $\Delta G_{Cd,R}$ in the reduction reaction (R) that occurs at the cadmium electrode during charging can be found in the same way as for (4.10b); see also (4.37).

$$
\begin{aligned}
\Delta G_{Cd,R} &= \mu_{Cd} + 2\mu_{OH^-} - \mu_{Cd(OH)_2} - 2\mu_{e^-} \\
&= \mu_{Cd}^o + 2\mu_{OH^-}^o - \mu_{Cd(OH)_2}^o - 2\mu_{e^-}^o \\
&\quad + RT \ln\left(\frac{a_{Cd}\left(a_{OH^-}\right)^2 a_{Cd(OH)_2}^{ref}}{a_{Cd}^{ref}\left(a_{OH^-}^{ref}\right)^2 a_{Cd(OH)_2}} \right) \\
&= \Delta G_{Cd,R}^o + RT \ln\left(\frac{a_{Cd}\left(a_{OH^-}\right)^2 a_{Cd(OH)_2}^{ref}}{a_{Cd}^{ref}\left(a_{OH^-}^{ref}\right)^2 a_{Cd(OH)_2}} \right)
\end{aligned}
\tag{4.51}
$$

The a_{OH^-} activity will be considered to be constant, as described in section 4.2.2. In addition, the activities of the solids $Cd(OH)_2$ and Cd are equal to the corresponding reference activities by definition. The ln term containing a_{OH^-} can now be added to $\Delta G^o_{Cd,R}$, while the ln terms that contain the activities of the $Cd(OH)_2$ and Cd species are zero. The following expression is then found for the equilibrium potential of the cadmium reaction:

$$
E_{Cd}^{eq} = \frac{-\Delta G_{Cd,R}}{n_{Cd}F} = \frac{-\Delta G_{Cd,R}^o - RT\ln\left(\frac{a_{OH^-}}{a_{OH^-}^{ref}}\right)^2}{n_{Cd}F}
\tag{4.52}
$$

$$
= \frac{-\Delta G_{Cd,R}^{o'}}{n_{Cd}F} = E_{Cd}^o
$$

where $n_{Cd}=2$, according to (4.37). Again, the same expression can be derived from the oxidation or discharge reaction in (4.37). Note that (4.52) expresses the true equilibrium potential, which means that the bulk activity a_{OH^-} should be used. The equilibrium potential for the cadmium reaction is constant, irrespective of the state of the reaction, because all activities are constant. The value in a strong alkaline solution should be used for $E^o{}_{Cd}$, which is -0.81 V vs SHE at 298 K [31], because the concentration of the electrolyte is present in $E^o{}_{Cd}$ through a_{OH^-}.

Kinetics of the cadmium reactions

The kinetics of the cadmium reaction are described by the Butler-Volmer equation. The exchange current value for the cadmium reaction is determined by:

$$I_{Cd}^o = n_{Cd} F A_{Cd} \left(k_{a,Cd} \right)^{1-\alpha_{Cd}} \left(k_{c,Cd} \right)^{\alpha_{Cd}} \cdot$$
$$(a_{Cd(OH)_2}^b)^{\alpha_{Cd} x_{Cd}} (a_{OH^-}^b)^{(1-\alpha_{Cd}) y_{Cd}} (a_{Cd}^b)^{(1-\alpha_{Cd}) z_{Cd}}$$
(4.53)

where $k_{a,Cd}$ and $k_{c,Cd}$ are the anodic and cathodic reaction rate constants of the cadmium reaction, respectively, x_{Cd}, y_{Cd} and z_{Cd} are the reaction orders of the corresponding species and A_{Cd} refers to the free metallic cadmium surface at which the charge transfer takes place. All bulk activities in (4.53) are constant. It is assumed that during charging, the cadmium is first formed in the form of single nuclei, which grow further into metallic particles, which eventually form a layer. The reverse process takes place during discharging. As a result, the free metallic cadmium surface area A_{Cd} is a function of the amount of cadmium in the cadmium electrode and hence of the SoC. It is further assumed that the time interval during which the nuclei are formed is sufficiently small to allow us to consider the nucleation process to be instantaneous. This means that all nuclei are formed when the charging starts at time t_o. During normal battery operation, Cd will already be present in the cadmium electrode at the beginning of the charging due to the overdischarge reserve $Q_{Cd,Cd}$; see Figure 4.16. This means that nuclei will indeed already be present at time t_o.

Hemispherical nuclei are assumed, which grow or shrink at the rate at which the reaction of (4.37) takes place. Suppose that the formation of metallic cadmium during the charging starts with N nuclei at $t=t_o$. The 'extended' electrode surface area θ_{ex} of N hemispherical particles at any time $t>t_o$ can be inferred from $\theta_{ex}=N\pi$ $r^2(t)$, where the radius $r(t)$ of each hemispherical particle can be inferred from the particle's volume. This is the area of the particle that is in direct contact with the surface. It is assumed that the number of particles remains constant. Only the size of the particles increases during charging or decreases during discharging. In addition, it is assumed that all the particles have the same volume at each time t. The volume of a hemispherical particle is half that of a sphere, or $(2/3)\pi r^3$. The total volume $V(t)$ of the deposited material can be obtained by taking N times the volume of a hemisphere, i.e.

$$V(t) = N \frac{2}{3} \pi r^3(t) = V^o + \frac{M_{Cd}}{\rho_{Cd}} \int_{t_o}^{t} J_{Cd}(t) dt$$
(4.54)

where $V(t)$ is the total volume of the N hemispherical particles in [m³], N the number of hemispherical particles, $r(t)$ the radius of hemispherical particles in [m], V^o the initial volume of the N hemispherical particles at $t=t_o$ in [m³], M_{Cd} the molecular weight of cadmium in [kg/mol], ρ_{Cd} the gravimetric density of cadmium in [kg/m³] and J_{Cd} is the chemical flow associated with the cadmium reaction of (4.37) in [mol/s]. J_{Cd} is positive during charging and the volume of the N hemispherical particles increases. During discharging, J_{Cd} is negative and the volume of the N particles decreases.

An expression for the extended surface area θ_{ex} can be inferred from (4.54). First, an expression for the radius $r(t)$ will be derived:

$$
r(t) = \left(\frac{3}{2} \frac{1}{N\pi} \left(V^o + \frac{M_{Cd}}{\rho_{Cd}} \int_{t_o}^{t} J_{Cd}(t)dt \right) \right)^{\frac{1}{3}}
$$

$$
= \left(\frac{3}{2} \frac{1}{N\pi} \frac{M_{Cd}}{\rho_{Cd}} \left(m_{Cd}^o + \int_{t_o}^{t} J_{Cd}(t)dt \right) \right)^{\frac{1}{3}}
$$

(4.55a)

where $m^o{}_{Cd}$ denotes the molar amount of the cadmium nuclei at $t=t_o$. This amount equals the molar amount of the overdischarge reserve $Q_{Cd,Cd}$ (see Figure 4.16) during normal battery operation, which means without the occurrence of deep discharges, where part of the overdischarge reserve has been consumed. Surface area θ_{ex} is now inferred from:

$$
\theta_{ex} = N\pi r^2(t) = \frac{N\pi}{(N\pi)^{\frac{2}{3}}} \left(\frac{3}{2} \frac{M_{Cd}}{\rho_{Cd}} \left(m_{Cd}^o + \int_{t_o}^{t} J_{Cd}(t)dt \right) \right)^{\frac{2}{3}}
$$

$$
= (N\pi)^{\frac{1}{3}} \left(\frac{3}{2} \frac{M_{Cd}}{\rho_{Cd}} \left(m_{Cd}^o + \int_{t_o}^{t} J_{Cd}(t)dt \right) \right)^{\frac{2}{3}}
$$

(4.55b)

The actual surface area covered by metallic cadmium A_{Cd} becomes smaller than θ_{ex} when the growing cadmium particles start to merge, because a correction has to be made for the overlapping areas. The occurrence of overlapping areas is clarified in Figure 4.18.

The relation between the actual surface area covered by metallic cadmium and θ_{ex} can be inferred from Avrami's theorem [34]:

$$
A_{Cd} = A_{Cd}^{max}(1 - e^{-\frac{\theta_{ex}}{A_{Cd}^{max}}})
$$

(4.56)

where A_{Cd}^{max} denotes the maximum surface area that can be covered by cadmium particles, i.e. the surface area of the cadmium electrode in $[m^3]$.

Diffusion limitation of the cadmium reactions
The only species that can be depleted at the electrode/electrolyte interface in the case of the cadmium reaction are OH^- ions. It was assumed for the nickel reaction that the surface activity of OH^- ions equals the bulk activity. This means that diffusion limitation of the nickel reaction due to depletion of OH^- ions was assumed to be negligible. This is justified, because the diffusion of H^+ ions is dominant in determining the diffusion overpotential for the nickel reaction. This is explained by the diffusion constant of H^+ ions, which is four orders of magnitude smaller than the

diffusion constant of OH⁻ ions [35],[36]. Diffusion of OH⁻ ions has been considered for the cadmium reaction for completeness, because OH⁻ ions are the only species that can be depleted from the electrode/electrolyte surface in this case. This means that the surface activity of OH⁻ ions may differ from the bulk activity. As was shown in (4.24), the apparent equilibrium potential $E_{Cd}^{eq^*}$ can be written as the sum of the true equilibrium potential of (4.52) and a diffusion overpotential η_{Cd}^{d}, which depends on the difference between the bulk and surface activities of OH⁻ ions.

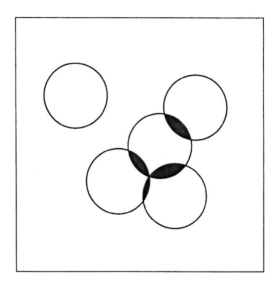

Figure 4.18: Overlapping metallic cadmium nuclei growing on the electrode surface

$$
\begin{aligned}
E_{Cd}^{eq^*} &= \frac{-\Delta G_{Cd,R}^{o} - RT \ln\left(\dfrac{a_{OH^-}^{s}}{a_{OH^-}^{ref}}\right)^2}{n_{Cd}F} \\[2em]
&= \frac{-\Delta G_{Cd,R}^{o} - RT \ln\left(\dfrac{a_{OH^-}^{b}}{a_{OH^-}^{ref}}\right)^2}{n_{Cd}F} + \frac{RT}{n_{Cd}F}\ln\left(\frac{\left(a_{OH^-}^{b}\right)^2}{\left(a_{OH^-}^{s}\right)^2}\right) \\[2em]
&= E_{Cd}^{eq} + \eta_{Cd}^{d} = E_{Cd}^{o} + \eta_{Cd}^{d}
\end{aligned}
\tag{4.57}
$$

The mass transport through diffusion of OH⁻ ions can be similarly described by means of an RC-ladder network, as described above for the mass transport of H⁺ ions in the nickel electrode. However, a simplification can be made for OH⁻ ions by considering stationary diffusion with a linear concentration gradient across a thin diffusion layer with a fixed thickness d_{OH^-}. This assumption is allowed when the time necessary to build the diffusion layer is much smaller than a typical battery charge/discharge time. The diffusion layer thickness can be assumed to be in the order of 1 μm for OH^- ions, while the diffusion constant D_{OH^-} is in the order of 10^{-10} m²/s [36]. A general law relates the thickness of the diffusion layer to the diffusion constant and the time needed to build the diffusion layer according to $d^2_{OH^-}/(\pi D_{OH^-})$ [37]. The time needed to build the diffusion layer in the case of OH^- ions is 0.3 ms, using the numbers mentioned, which is indeed a lot smaller than normal charge/discharge times. On the other hand, the diffusion constant for H^+ ions D_{H^+} is in the order of 10^{-14} m²/s [35]. The time needed to build a diffusion layer of 1 μm is indeed 3 s. Therefore, stationary diffusion cannot be assumed for H^+ ions and an RC-ladder network should be used.

In the case of stationary diffusion, the diffusion current $I_{OH^-}^{dif}$ can be inferred from Fick's first law, as described in chapter 3, with $I = nFJ$ and $a = \gamma c$. Hence,

$$I_{OH^-}^{dif} = nFJ_{OH^-}^{dif} = \frac{nFD_{OH^-}A_{Cd}a_{OH^-}^b}{\gamma_{OH^-}d_{OH^-}} \qquad (4.58)$$

where $n=2$ and the activity coefficient γ_{OH^-} will again be considered to be unity. Area A_{Cd} defined by (4.56) has to be used, because the charge transfer reaction takes place at the cadmium surface only. The kinetic current I_{Cd}^{kin} is described by the Butler-Volmer equation for the cadmium reaction of (4.37), with the exchange current I^o_{Cd} being given by (4.53). A simple way of including the effect of OH^- diffusion on the overpotential of the cadmium reaction is to find an expression for the overall reaction current. This current is determined by the kinetic current for small overpotentials and by the diffusion-limited current $I_{OH^-}^{dif}$ for large overpotentials; see Figure 3.4. The overall reaction current can be found in [15], assuming a mixed kinetic-diffusion controlled reaction:

$$I_{Cd} = \frac{I_{Cd}^{kin}I_{OH^-}^{dif}}{I_{Cd}^{kin} + I_{OH^-}^{dif}} \qquad (4.59)$$

Of course, the effect of OH^- ion diffusion on the cadmium electrode potential can also be modelled by means of an RC-ladder network. This has not been considered in the present NiCd model for simplicity.

4.2.4   The oxygen reactions

The reaction equation for oxygen production was given above, by (4.38), whereas that for oxygen recombination was given by (4.39). When oxygen production occurs at one electrode, oxygen recombination will occur at the other electrode and *vice versa*, as described in section 4.2.1. Where applicable, the electrode at which the reaction takes place will be mentioned in the text.

Thermodynamics of the oxygen reactions

The change in free energy $\Delta G_{O_2,o}$ in the oxidation reaction given by (4.38), which occurs, for example, at the nickel electrode during overcharging, is given by:

$$\Delta G_{O_2,0} = \mu_{O_2} + 2\mu_{H_2O} + 4\mu_{e^-} - 4\mu_{OH^-}$$

$$= \mu_{O_2}^o + 2\mu_{H_2O}^o + 4\mu_{e^-}^o - 4\mu_{OH^-}^o$$

$$+ RT \ln\left(\frac{a_{O_2}\left(a_{H_2O}\right)^2 \left(a_{OH^-}^{ref}\right)^4}{a_{O_2}^{ref}\left(a_{H_2O}^{ref}\right)^2 \left(a_{OH^-}\right)^4} \right)$$

$$= \Delta G_{O_2,0}^o + RT \ln\left(\frac{\left(a_{H_2O}\right)^2 \left(a_{OH^-}^{ref}\right)^4}{\left(a_{H_2O}^{ref}\right)^2 \left(a_{OH^-}\right)^4} \right) + RT \ln\left(\frac{a_{O_2}}{a_{O_2}^{ref}} \right)$$

$$= \Delta G_{O_2,0}^{o'} + RT \ln\left(\frac{a_{O_2}}{a_{O_2}^{ref}} \right)$$

(4.60)

The a_{OH^-} and a_{H_2O} activities have again been considered to be constant for simplicity and their reference activities have been taken the same. Hence, the ln term that contains these constant activities is added to $\Delta G^o_{O_2,0}$ to yield the term $\Delta G^{o'}_{O_2,0}$, while the ln term that contains the reference activities is zero. The following expression for the equilibrium potential of the oxygen reaction is inferred from (4.60):

$$E_{O_2}^{eq} = \frac{\Delta G_{O_2,0}}{n_{O_2}F} = \frac{\Delta G_{O_2,0}^{o'}}{n_{O_2}F} + \frac{RT}{n_{O_2}F} \ln\left(\frac{a_{O_2}}{a_{O_2}^{ref}} \right)$$

$$= E_{O_2}^o + \frac{RT}{n_{O_2}F} \ln\left(\frac{a_{O_2}}{a_{O_2}^{ref}} \right)$$

(4.61)

where $n_{O_2} = 4$, according to (4.38). Again, the same expression can be derived by considering the reduction reaction of (4.39). The bulk activity of O_2 should be used, which equals the surface activity in a state of true equilibrium, because (4.61) expresses the true equilibrium potential. The value of $E^o_{O_2}$ in a strong alkaline is 0.40 V vs SHE at 298 K [31].

The oxygen inside the battery is present in either the gas phase (g) in the free volume inside the battery or dissolved in the electrolyte (l). The oxygen dissolution reaction can be represented by:

$$O_2(l) \underset{k_2}{\overset{k_1}{\rightleftharpoons}} O_2(g)$$

(4.62)

where k_1 and k_2 are the corresponding reaction rate constants. Henri's law applies in the case of an ideal gas [21]. This law relates the activity of oxygen dissolved in the electrolyte to the oxygen pressure in the gas phase through:

$$a_{O_2} = \frac{\gamma_g k_2}{k_1} P_{O_2} = K_{O_2} P_{O_2} \tag{4.63}$$

where (4.63) also holds for the activity in the reference state $a_{O_2}^{ref}$ and the reference oxygen pressure $P_{O_2}^{ref}$, which is taken to be 1 atmosphere ($\approx 10^5$ Pa). Moreover, γ_g denotes the fugacity coefficient in [mol/(m³.Pa)] and K_{O_2} the oxygen solubility constant in [mol/(m³.Pa)]. Using (4.63), the equilibrium potential $E_{O_2}^{eq}$ of (4.61) can be rewritten as:

$$E_{O_2}^{eq} = E_{O_2}^o + \frac{RT}{n_{O_2} F} \ln(\frac{P_{O_2}}{P_{O_2}^{ref}}) \tag{4.64}$$

Kinetics of the oxygen reactions
The exchange current $(I_{O_2}^o)_{Ni}^0$ for the oxygen production reaction that occurs at the nickel electrode surface A_{Ni} during overcharging is given by:

$$(I_{O_2}^o)_{Ni}^0 = n_{O_2} F A_{Ni} \left(k_{a,O_2}^{Ni} \right)^{-\alpha_{O_2}} \left(k_{c,O_2}^{Ni} \right)^{\alpha_{O_2}} \cdot$$
$$(a_{O_2}^b)^{\alpha_{O_2} x_{O_2}} (a_{H_2O}^b)^{\alpha_{O_2} y_{O_2}} (a_{OH^-}^b)^{(1-\alpha_{O_2}) k_{O_2}} \tag{4.65}$$

where the bulk activities of OH⁻ ions and water are again considered to be constant for simplicity, k_{a,O_2}^{Ni} and k_{c,O_2}^{Ni} are the anodic and cathodic reaction rate constants of the oxygen production reaction at the nickel electrode, respectively, and x_{O_2}, y_{O_2} and z_{O_2} again denote the reaction orders. In order to reduce the complexity of the mathematical treatment, in the simple expression for the anodic branch of the Butler-Volmer equation, which describes the oxygen evolution, it is assumed that the surface coverage of oxygen molecules remains negligibly low and that the concentrated KOH electrolyte behaves like an ideal solution. An expression similar to (4.65), but in which A_{Ni} has been replaced by A_{Cd} and with different values for the reaction rate constants, holds for the oxygen production reaction that occurs at the cadmium electrode during the second overdischarging step. This was described in section 4.2.1.

The oxygen that has been formed at the nickel electrode during overcharging will eventually be reduced at the cadmium electrode, as was explained in section 4.2.1. The oxygen can be transported to the cadmium electrode through the electrolyte or through the gas phase. It is most likely that the oxygen will be transported mainly through the free gas volume inside the battery [38], since the solubility of oxygen in the highly concentrated KOH electrolyte solution is relatively low, i.e. 4.10^{-5} mol/l in 8 M KOH [39]. The presence of water is essential for the recombination reaction, as can be inferred from the oxygen recombination reaction of (4.39). Therefore, the dissolution of oxygen in the electrolyte and its transport to the electrode surface through the electrolyte is an important step in the oxygen reduction reaction. The kinetics of the oxygen reduction reaction are

described by the cathodic branch of the Butler-Volmer equation. The exchange
current of the oxygen reduction reaction that takes place at the free metallic
cadmium surface A_{Cd} during overcharging is inferred from:

$$
(I_{O_2}^o)_{Cd}^R = n_{O_2} FA_{Cd} \left(k_{a,O_2}^{Cd}\right)^{1-\alpha_{O_2}^R} \left(k_{c,O_2}^{Cd}\right)^{\alpha_{O_2}^R} \cdot
$$
$$
(a_{O_2}^b)^{\alpha_{O_2}^R x_{O_2}^R} (a_{H_2O}^b)^{\alpha_{O_2}^R y_{O_2}^R} (a_{OH^-}^b)^{(1-\alpha_{O_2}^R) k_{O_2}^R}
$$

(4.66)

where the charge transfer coefficient $\alpha_{O_2}^R$ may have a different value than the charge
transfer coefficient α_{O_2} for the oxygen production reaction and the same remark
holds for the anodic and cathodic reaction rate constants k_{a,O_2}^{Cd} and k_{c,O_2}^{Cd} in relation
to those in (4.65). An expression similar to (4.66) can be used for the oxygen
reduction at the nickel electrode during the second overdischarging step. This
reaction then competes with the cadmium reduction reaction of (4.40) that takes
place at the surface of the exposed cadmium at the nickel electrode A_{Cd}^{Ni}. Hence, the
area of this cadmium should be used in the equation for the exchange current of the
oxygen reduction reaction that takes place during the second overdischarging step.

Diffusion limitation of the oxygen reactions
For the oxygen reactions, water is assumed to be present in abundance. Therefore,
mass transport limitation can only occur as a result of depletion of either O_2
molecules or OH^- ions. The diffusion layer thickness is assumed to be equal for both
species and in the order of 1 μm. The diffusion constant for both species is in the
order of 10^{-10} m^2/s [36],[39]. Therefore, stationary diffusion can be assumed for both
species, as discussed in the previous section. This leads to a diffusion limitation
current for OH^- ions, defined by (4.58), for which the appropriate area A_i should be
used, which will depend on the electrode at which the O_2 reaction occurs. A similar
expression can be found for stationary diffusion of O_2 molecules through a diffusion
layer with a thickness d_{O_2}. This diffusion occurs for the reduction reaction of (4.39):

$$
I_{O_2}^{dif} = \frac{nFA_i D_{O_2} a_{O_2}^b}{\gamma_{O_2} d_{O_2}}
$$

(4.67)

where $n=4$, D_{O_2} is the diffusion coefficient of oxygen in [m^2/s] and the activity
coefficient γ_{O_2} will again be considered to be unity. The appropriate area should be
used for A_i.

(4.58) and (4.67) show that the diffusion limitation currents of OH^- ions and O_2
molecules depend on the bulk concentrations of OH^- ions and O_2, respectively. As
the bulk activity of OH^- ions will be a lot higher than that of O_2 molecules in the
electrolyte due to the strong alkaline electrolyte, only diffusion limitation of O_2
molecules is considered for the oxygen reactions in the NiCd model. A difference
between the oxygen bulk activity and the oxygen activity at the electrode surface
gives rise to a diffusion overpotential $\eta_{O_2}^d$. The apparent equilibrium potential $E_{O_2}^{eq*}$
can be written as the sum of the true equilibrium potential $E_{O_2}^{eq}$ of (4.61) and $\eta_{O_2}^d$,
in accordance with (4.24):

$$
\begin{aligned}
E_{O_2}^{eq\,*} &= E_{O_2}^{o} + \frac{RT}{n_{O_2}F} \ln\left(\frac{a_{O_2}^{s}}{a_{O_2}^{ref}}\right) \\
&= E_{O_2}^{o} + \frac{RT}{n_{O_2}F} \ln\left(\frac{a_{O_2}^{b}}{a_{O_2}^{ref}}\right) + \frac{RT}{n_{O_2}F} \ln\left(\frac{a_{O_2}^{s}}{a_{O_2}^{b}}\right) \\
&= E_{O_2}^{eq} + \eta_{O_2}^{d}
\end{aligned}
\tag{4.68}
$$

Whether or not diffusion limitation plays a role in the reaction will depend on the value of the overpotential. This was shown in Figure 3.4, in which it was illustrated that the electrode current levels off to the diffusion-limited current for large overpotentials. The electrode current is however determined mainly by the reaction kinetics at relatively low overpotentials. A mixed kinetic-diffusion-controlled reaction takes place in the intermediate region, as considered before in section 4.2.3. The values of the overpotentials for the oxygen reactions which take place at the nickel and cadmium electrodes during overcharging and overdischarging differ substantially.

Oxygen is produced at the nickel electrode during overcharging, as explained in section 4.2.1. The reaction takes place at an electrode potential that is at least 120 mV more positive than the equilibrium potential for the oxygen reaction. This can be understood by considering the difference between E^{o}_{Ni} of 0.52 V vs SHE and $E^{o}_{O_2}$ of 0.40 V vs SHE. In this case it is assumed that the kinetics of the reaction can be simply described by the anodic branch of the Butler-Volmer relationship. It is hence assumed that the overpotential is generally not large enough for diffusion limitation of oxygen to be a limiting factor. This means that the reaction current does not level off to the diffusion-limited current at the usual values of the overpotential.

The oxygen reduction reaction during overcharging takes place at a very high overpotential because the cadmium electrode potential is very negative with respect to $E^{o}_{O_2}$. Here, it is reasonable to assume that mass transport of oxygen through the electrolyte to the free metallic cadmium surface becomes a limiting factor. Whether the electrode current will equal the diffusion-limited current or be less will depend on the parameter values of the oxygen reaction and the oxygen gas pressure through the bulk activity of oxygen; see (4.67). Therefore, as for the cadmium reaction, a mixed kinetic-diffusion-controlled reaction is assumed for the oxygen reduction reaction. Hence, the reaction current for the oxygen reduction reaction is given by:

$$
I_{O_2}^{R} = \frac{I_{O_2}^{kin} I_{O_2}^{dif}}{I_{O_2}^{kin} + I_{O_2}^{dif}}
\tag{4.69}
$$

Again, the diffusion of oxygen can also be modelled by an RC-ladder network. This has not been considered in the present NiCd model for simplicity.

Oxygen is produced at the cadmium electrode and reduced at the nickel electrode during the second overdischarging step, as explained in section 4.2.1. The situation described for overcharging is now reversed. The oxygen evolution at the cadmium electrode takes place at a relatively low overpotential. Therefore, only the anodic branch of the Butler-Volmer equation is considered in the NiCd model. The reduction reaction takes place at the cadmium surface at the nickel electrode at relatively high overpotentials, like the oxygen reduction during overcharging. This again leads to the use of a mixed kinetic-diffusion-controlled reaction in the model, as described by (4.69).

Oxygen gas pressure calculation
The relation between the oxygen gas pressure inside the battery and the molar amount of oxygen is described by the general gas law [21]:

$$P_{O_2} = \frac{RTm_{O_2}}{V_g}$$
(4.70)

where m_{O_2} denotes the molar amount of oxygen in the gas phase in [mol] and V_g is the free gas volume inside the battery in [m^3]. The solubility of oxygen in the highly concentrated KOH electrolyte solution is poor, as described above. Therefore, the contribution to m_{O_2} of oxygen dissolved in the electrolyte is neglected [38]. The amount of oxygen in the gas phase will depend on the difference between the amount of oxygen produced at one electrode and the amount reduced at the other. Therefore, the net molar amount of oxygen inside the battery can be found by integrating the net chemical flow of oxygen inside the battery:

$$\frac{dm_{O_2}}{dt} = J_{O_2}^{Ni} + J_{O_2}^{Cd} - J_{O_2}^{R,Ni} - J_{O_2}^{R,Cd}$$
$$= \frac{I_{O_2}^{Ni} + I_{O_2}^{Cd} - I_{O_2}^{R,Ni} - I_{O_2}^{R,Cd}}{n_{O_2}F}$$
(4.71)

where the latter term is inferred from the relation between the chemical and electrical domains, as explained in section 4.1.

4.2.5   Temperature dependence of the reactions

The temperature dependence of the reaction rate constants k_i and diffusion coefficients D_i in the equations is described by the Arrhenius relation for all reactions that take place inside the NiCd battery, which is given in a general form in [15]:

$$par(T) = par^o \exp(\frac{-E_{par}^a}{RT})$$
(4.72)

where $par(T)$ denotes a temperature-dependent parameter, which can be any reaction rate constant k_i or diffusion coefficient D_i, par^o denotes the pre-exponential factor and E_{par}^a is the activation energy of par in [J/mol]. The temperature dependence of the standard redox potentials E_i^o of the reactions i is given by [15],[31]:

$$E_i^o(T) = E_i^o(298[K]) + (T - 298[K])\frac{dE_i^o}{dT}$$

$$= E_i^o(298[K]) + (T - 298[K])\frac{\Delta S_i}{n_i F}$$

(4.73)

where ΔS_i denotes the entropy change associated with reaction i in [J/(mol.K)]. It is assumed that ΔS_i is constant within the limited temperature range in which batteries usually operate. The sign of ΔS_i will depend on whether reaction i is endothermic or exothermic, as explained in section 4.1.4.

4.2.6 The model

The complete simulation model for a sealed rechargeable NiCd battery based on the schematic representation given in Figure 4.16 is shown in Figure 4.19. This model is based on the general principles described in section 4.1. Figure 4.19a represents the equivalence of the electrochemical behaviour and Figure 4.19b that of the thermal behaviour. In Figure 4.19a, the nickel electrode model is shown on the left side and the cadmium electrode model on the right side. The electrodes are modelled by a series resistance R_i that represents the electronic conductivity of the electrode material, in series with a parallel connection of several reaction paths and a capacitance C_i^{dl} that represents the double-layer capacitance.

The electrodes are separated by a resistance R_e, which models the ionic conductivity of the electrolyte and the separator. This is a simplification of the model describing mass transport as a result of diffusion and migration in section 4.1.3 (see Figure 4.13) in which OH⁻ ions have replaced the M^{n+} ions, because both ion types react at the electrode surfaces. K^+ ions have replaced the X^{n-} ions, because neither ions react at the electrode surfaces. The bulk activity of the OH⁻ ions in the electrolyte is assumed to be constant and uniformly distributed, as described above. This means that no diffusion profile is present for the OH⁻ ions and, because of charge neutrality, the same remark holds for the K^+ ions. The bulk and surface activities of the OH⁻ ions are assumed to be the same for the nickel and oxygen reactions, as described above. Only for the cadmium reactions is a difference between the surface and bulk activities of the OH⁻ ions considered. This will be modelled in the anti-parallel O and R diodes, as will be described below, and not by two chemical capacitances describing the surface and bulk activities of the OH⁻ ions, respectively. This means that the molar amounts of the OH⁻ and K^+ ions are constant in each of the spatial elements $1,2,..p$ in the network shown in Figure 4.13. Consequently all these chemical capacitances can just as well be omitted from the network. The resistance string for K^+ ions will then become floating and can also be omitted. No current will flow through the geometric capacitances when the electrical field is assumed to be stationary. Hence, these capacitances can also be omitted. This means that transient effects in the electrolyte will not be considered. What remains is a string of resistances that describes the migration of OH⁻ ions from one electrode to another. This string of resistances can be transferred to the electrical domain and can be simply modelled by a resistance R_e with a constant relatively low value in the mΩ-range. This gives rise to a relatively small voltage drop across the electrolyte. This low voltage drop is due to the high concentration of the electrolyte.

In section 4.3 it will be shown for a Li-ion battery model how the electrolyte can be modelled taking into account mass transport as a result of diffusion and migration.

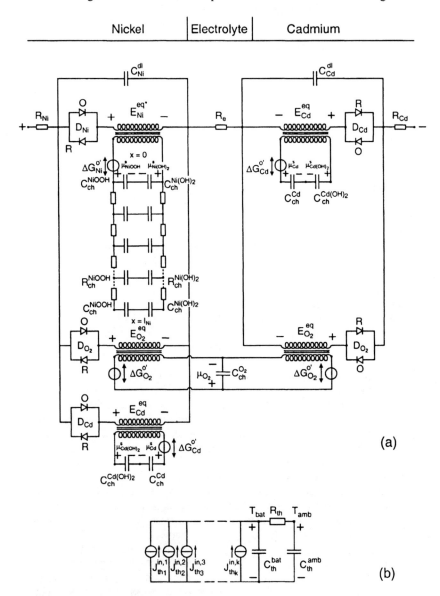

Figure 4.19: Simulation model in the electrical, chemical and thermal domains of a rechargeable NiCd battery based on the schematic representation of Figure 4.16. (a) Equivalence of electrochemical behaviour, (b) Equivalence of thermal behaviour

Three parallel reaction paths can be distinguished in the part of the model representing the nickel electrode, which from top to bottom denote: (i) the nickel storage reaction, (ii) the oxygen side-reaction and (iii) the cadmium side-reaction.

Two parallel reaction paths can be distinguished in the part of the model representing the cadmium electrode, which from top to bottom denote: (i) the cadmium storage reaction and (ii) the oxygen side-reaction.

Two domains can be distinguished in Figure 4.19a, notably the electrical and the chemical domain. Energy storage in the electrical domain is modelled by the double-layer capacitances C^{dl}, that in the chemical domain by chemical capacitances C_{ch}^i, where i denotes the reacting species. The coupling between the electrical and chemical domains is represented by ideal transformers, as described in section 4.1. Each reaction path j is modelled as an ideal transformer in series with two antiparallel diodes D_j, with one diode representing the oxidation (O) reaction and the other the reversed reduction (R) reaction, as indicated in the figure.

Only changes in the activities of NiOOH and Ni(OH)$_2$ are considered for the nickel reaction, as discussed in section 4.2.2. Therefore, only the capacitances C_{ch}^{NiOOH} and $C_{ch}^{Ni(OH)2}$ are present in the chemical domain. The source $\Delta G^{o'}_{Ni}$ has been defined in (4.45) for an oxidation reaction. This means that this source includes all μ^o values and hence the μ^o term is omitted in the electrochemical potentials μ across capacitances C_{ch}^{NiOOH} and $C_{ch}^{Ni(OH)2}$. $\Delta G^{o'}_{Ni}$ also includes the constant activities of water and OH$^-$ ions; see (4.45).

The diffusion process of H$^+$ ions in the x-direction inside the electrode is modelled by an RC-ladder network, as can be seen in Figure 4.19a. The electrode has been divided into spatial elements as shown in Figure 4.5. A network similar to that shown in Figure 4.8 is used, because no electrical field is present; see section 4.2.2. The capacitances on the left-hand side contain the molar amounts of the NiOOH species in spatial elements i and the capacitances on the right-hand side contain the molar amounts of the Ni(OH)$_2$ species in spatial elements i. The top of the RC-ladder network denotes the electrode/electrolyte interface at $x=0$, where the actual electrochemical reaction takes place, whereas the bottom denotes the electrode/current collector interface at $x=l_{Ni}$. The diffusing H$^+$ ions can only cross the electrode/electrolyte boundary due to the electrochemical reaction, whereas the electrode/current collector boundary cannot be crossed. The chemical resistances $R_{ch}^{Ni(OH)2}$ and R_{ch}^{NiOOH} are given by (4.21c), in which the diffusion coefficient of H$^+$ ions in the nickel electrode (D_{H^+}) is used for both resistances. The resistances have a value which depends on the position x (see Figure 4.8) because the value of the resistances depends on the molar amount m of the corresponding species at position x, according to (4.21c).

A concentration profile of Ni(OH)$_2$ and NiOOH species throughout the electrode material leads to a change in chemical potential of both species throughout the electrode. However, this change in chemical potential can only be measured in electrical terms as an extra voltage drop η^d across the electrode/electrolyte interface, as described by (4.24). Hence, the apparent equilibrium potential $E_{Ni}^{eq^*}$ for the nickel reaction is present across the electrical port of the transformer. The current-overpotential relation for the diodes D_{Ni} is described by the Butler-Volmer equation, with the exchange current I^o_{Ni} being given by (4.46).

Only the change in the activity of oxygen is considered for the oxygen reaction, as described in section 4.2.4. This is expressed by the capacitance C_{ch}^{O2}, which contains the total molar amount of oxygen present inside the battery. The contribution of oxygen dissolved in the electrolyte to the molar amount of oxygen is neglected; see section 4.2.4. Therefore, capacitance C_{ch}^{O2} contains the molar amount of oxygen present in the gas phase, which is directly linked to the internal oxygen pressure through (4.70). The sources $\Delta G^{o'}_{O_2}$ are defined by (4.60). The net chemical flow of oxygen production (see (4.71)) is into C_{ch}^{O2}, whereas that of oxygen reduction is out of C_{ch}^{O2}. This has been realized in the model shown in Figure 4.19 by connecting C_{ch}^{O2} to the oxygen reaction path at both electrodes with the signs indicated.

Only one capacitance describing the bulk molar amount of oxygen is present in the model. Hence, no distinction is made between the surface and bulk activities of oxygen in the chemical domain. This means that the voltage across the electrical port of the ideal transformer is the true equilibrium potential; see (4.61). The diffusion overpotential η^d is taken into account in the current-voltage relation for the anti-parallel diodes D_{O_2}. Hence, the overpotential η^{ft} across the anti-parallel diodes equals $\eta^k + \eta^d$. No diffusion limitation of oxygen is to be expected for the oxygen evolution reaction at either electrode; see section 4.2.4. Therefore, the anodic branch of the Butler-Volmer equation is used for the oxidation diodes. Here, (4.65) will be used for the nickel electrode and a similar expression with different parameters for the cadmium electrode. Diffusion limitation is to be expected in the oxygen reduction reactions. Therefore, (4.69) will be used to describe the current-voltage relationship for the reduction diodes in the nickel and cadmium electrode models, using the appropriate parameter values.

The activities of all the reacting species except OH⁻ can be considered to be constant in the cadmium reaction, as mentioned above. The molar amounts of both Cd and Cd(OH)₂ have been modelled by a chemical capacitance in Figure 4.19 in consistence with the rest of the model. However, it should be noted that in this case the chemical potential across these capacitances is actually constant. Therefore, these capacitances are actually sources with a constant chemical potential. The molar amount of Cd and Cd(OH)₂ can be found by integrating the chemical flow through these sources.

No capacitance that describes the molar amount of OH⁻ ions is to be found in the cadmium electrode model. The bulk activity of OH⁻ ions has been taken into account in source $\Delta G^{o'}_{Cd}$; see (4.52). The voltage across the electrical port of the transformer is the true equilibrium potential for the cadmium reaction. Diffusion limitation is taken into account in the current-voltage relationship for the anti-parallel diodes D_{Cd}, in a similar way as described for the oxygen reaction. The current-voltage relationship is described by (4.59) for both the oxidation and the reduction diodes in the nickel and cadmium electrodes. The same value is used for k_{Cd} in the kinetic current I_{Cd}^{kin} for both electrodes. However, the maximum surface area A_{Cd}^{max} in (4.56) is taken to be lower for the nickel electrode than for the cadmium electrode for the dependency of A_{Cd} on the amount of cadmium, because $Q_{Cd(OH)_2,Ni}$ is much smaller than $Q_{Cd}^{max} + Q_{Cd,Cd}$ in Figure 4.16.

The thermal behaviour outlined in Figure 4.19b is modelled in the same way as discussed in section 4.1.4. The contributions to the internally generated heat flow are described by (4.32). The calculated battery temperature is used in the model equations to model the temperature dependence.

During current flow, the current divides itself over the available reaction paths in each electrode. Which reactions will occur will depend on the value of the equilibrium potential of each reaction and the rate at which each reaction can take place. In principle, the redox reaction with the lowest value of the equilibrium potential will thermodynamically be more favourable than a redox reaction with a higher value of the equilibrium potential. Applying Kirchhoff's voltage law to two parallel reaction paths reveals that the thermodynamically favourable reaction has a higher overpotential η across the antiparallel diodes. However, this does not necessarily mean that this reaction will actually take place, because the reaction rate might be too low, for example due to a low exchange current or diffusion limitation. More information on the parameter values used in the NiCd model will be given in section 4.4.

4.3   A simulation model of a rechargeable Li-ion battery

4.3.1   Introduction

The general principles described in section 4.1 will be used in this section to construct a network model for a rechargeable Li-ion battery [40],[41]. Again, there are many different types of Li-ion batteries. The CGR17500 Li-ion battery developed by Panasonic will be modelled in this section. This battery consists of a $LiCoO_2$ positive electrode and a graphite LiC_6 negative electrode. The electrolyte of a Li-ion battery consists of a salt dissolved in an organic solvent; see chapter 3. The dissolved salt is $LiPF_6$ in the case of the CGR17500 Li-ion battery considered in this section, which yields Li^+ and PF_6^- ions in the organic solvent. This solvent is a 1:3 mixture of ethylene carbonate (EC) and ethyl methyl carbonate (EMC). The use of EC as a co-solvent is of major importance for the performance of the Li-ion battery, because it forms a protective layer on the graphite electrode [40]. This protective layer is denoted as a Solid Electrolyte Interface (SEI) and it allows only Li^+ ions to be transported to and from the graphite electrode. It prevents the risk of solvent molecules entering the graphite electrode together with these ions [43]. The SEI is electrically insulating and, when sufficiently thick, it suppresses the decomposition of the electrolyte at the electrode surface. Besides the two electrodes and the electrolyte solution, a separator is present inside the battery. A Positive-Temperature Coefficient resistor (PTC) is also present inside the battery for safety reasons; see chapter 3. The SEI will not be considered in the Li-ion model described in this section. A PTC can be modelled by a series resistance. The CGR17500 is a cylindrical battery, with a diameter of 17 mm and a height of 50 mm. An external series safety switch has to be added in applications, as described in chapters 2 and 3.

The overall reaction during charging and discharging of the considered Li-ion battery, in which x Li^+ ions are involved, is given by:

$$xLiCoO_2 + xC_6 \overset{ch}{\underset{d}{\rightleftarrows}} xCoO_2 + xLiC_6 \qquad (4.74)$$

Li^+ ions are moved back and forth between the electrodes, as expressed in (4.74), which explains the frequently encountered term 'rocking-chair battery'. The Li^+ ions move from the $LiCoO_2$ electrode to the graphite electrode during charging. They move in the other direction during discharging. Both electrodes are so-called intercalation electrodes, with the Li^+ ions being guests that can be inserted into or extracted from the host lattice. In an intercalation electrode, the host lattice does not change significantly upon insertion of the guest atoms [44].

Unlike in the NiCd battery model presented in the previous section, no side-reactions have been modelled. Side-reactions that could take place inside a Li-ion battery include Li metal deposition and electrolyte decomposition. These side-reactions could be responsible for aging effects leading to gradual capacity decrease during the battery's lifetime. So these side-reactions are important. However, they will be left out of the model presented in this book, because of insufficient knowledge of their characteristics. Hence, in each of the electrodes only one reaction is included in the model. The two electrode reactions will be described separately in the next sections. Moreover, the behaviour of the electrolyte will be described.

4.3.2   The $LiCoO_2$ electrode reaction

Li^+ ions are extracted from the $LiCoO_2$ electrode during charging and inserted into it during discharging. The extraction and insertion of Li^+ ions occurs through oxidation and reduction, respectively, as described by the following basic reaction equation for one Li^+ ion:

$$LiCoO_2 \underset{k_{c,pos}}{\overset{k_{a,pos}}{\rightleftharpoons}} CoO_2 + Li^+ + e^- \tag{4.75}$$

where $k_{a,pos}$ and $k_{c,pos}$ are the anodic and cathodic rate constants of the $LiCoO_2$ (pos=positive) electrode reaction, respectively. The electrode is modelled in a way comparable with that used to model of the nickel electrode in section 4.2. The electrode is considered to be a solid-solution electrode, with $LiCoO_2$ being the *Red* species and CoO_2 the *Ox* species. The *Ox* species actually denotes the portion of the electrode material from which Li^+ ions have been extracted. It is assumed that $LiCoO_2$ and CoO_2 are the only species present in the electrode. As a result, the sum of the mol fractions of $LiCoO_2$ and CoO_2 is unity. As in the NiCd model, volume changes in the active material during charging and discharging are neglected. The positions of the CoO_2 centres in the electrode lattice are fixed and the Li^+ ions are transported through the electrode. Hence, the $LiCoO_2$ species can be compared with the $Ni(OH)_2$ species in the NiCd battery and the CoO_2 species with the $NiOOH$ species. The Li^+ ions can be compared with the H^+ ions in the NiCd battery. The mol fraction of CoO_2 corresponds to the SoC of the Li-ion battery, because the CoO_2 species is formed during charging. The mol fraction of $LiCoO_2$ equals the mol fraction of Li^+ ions inside the positive electrode, i.e. ($\gamma=1$ assumed):

$$x_{Li}^{pos} = x_{LiCoO_2}^{pos} = \frac{a_{LiCoO_2}}{a_{LiCoO_2} + a_{CoO_2}} \tag{4.76}$$

In theory, the mol fraction x_{Li}^{pos} can vary between 0, when all the Li^+ ions have been extracted, and 1, when all the lattice sites have been filled. In the latter case there are no longer any vacancies for Li^+ ions. In practice, only part of this full capacity can be used reversibly under normal operating conditions. How large this part will be will depend on the material used for the positive electrode. For example, the reversible range of x_{Li}^{pos} is less than or equal to 0.5 in the case of the material used in the CGR17500 Li-ion battery [18]. For the CGR17500 battery this means that the $LiCoO_2$ electrode will only cycle between $x_{Li}^{pos} = 0.5$ and $x_{Li}^{pos} = 0.95$ under normal operating conditions. This means that the rated nominal capacity, derived from the range $0.5 < x_{Li}^{pos} < 0.95$, will be only a little less than half of the full capacity of the $LiCoO_2$ electrode in this case. The mol fraction of the CoO_2 species can be derived in a similar way as shown in (4.76). This mol fraction equals $1 - x_{Li}^{pos}$.

Thermodynamics of the $LiCoO_2$ electrode reaction

The change in free energy $\Delta G_{LiCoO_2,0}$ in the oxidation reaction that occurs at the $LiCoO_2$ electrode during charging can be found in a similar way as shown in (4.10).

As discussed above, the LiCoO$_2$ electrode is modelled in a similar way as the nickel electrode. The equilibrium potential of the nickel electrode can be described by the Nernst equation; see (4.45). However, in the case of the extraction and insertion of Li$^+$ ions from and into an intercalation electrode, such as the LiCoO$_2$ electrode, the electrical and elastic interaction between the intercalated Li$^+$ ions and between Li$^+$ ions and the host atoms has to be accounted for [44],[45]. This interaction can be described with the aid of an extra interaction energy term. The change in free energy can be inferred from (4.75) as follows:

$$
\Delta G_{LiCoO_2,0} = \mu_{CoO_2} + \mu_{Li^+} + \mu_{e^-} - \mu_{LiCoO_2}
$$
$$
- RTU_{pos}x_{Li}^{pos} + RT\zeta_{pos}
$$

$$
= \mu_{CoO_2}^o + \mu_{Li^+}^o + \mu_{e^-}^o - \mu_{LiCoO_2}^o + RT\ln\left(\frac{a_{CoO_2}a_{Li^+}a_{LiCoO_2}^{ref}}{a_{LiCoO_2}a_{CoO_2}^{ref}a_{Li^+}^{ref}}\right) \qquad (4.77)
$$
$$
- RTU_{pos}x_{Li}^{pos} + RT\zeta_{pos}
$$

$$
= \Delta G_{LiCoO_2,0}^o + RT\ln\left(\frac{a_{CoO_2}}{a_{LiCoO_2}}\right) - RTU_{pos}x_{Li}^{pos} + RT\zeta_{pos}
$$

where U_{pos} denotes the dimensionless interaction energy coefficient in the LiCoO$_2$ electrode, RTU_{pos} is the interaction energy in [J/mol], x_{Li}^{pos} is defined in (4.76), and ζ_{pos} denotes a dimensionless constant in the LiCoO$_2$ electrode, where $RT\zeta_{pos}$ is an interaction energy term which does not depend on x_{Li}^{pos}. As before, bulk activities have been used in (4.77) because the equilibrium condition is valid. All reference activities a^{ref} have been assumed the same in the last line, as in earlier derivations, and the activity a_{Li^+} of Li$^+$ ions has been assumed to be the same as its reference activity for simplicity. This has been done because the electrolyte concentration is approximately 1 M and no concentration profile is present in the electrolyte in a state of equilibrium. As before, the activity of electrons has been taken the same as the reference activity a_e^{ref}. The following expression is then found for the equilibrium potential of the LiCoO$_2$ electrode:

$$
E_{LiCoO_2}^{eq} = \frac{\Delta G_{LiCoO_2,0}}{n_{Li}F}
$$

$$
= \frac{\Delta G_{LiCoO_2,0}^o}{n_{Li}F} + \frac{RT}{n_{Li}F}\left[\ln\left(\frac{a_{CoO_2}}{a_{LiCoO_2}}\right) - U_{pos}x_{Li}^{pos} + \zeta_{pos}\right] \qquad (4.78)
$$

$$
= E_{LiCoO_2}^o + \frac{RT}{n_{Li}F}\left[\ln\left(\frac{1-x_{Li}^{pos}}{x_{Li}^{pos}}\right) - U_{pos}x_{Li}^{pos} + \zeta_{pos}\right]
$$

where $n_{Li} = 1$ (see (4.75)) and the extra terms due to the interaction between the intercalated Li$^+$ ions and between these ions and the host material are clearly identifiable. The same expression can be derived by considering the reduction reaction of (4.75). This was shown in section 4.1.2. Both activities in the ln term have been replaced by the corresponding mol fractions, as defined in (4.76).

The expression for $E^{eq}_{LiCoO_2}$ in (4.78) is in agreement with an expression for the equilibrium potential for the $LiCoO_2$ electrode given in [45]. The expression for the equilibrium potential is similar to the Nernst-like equilibrium potentials derived in previous sections in this chapter when the interaction energy between the intercalated Li^+ ions is zero, or $U_{pos} = 0$.

The equilibrium potential drops more steeply with increasing x_{Li}^{pos} in the case of repulsive interaction, which occurs when $U_{pos} > 0$. Repulsive interaction means that if we place a Li^+ ion and an electron at one site in the lattice, the nearest neighbour sites will be forbidden for the next ion that is inserted [45]. The repulsive interaction can lead to the formation of a 'superlattice' inside the electrode, in which the intercalated Li^+ ions are ordered at fixed distances, so that interactions are avoided as much as possible, which means that maximum distance between nearest neighbours is achieved [44]. One can easily imagine a repulsive electrical interaction between ions of the same charge. If Li^+ ions are intercalated in layers, one can imagine that repulsive elastic interaction will result when a Li^+ ion is present in layer n, thereby leaving less space for intercalation directly above and below the ion in layers $n-1$ and $n+1$.

The equilibrium potential will drop less steeply than with pure Nernst-like behaviour with an increasing x_{Li}^{pos} for attractive interaction, where $U_{pos} < 0$. Attractive interaction means that if we place a Li^+ ion and an electron at one site in the lattice, the nearest neighbour sites will be preferred for the next insertion [45]. In an extreme case, the equilibrium potential can become constant for a certain range of x_{Li}^{pos}. Phase separation will then occur and the intercalated Li^+ ions will condense in two separate phases [44],[45]. A phase is characterized by a certain lattice structure. The energy of interaction between the phases is negligible in comparison with the interaction energy within each phase [44]. In simple words, the Li^+ ions stick together in two separate phases. Screening of the Li^+ ions by electrons present in the lattice will occur, which prevents that the Li^+ ions show repulsive behaviour. Therefore, attractive electrical interaction can take place. One can imagine attractive elastic interaction in the above example in which the Li^+ ion was intercalated in layer n in the lattice. As a result, more space will be created for the sites directly next to this ion in the same layer.

When Li^+ ions are intercalated in a host lattice, phase transitions between possible lattice structures may occur depending on the number of Li^+ ions present in the host lattice [44]. Each possible lattice structure is characterized by a certain amount of free energy, which will vary during intercalation. Depending on the free energy values of each of the possible lattice structures, a first-order phase transition will take place to the lattice structure with the smallest amount of free energy. The two phases may exist together over a certain range of x_{Li}^{pos} when attractive interaction occurs. This was discussed above.

Two first-order phase transitions occur within the $LiCoO_2$ electrode at approximately $x_{Li}^{pos} = 0.25$ and $x_{Li}^{pos} = 0.75$ [46]. The phase transition at $x_{Li}^{pos} = 0.25$ is not noticed during normal battery operation, because the battery is not cycled in this region, as previously discussed. The other phase transition is noticed as a change in the slope of the true equilibrium potential as a function of x_{Li}^{pos}. This change in the slope of the equilibrium potential is realized in the present Li-ion battery model by a change in the interaction energy between the intercalated Li^+ ions from a value $U_{pos,1}$ in phase 1 to a value $U_{pos,2}$ in phase 2. The values of the dimensionless constants $\zeta_{pos,1}$ and $\zeta_{pos,2}$ in phases 1 and 2, respectively, are chosen so that a smooth transition is achieved from the equilibrium potential in phase 1 to that in phase 2. This modelling approach is similar to the one described in [45]. The two equilibrium potentials in the different phases for the considered CGR17500 Li-ion battery are described by:

$$\left(E^{eq}_{LiCoO_2}\right)_{phase1} = E^o_{LiCoO_2} + \frac{RT}{n_{Li}F}\left[\ln\left(\frac{1-x^{pos}_{Li}}{x^{pos}_{Li}}\right) - U_{pos,1}x^{pos}_{Li} + \zeta_{pos,1}\right] \quad (4.79)$$

for $0.75 \le x_{Li}^{pos} \le 0.95$, and

$$\left(E^{eq}_{LiCoO_2}\right)_{phase2} = E^o_{LiCoO_2} + \frac{RT}{n_{Li}F}\left[\ln\left(\frac{1-x^{pos}_{Li}}{x^{pos}_{Li}}\right) - U_{pos,2}x^{pos}_{Li} + \zeta_{pos,2}\right] \quad (4.80)$$

for $x_{Li}^{pos} < 0.75$. The equilibrium potential is found to be independent of x_{Li}^{pos} in phase 1, which means that this is a two-phase region and the interaction is attractive [46]. Phase 1 and phase 2 actually co-exist in this region of x_{Li}^{pos}. The slope of the equilibrium potential is steeper than predicted by pure Nernst behaviour in phase 2. Hence, the interaction becomes repulsive in phase 2. In order to obtain a smooth transition of the equilibrium potential at the phase transition, the relation between $\zeta_{pos,1}$ and $\zeta_{pos,2}$ is defined as:

$$\begin{aligned}\zeta_{pos2} &= (U_{pos,2} - U_{pos,1})x^{pos}_{phasetransition} + \zeta_{pos,1} \\ &= (U_{pos,2} - U_{pos,1})\cdot 0.75 + \zeta_{pos,1}\end{aligned} \quad (4.81)$$

It should be noted that a smooth transition can also be obtained by assigning a different value to the standard redox potential $E^o_{LiCoO_2}$ in each of the two phases. The terms $\zeta_{pos,1}$ and $\zeta_{pos,2}$ can then of course be omitted in (4.79) and (4.80), respectively [45].

Kinetics of the LiCoO_2 electrode reaction

The reaction kinetics of the charge-transfer reaction that takes place at the surface of the LiCoO_2 electrode are described by the Butler-Volmer relationship. The exchange current value for this reaction is determined by:

$$\begin{aligned}I^o_{LiCoO_2} &= n_{Li}FA_{LiCoO_2}\left(k_{a,pos}\right)^{-\alpha_{LiCoO_2}}\left(k_{c,pos}\right)^{\alpha_{LiCoO_2}} \cdot \\ &\quad (a^b_{CoO_2})^{\alpha_{LiCoO_2}x_{LiCoO_2}}(a^b_{LiCoO_2})^{(1-\alpha_{LiCoO_2})y_{LiCoO_2}}\end{aligned} \quad (4.82)$$

where A_{LiCoO_2} denotes the surface area of the LiCoO_2 electrode and x_{LiCoO_2} and y_{LiCoO_2} the reaction orders. Again, bulk activities have been used, because an expression for I^o is always derived under equilibrium circumstances. As a simplification, it is in the present model assumed that the kinetics of the reaction can be described by the same parameter values in both phases.

Diffusion limitation of the LiCoO₂ electrode reaction

Diffusion of Li^+ ions is important in determining the rate at which the reaction of (4.75) occurs. Diffusion of Li^+ ions inside the electrode is important in determining the reaction rate. The movement of electrons is again assumed to be much faster than the movement of Li^+ ions and mass transport limitations inside the electrode are therefore attributed to Li^+ ions only. Moreover, a negligible electrical field is assumed to be present inside the electrode. Mass transport through convection is neglected. Therefore, only mass transport due to diffusion of Li^+ ions inside the electrode material is considered. Concentration profiles of the $LiCoO_2$ and the CoO_2 species arise because the Li^+ ions diffuse through the electrode material. This can be modelled in the same way as described above for the nickel electrode.

Li⁺ ions enter or are retrieved from the electrolyte in the same form, as opposed to the situation at the nickel electrode described in section 4.2.2, where H^+ ions form water molecules or are obtained from water molecules at the electrode surface. The Li^+ ions are transported further in the electrolyte. Accumulation or depletion of Li^+ ions can occur at the electrode surface, because Li^+ ions are not present in abundance in the electrolyte, and a concentration profile can consequently be formed. So, mass transport limitation of Li^+ ions in the electrolyte must be considered. This influences the surface activity $a^s_{Li^+}$ of Li^+ ions and hence also the potential across the electrode/electrolyte interface. Rearrangement of the terms in (4.78) yields an expression for the true equilibrium potential $E^{eq}_{LiCoO_2}$ and the diffusion overpotential $\eta^d_{LiCoO_2}$, which in summation yield the apparent equilibrium potential $E^{eq^*}_{LiCoO_2}$. The latter potential can be inferred from (4.78) by using surface activities instead of bulk activities. The activity a_{Li^+} of the Li^+ ions must now be included, which was neglected in the equilibrium situation in (4.78); see (4.77).

The second term in the last line of (4.83) describes the overpotential across the electrode/electrolyte interface due to the mass transport of Li^+ ions through the electrode material. The third term in the last line of (4.83) expresses the effect of a concentration profile of Li^+ ions in the electrolyte solution.

$$
E_{LiCoO_2}^{eq^\bullet} = E_{LiCoO_2}^o + \frac{RT}{n_{Li}F}\left[\ln\left(\frac{a_{CoO_2}^s}{a_{LiCoO_2}^s}\right) - U_{pos}x_{Li}^{pos} + \zeta_{pos}\right]
$$
$$
+ \frac{RT}{n_{Li}F}\ln\left(\frac{a_{Li^+}^s}{a_{Li^+}^{ref}}\right)
$$
$$
= E_{LiCoO_2}^o + \frac{RT}{n_{Li}F}\left[\ln\left(\frac{a_{CoO_2}^s a_{CoO_2}^b a_{LiCoO_2}^b}{a_{LiCoO_2}^s a_{CoO_2}^b a_{LiCoO_2}^b}\right) - U_{pos}x_{Li}^{pos} + \zeta_{pos}\right]
$$
$$
+ \frac{RT}{n_{Li}F}\ln\left(\frac{a_{Li^+}^s}{a_{Li^+}^{ref}}\right)
$$
$$
= E_{LiCoO_2}^o + \frac{RT}{n_{Li}F}\left[\ln\left(\frac{a_{CoO_2}^b}{a_{LiCoO_2}^b}\right) - U_{pos}x_{Li}^{pos} + \zeta_{pos}\right] \tag{4.83}
$$
$$
+ \frac{RT}{n_{Li}F}\ln\left(\frac{a_{CoO_2}^s a_{LiCoO_2}^b}{a_{LiCoO_2}^s a_{CoO_2}^b}\right) + \frac{RT}{n_{Li}F}\ln\left(\frac{a_{Li^+}^s}{a_{Li^+}^{ref}}\right)
$$
$$
= E_{LiCoO_2}^o + \frac{RT}{n_{Li}F}\left[\ln\left(\frac{1 - x_{Li}^{pos}}{x_{Li}^{pos}}\right) - U_{pos}x_{Li}^{pos} + \zeta_{pos}\right]
$$
$$
+ \frac{RT}{n_{Li}F}\ln\left(\frac{a_{CoO_2}^s a_{LiCoO_2}^b}{a_{LiCoO_2}^s a_{CoO_2}^b}\right) + \frac{RT}{n_{Li}F}\ln\left(\frac{a_{Li^+}^s}{a_{Li^+}^{ref}}\right)
$$
$$
= E_{LiCoO_2}^{eq} + \frac{RT}{n_{Li}F}\ln\left(\frac{a_{CoO_2}^s a_{LiCoO_2}^b}{a_{LiCoO_2}^s a_{CoO_2}^b}\right) + \frac{RT}{n_{Li}F}\ln\left(\frac{a_{Li^+}^s}{a_{Li^+}^{ref}}\right)
$$
$$
= E_{LiCoO_2}^{eq} + \eta_{LiCoO_2}^d + \frac{RT}{n_{Li}F}\ln\left(\frac{a_{Li^+}^s}{a_{Li^+}^{ref}}\right)
$$

4.3.3 The LiC$_6$ electrode reaction

Li$^+$ ions are inserted into the LiC$_6$ electrode during charging and extracted from it during discharging. The insertion and extraction of Li$^+$ ions occurs through reduction and oxidation, respectively, as described by the following basic reaction equation for one Li$^+$ ion:

$$
C_6 + Li^+ + e^- \underset{k_{a,neg}}{\overset{k_{c,neg}}{\rightleftharpoons}} LiC_6 \tag{4.84}
$$

where $k_{a,neg}$ and $k_{c,neg}$ are the anodic and cathodic reaction rate constants of the LiC$_6$ (neg=negative) electrode reaction, respectively. The electrode is modelled in the same way as in the model of the LiCoO$_2$ electrode outlined in the previous section. The electrode is considered to be a solid-solution electrode, with LiC$_6$ being the *Red* and C$_6$ the *Ox* species. It is assumed that LiC$_6$ and C$_6$ are the only species present in the electrode and that the sum of their mol fractions is unity. The positions of the C$_6$ centres in the electrode host lattice are fixed and the Li$^+$ ions are transported through the electrode and can fill the sites intended for guest atoms. The mol fraction of LiC_6 in this case corresponds to the SoC of the Li-ion battery, which equals the mol fraction of Li^+ ions inside the electrode, or ($\gamma = 1$ assumed):

$$x_{Li}^{neg} = x_{LiC_6}^{neg} = \frac{a_{LiC_6}}{a_{LiC_6} + a_{C_6}}$$

(4.85)

The mol fraction of the C_6 species can be derived in a similar way and equals $1 - x_{Li}^{neg}$. Unlike with the $LiCoO_2$ electrode, x_{Li}^{neg} can be cycled from nearly zero to 1 in the LiC_6 electrode, which means that the electrode can be completely filled by Li^+ ions. The value of x_{Li}^{neg} will remain somewhat larger than zero after the first charge/discharge cycle. This is because the SEI layer is formed during the first charge/discharge cycle, as described in section 4.3.1. This means that the LiC_6 electrode in the CGR17500 battery is cycled between $x_{Li}^{neg} = 0.06$ and $x_{Li}^{neg} = 1$ under normal operating conditions.

Thermodynamics of the LiC_6 electrode reaction

As the LiC_6 electrode is an intercalation electrode, just like the $LiCoO_2$ electrode, the interaction between the intercalated Li^+ ions and between these ions and the host atoms has to be accounted for. Therefore, a similar expression holds for the change in free energy $\Delta G_{LiC_6,R}$ of the reduction reaction that occurs at the LiC_6 electrode during charging:

$$\Delta G_{LiC_6,R} = \mu_{LiC_6} - \mu_{C_6} - \mu_{Li^+} - \mu_{e^-} + RTU_{neg} x_{Li}^{neg} - RT\zeta_{neg}$$

$$= \mu_{LiC_6}^o - \mu_{C_6}^o - \mu_{Li^+}^o - \mu_{e^-}^o + RT \ln\left(\frac{a_{LiC_6} a_{C_6}^{ref} a_{Li^+}^{ref}}{a_{C_6} a_{Li^+} a_{LiC_6}^{ref}}\right)$$

$$+ RTU_{neg} x_{Li}^{neg} - RT\zeta_{neg}$$

(4.86)

$$= \Delta G_{LiC_6,R}^o + RT \ln\left(\frac{a_{LiC_6}}{a_{C_6}}\right) + RTU_{neg} x_{Li}^{neg} - RT\zeta_{neg}$$

where U_{neg} denotes the dimensionless interaction energy coefficient in the LiC_6 electrode, RTU_{neg} the interaction energy in [J/mol], x_{Li}^{neg} was defined by (4.85), and ζ_{neg} denotes a dimensionless constant in the LiC_6 electrode. Bulk activities have been used, reference activities have been taken the same, the activity a_{Li^+} of Li^+ ions in the electrolyte has been assumed to be the same as the reference activity in a state of equilibrium and the electron activity has been taken to be the same as its reference activity, as before. This yields the following expression for the equilibrium potential of the LiC_6 electrode reaction:

$$E^{eq}_{LiC_6} = \frac{-\Delta G_{LiC_6,R}}{n_{Li}F} = \frac{-\Delta G^o_{LiC_6,Red}}{n_{Li}F}$$

$$+ \frac{RT}{n_{Li}F}\left[\ln\left(\frac{a_{C_6}}{a_{LiC_6}}\right) - U_{neg}x^{neg}_{Li} + \zeta_{neg}\right] \quad (4.87)$$

$$= E^o_{LiC_6} + \frac{RT}{n_{Li}F}\left[\ln\left(\frac{1-x^{neg}_{Li}}{x^{neg}_{Li}}\right) - U_{neg}x^{neg}_{Li} + \zeta_{neg}\right]$$

The same expression can be derived starting from the oxidation reaction in (4.84). Mol fractions have been used instead of activities in the last line of (4.87); see (4.85).

The so-called staging phenomenon occurs in the LiC_6 electrode [47],[48]. This phenomenon implies that the Li^+ ions do not occupy all the available layers at the beginning of the intercalation process. A number of $k-1$ empty layers exists between layers in which Li^+ ions are present when the electrode is in the kth stage. The phase of the electrode is characterized by the stage or stages in which the electrode resides. Only high stages are present when charging starts. A phase transitions will occur each time a new stage arises. This is caused by empty layers that also start to be filled. Four phase transitions that take place between five phases in the LiC_6 electrode have been reported in the literature [47]. Two stages co-exist in most of these five phases. The equilibrium potential is constant in these regions, as discussed above. One stage might be present or two other stages might co-exist after a phase transition and the equilibrium potential changes to a new value. For example, a phase transition may occur from the co-existence of stages 2 and 3 to the co-existence of stages 1 and 2. The magnitude of such a potential change depends strongly on the inter- and intra-layer interaction energies [48].

A good example of a lattice-gas model for the staging phenomenon in the graphite electrode can be found in [48]. However, only two phases and one phase transition at $x_{Li}^{neg} = 0.25$ will be considered in the present Li-ion model as a simplification. The reason is that measurements have shown that phase transitions at other values of x_{Li}^{neg} have only a negligible effect on the equilibrium potential [40],[41]. The two phases have been modelled in the same way, as described for the $LiCoO_2$ electrode. The two equilibrium potentials in the two different phases in the considered CGR17500 Li-ion battery are described by:

$$\left(E^{eq}_{LiC_6}\right)_{phase1} = E^o_{LiC_6} + \frac{RT}{n_{Li}F}\left[\ln\left(\frac{1-x^{neg}_{Li}}{x^{neg}_{Li}}\right) - U_{neg,1}x^{neg}_{Li} + \zeta_{neg,1}\right] \quad (4.88)$$

for $x_{Li}^{neg} < 0.25$ and

$$\left(E^{eq}_{LiC_6}\right)_{phase2} = E^o_{LiC_6} + \frac{RT}{n_{Li}F}\left[\ln\left(\frac{1-x_{Li}^{neg}}{x_{Li}^{neg}}\right) - U_{neg,2}x_{Li}^{neg} + \zeta_{neg,2}\right] \quad (4.89)$$

for $x_{Li}^{neg} \geq 0.25$. The dimensionless constants $\zeta_{neg,1}$ and $\zeta_{neg,2}$ are defined in the same way as for (4.81).

Kinetics of the LiC$_6$ electrode reaction

The reaction kinetics of the reaction that takes place at the surface of the LiC$_6$ electrode are described by the Butler-Volmer relationship. The exchange current value for the reaction is determined by:

$$I^o_{LiC_6} = n_{Li}FA_{LiC_6}\left(k_{a,neg}\right)^{1-\alpha_{LiC_6}}\left(k_{c,neg}\right)^{\alpha_{LiC_6}} \cdot$$
$$(a^b_{C_6})^{\alpha_{LiC_6}x_{LiC_6}}(a^b_{LiC_6})^{(1-\alpha_{LiC_6})y_{LiC_6}} \quad (4.90)$$

where A_{LiC_6} denotes the electrode surface area and x_{LiC_6} and y_{LiC_6} the reaction orders. As a simplification, it is in the present model assumed that the kinetics of the reaction can be described by the same parameter values in both phases.

Diffusion limitation of the LiC$_6$ electrode reaction

As with the LiCoO$_2$ electrode, diffusion of Li$^+$ ions both inside the electrode and in the electrolyte plays a role in determining the rate of the reaction. An expression for the apparent equilibrium potential $E^{eq^*}_{LiC_6}$ can be found in the same way as shown in the derivation of (4.83), using (4.86) and (4.87). This potential is the summation of the true equilibrium potential $E^{eq}_{LiC_6}$ and the diffusion overpotential due to diffusion limitation of the electrode reaction of Li$^+$ ions inside the electrode and inside the electrolyte. For the latter diffusion, the activity of Li$^+$ ions in (4.86) has to be taken into account. This is shown in the following equation.

$$E^{eq^*}_{LiC_6} = E^o_{LiC_6} + \frac{RT}{n_{Li}F}\left[\ln\left(\frac{1-x_{Li}^{neg}}{x_{Li}^{neg}}\right) - U_{neg}x_{Li}^{neg} + \zeta_{neg}\right]$$
$$+ \frac{RT}{n_{Li}F}\ln\left(\frac{a^s_{C_6}a^b_{LiC_6}}{a^s_{LiC_6}a^b_{C_6}}\right) + \frac{RT}{n_{Li}F}\ln\left(\frac{a^s_{Li^+}}{a^{ref}_{Li^+}}\right)$$
$$= E^{eq}_{LiC_6} + \eta^d_{LiC_6} + \frac{RT}{n_{Li}F}\ln\left(\frac{a^s_{Li^+}}{a^{ref}_{Li^+}}\right) \quad (4.91)$$

As in (4.83), the second term in the last line expresses the overpotential due to mass transport of Li$^+$ ions inside the electrode. The last term expresses the overpotential due to mass transport limitation of Li$^+$ ions in the electrolyte.

4.3.4   The electrolyte solution

Li^+ ions are formed at one electrode in the considered Li-ion battery and injected into the electrolyte at the electrode/electrolyte interface. These Li^+ ions are removed from the electrolyte at the other electrode and allowed to enter the electrode through the electrolyte/electrode interface. The electrolyte of the considered CGR17500 Li-ion battery consists of a $LiPF_6$ salt dissolved in an organic solvent, as discussed in section 4.3.1. As a result, Li^+ ions and PF_6 ions co-exist in the electrolyte solution. Both ions are transported inside the electrolyte under the influence of concentration gradients through diffusion and under the influence of an electrical field through migration. The influence of convection is neglected. The ionic conductivity of the non-aqueous electrolyte used in Li-ion batteries is relatively low, as opposed to the electrolyte in the NiCd battery considered in section 4.2. The conductivities of typical non-aqueous electrolytes are in the range of 10^{-3} to 10^{-2} $(\Omega cm)^{-1}$, whereas typical conductivity values of aqueous electrolytes are in the range of 10^{-1} to 1 $(\Omega cm)^{-1}$ [43]. Therefore, the electrical field inside the electrolyte cannot be neglected in the Li-ion model, as was done in the NiCd battery model.

The total current that flows through the electrolyte is the sum of the diffusion and migration contributions of both ion types, as defined by (4.19) for the chemical domain. The diffusion coefficient $D_{Li^+}{}^{elyt}$ of the Li^+ ions in the electrolyte should be used for the Li^+ ions and $z_{Li^+}= 1$. Likewise, the diffusion coefficient of PF_6 ions in the electrolyte $D_{PF_6^-}{}^{elyt}$ should be used for the PF_6 ions and $z_{PF_6} = -1$. The movement of the ions is such that the solution is electrically neutral anywhere inside the electrolyte, except in the very thin Gouy-Chapman layers at the electrode surfaces. Figure 4.20 illustrates the movement of ions in the electrolyte during the charging of the Li-ion battery considered in this section.

Figure 4.20 illustrates the injection of Li^+ ions into the electrolyte at the positive electrode and their extraction from the electrolyte at the negative electrode. The injection occurs at the rate of the $LiCoO_2$ electrode reaction, as discussed in section 4.3.2, whereas the extraction occurs at the rate of the LiC_6 electrode reaction; see section 4.3.3. The PF_6 ions are not allowed to cross either of the electrode/electrolyte interfaces. The net current supported by these ions can become zero because the contributions to the total current of the PF_6 ions by diffusion and migration have opposite signs. In that case, a steady-state situation is reached and all current in the electrolyte is supported by Li^+ ions.

The self-discharge rate of Li-ion batteries is considerably lower than that of NiCd and NiMH batteries; see chapter 3. However, it cannot be neglected. The origin of self-discharge in Li-ion batteries is not exactly known. It consists of a reversible part, which implies a temporary capacity loss that can be recovered, and an irreversible part, which is a permanent capacity loss that cannot be recovered. The reversible part could be attributable to electronic conductivity of the electrolyte solution, whereas the irreversible part could be attributable to electrolyte decomposition.

In the ideal case, the electrolyte solution would be a perfect electronic insulator and current flow inside the electrolyte would be supported only by ions. However, the electrical resistivity of the electrolyte is not infinitively large in practice. On the basis of an average self-discharge rate of 5% per month at room temperature (see chapter 3), a rated nominal capacity of 720 mAh and a battery voltage of 4 V, an equivalent resistor R_{leak} of 80 kΩ is found to be present between the positive and negative electrodes of the CGR17500 battery.

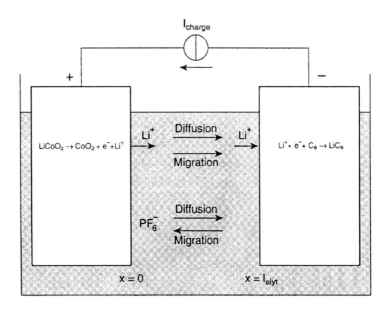

Figure 4.20: Schematic representation of the charging of a Li-ion battery, in which Li^+ and PF_6^- ions are transported through the electrolyte by means of combined diffusion and migration

4.3.5   Temperature dependence of the reactions

The temperature dependence according to the Arrhenius relation of (4.72) should be taken into account for the reaction rate constants k_i and the diffusion coefficients D_i. The temperature dependence given by (4.73) should be considered for the standard redox potentials E^o. Moreover, the temperature dependence of self-discharge should be considered. The temperature dependencies mentioned above have not yet been taken into account in the present Li-ion model version.

4.3.6   The model

The simulation model for a CGR17500 Li-ion battery is shown in Figure 4.21. This model is based on the general principles given in section 4.1. No thermal network is shown. A network similar to the one in Figure 4.19b can be easily included, as soon as the temperature dependencies have been added to the model equations. The $LiCoO_2$ electrode model is shown on the left-hand side in Figure 4.21 and the LiC_6 electrode model on the right-hand side. The electrolyte model is shown in the middle connecting the two electrodes. The self-discharge of the battery is modelled by R_{leak}, as described above. This resistor is connected between the positive electrode and the negative electrode in the electrical domain.

The model of the CGR17500 battery shown in Figure 4.21 is built up almost entirely in the chemical domain. The only components present in the electrical domain are the ohmic series resistances R_{LiCoO_2} and R_{LiC_6} of the electrodes and the self-discharge resistor R_{leak}. This has been done to clarify how the mass transport model of Figure 4.13 has been included in a battery model. No side-reactions have been modelled, as stated in section 4.3.1, and hence only one reaction path is present in each electrode.

Energy storage in the chemical domain is modelled by chemical capacitances C_{ch}^i, with i denoting the reacting species. The coupling between the electrical and chemical domains is represented by ideal transformers, as described in section 4.1. Both electrode reactions have been modelled in the same way. The kinetics of both reactions have been modelled in the chemical domain. This means that the change in free electrochemical energy $\Delta \bar{G}$ is now present across the anti-parallel diodes D_j; see Table 4.2. One of the anti-parallel diodes represents the oxidation reaction and the other represents the reversed reduction reaction. This has been indicated in the figure.

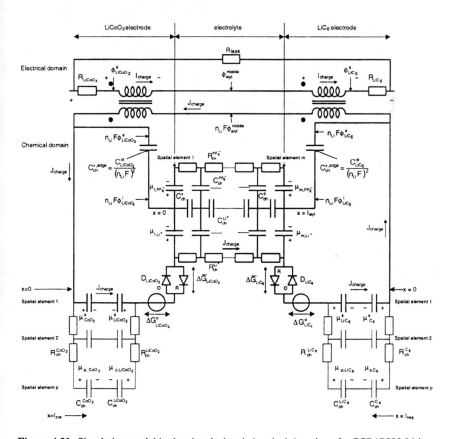

Figure 4.21: Simulation model in the electrical and chemical domains of a CGR17500 Li-ion battery

The species CoO_2, $LiCoO_2$ and Li^+ ions take part in the reaction at the $LiCoO_2$ electrode, as expressed by (4.75), and therefore the chemical capacitances that contain the molar amounts of these species are present in the $LiCoO_2$ electrode model. All μ^0 values have been omitted in the chemical potentials μ across the capacitances of the CoO_2 and $LiCoO_2$ species, because these values are already represented in source $\Delta G^{o}_{LiCoO_2}$; see (4.77). Moreover, the term $(-U_{pos} x_{Li}{}^{pos} + \zeta_{pos})$ has been added to the description of the capacitances of the CoO_2 and $LiCoO_2$ species to obtain (4.83).

The diffusion of Li^+ ions inside the electrode is described by an RC-ladder network. The electrode has again been divided into p spatial elements each with a thickness Δx. The thickness of the electrode is l_{pos}. The capacitances on the left-hand side contain the molar amounts of the CoO_2 species in the spatial elements and the capacitances on the right-hand side contain the molar amounts of the $LiCoO_2$ species in these elements. The top of the RC-ladder network denotes the electrode/electrolyte interface at $x=0$, where the actual electrochemical reaction takes place. The bottom denotes the electrode/current collector interface at $x=l_{pos}$. The diffusing Li^+ ions can pass the electrode/electrolyte boundary only in the electrochemical reaction; they cannot pass the electrode/current collector boundary. The chemical resistances $R_{ch}{}^{CoO_2}$ and $R_{ch}{}^{LiCoO_2}$ are given by (4.21c), with the diffusion coefficient $D_{Li^+}{}^{pos}$ of the Li^+ ions in the positive electrode being used for both resistances. Again, the value of the resistances depends on the molar amount m of the Li^+ ions in the spatial elements according to (4.21c), which equals the molar amount of the $LiCoO_2$ species. This means that resistances $R_{ch}{}^{CoO_2}$ and $R_{ch}{}^{LiCoO_2}$ have the same values in each spatial element. Different spatial elements will have different values for $R_{ch}{}^{CoO_2}=R_{ch}{}^{LiCoO_2}$ when a diffusion profile of Li^+ ions occurs. Such a diffusion profile will give rise to a diffusion overpotential $\eta^{d}_{LiCoO_2}$ across the electrode/electrolyte interface, as defined by (4.83). The relation between the change in free electrochemical potential $\Delta \overline{G}$ across the diodes D_{LiCoO_2} and chemical flow J_{ch} is described by the Butler-Volmer equation in the chemical domain; see (4.12). The exchange current $I^o_{LiCoO_2}$ is given by (4.82).

The LiC_6 electrode has been modelled in the same way as the $LiCoO_2$ electrode. The electrode has again been divided into p spatial elements for simplicity, with the electrode thickness having been defined as l_{neg}. The number of spatial elements does not have to be the same as in the positive electrode, because the electrode thicknesses or grain sizes do not have to be the same. The diffusion coefficient $D_{Li^+}{}^{neg}$ of the Li^+ ions in the negative electrode is used for the chemical resistances. The exchange current for the anti-parallel diodes is given by (4.90).

The electrolyte has been modelled in the chemical domain, as described in section 4.1.3 and shown in Figure 4.13. The electrolyte has been divided into m spatial elements, as shown in Figure 4.21, where the electrolyte thickness is l_{elyt}. The electrolyte model comprises capacitances $C_{ch}{}^{Li^+}$ and $C_{ch}{}^{PF_6}$, described by (4.3), as well as the geometric capacitances $C_{ch}{}^{\varepsilon}$ defined in Figure 4.11. The resistances $R_{ch}{}^{Li^+}$ and $R_{ch}{}^{PF_6}$ are described by (4.21c), with the resistance values again depending on the spatial element. The diffusion coefficients $D_{PF_6}{}^{elyt}$ and $D_{Li^+}{}^{elyt}$ of the PF_6 ions and Li^+ ions in the electrolyte are used in the resistance expressions, respectively. The resistances $R_{ch}{}^{Li^+}$ and $R_{ch}{}^{PF_6}$ have different values in the same spatial element when the diffusion coefficients of the ions differ. The top part of the network describes the mass transport of the PF_6 ions as a result of diffusion and migration, while the bottom part describes the mass transport of the Li^+ ions as a result of diffusion and migration. The Li^+ ions are allowed to cross the electrode/electrolyte interfaces at both the $LiCoO_2$ electrode at $x=0$ and the LiC_6 electrode at $x=l_{elyt}$. This is represented in Figure 4.21 by the connections between the capacitances $C_{ch}{}^{Li^+}$ and the electrode reactions at $x=0$ and $x=l_{elyt}$. The PF_6 ions are not allowed to cross these boundaries because they do not participate in the electrochemical reactions.

The geometrical capacitances $C_{ch}^{\varepsilon,edge}$ denote the double-layer capacitances C^{dl} in the chemical domain; see section 4.1.3. The chemical capacitance equals the electrical capacitance C^{dl} divided by $(n_{Li}F)^2$, as expressed by (4.17). This is indicated in Figure 4.21 for both electrodes. The electrostatic potentials at the electrode/electrolyte interfaces at $x=0$ and $x=l_{elyt}$, expressed in the chemical domain, are also indicated in the figure. A chemical electrostatic potential difference of $n_{Li}F(\phi^s_{LiCoO_2}-\phi_{LiCoO_2})$ is present across the chemical double-layer capacitance of the positive electrode. This chemical potential difference is $n_{Li}F(\phi^s_{LiC_6}-\phi_{LiC_6})$ for the negative electrode. The electrostatic potentials $\phi^s_{LiCoO_2}$ and $\phi^s_{LiC_6}$ at the electrical ports of the ideal transformers are present at the indicated nodes. The electrostatic potential at the node at which the two ideal transformers are connected equals the electrostatic potential ϕ_{elyt}^{middle} in the middle of the electrolyte, because the electrolyte model is completely symmetrical. This can be easily envisioned by splitting the battery in half. The voltage across the left-hand ideal transformer covers the positive electrode potential and half of the potential across the electrolyte. The right-hand ideal transformer covers the potential across the other half of the electrolyte and the negative electrode potential.

The direction of a charge current I_{charge} is indicated in the electrical domain in Figure 4.21. The currents through the electrical ports of both ideal transformers will always be the same because of the network's structure. The direction of the chemical flow J_{charge} in the chemical domain is shown for a steady-state situation. This means that the electrostatic potentials and the diffusion profiles remain unchanged. J_{charge} then flows through the chemical capacitances of the CoO_2 and $LiCoO_2$ species at the electrode/electrolyte interface at the positive electrode and the LiC_6 and C_6 species at the negative electrode. The molar amounts of CoO_2 and LiC_6 increase, as is reflected by the polarity of these capacitances, whereas the molar amounts of $LiCoO_2$ and C_6 decrease. The current flows only through the chemical resistors $R_{ch}^{Li^+}$ in the electrolyte because of the steady-state situation. No current flows through the chemical capacitances containing the molar amount of Li^+ ions in the electrolyte at the interfaces at $x=0$ and $x=l_{elyt}$ because the diffusion profile of Li^+ ions in the electrolyte remains unchanged in a steady-state situation.

As a further illustration of the theory presented in section 4.1, the model of Figure 4.21 will be represented in the electrical domain only. This can be done as described in section 4.1.2. In Figure 4.22, the network for the chemical domain has been transferred to the electrical domain by 'shifting' it through the ideal transformers. Therefore, all impedances have been multiplied by $N^2=1/F^2$, as $n_{Li}=1$.

Moreover, a distinction between the true equilibrium potential and the diffusion overpotential due to diffusion of the Li^+ ions in the electrodes, as described by (4.83) and (4.91), will be made for both electrodes. This was illustrated above in Figure 4.10. The results of this exercise are shown in Figure 4.23. A capacitance describing the bulk activities of the Ox and Red species has been added to the network in both electrodes (labelled 3 for $LiCoO_2$ electrode and 4 for LiC_6 electrode). In addition, the RC-ladder networks describing the diffusion of Li^+ ions inside the electrodes have been changed slightly by moving the chemical resistances to one side of the RC-ladder network. This has been done to show that the electrical potential inside the electrode hardly changes thanks to the good conductivity of the electrodes. This can be inferred from the direct connection between the electrode/current collector interfaces at $x=l_{pos}$ and $x=l_{neg}$, respectively, and the electrode/electrolyte interfaces at $x=0$.

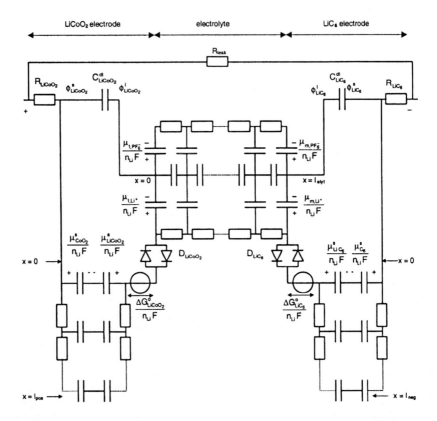

Figure 4.22: Simulation model of the CGR17500 Li-ion battery in the electrical domain only. The model is based on that shown in Figure 4.21, as indicated in the text

The networks shown in Figures 4.22 and 4.23 are identical, as can be understood by applying Kirchhoff's voltage law to the interface between two spatial elements in the RC-ladder network. The circular arrows indicate this. Figure 4.23 shows that the electrostatic potential inside the electrode is constant, which means that the same potential is valid at $x=0$ and $x=l_{pos}$ or $x=l_{neg}$, and that it hence equals ϕ^s. The electrochemical potential may vary throughout the electrode due to concentration gradients. The resulting diffusion overpotential can only be measured at the electrode/electrolyte interface.

The application of Kirchhoff's voltage law to both electrodes in Figure 4.23 illustrates that the voltage across the electrode/electrolyte interface, ϕ^s-ϕ^l, equals the sum of various contributions. They are the true equilibrium potential E^{eq}, the diffusion overpotential due to diffusion of Li$^+$ ions inside the electrodes η^d, the kinetic overpotential η^k across the diodes D_j and the voltage across the capacitance describing the activity of the Li$^+$ ions at the interface. This latter voltage is in agreement with the last term in (4.83) and (4.91). The total battery voltage can hence be obtained from:

$$V_{bat} = IR_{LiCoO_2} + IR_{LiC_6} + E_{LiCoO_2} + E_{elyt} - E_{LiC_6}$$

$$= IR_{LiCoO_2} + IR_{LiC_6} + \left(\phi^s_{LiCoO_2} - \phi^l_{LiCoO_2}\right)$$

$$+ (\phi^l_{LiCoO_2} - \phi^l_{LiC_6}) - \left(\phi^s_{LiC_6} - \phi^l_{LiC_6}\right) \qquad (4.92)$$

$$= IR_{LiCoO_2} + IR_{LiC_6} + (\eta^d_{LiCoO_2} + E^{eq}_{LiCoO_2} + \eta^k_{LiCoO_2} + \frac{\mu_{1,Li^+}}{n_{Li}F})$$

$$+ (\phi^l_{LiCoO_2} - \phi^l_{LiC_6}) - (\eta^d_{LiC_6} + E^{eq}_{LiC_6} + \eta^k_{LiC_6} + \frac{\mu_{m,Li^+}}{n_{Li}F})$$

where E_{elyt} is the electrostatic potential across the electrolyte.

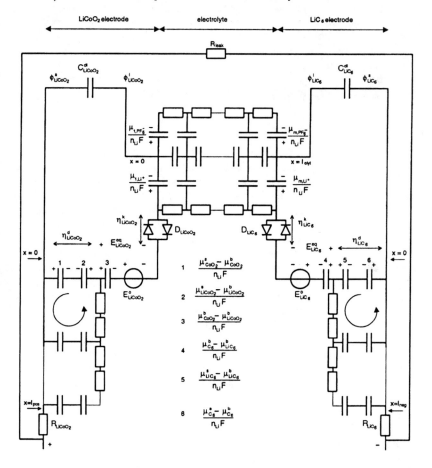

Figure 4.23: Alternative simulation model based on that shown in Figure 4.22 with modified definitions of capacitances to allow a distinction to be made between the true equilibrium potential and the diffusion overpotential due to diffusion of Li$^+$ ions in the electrodes. Moreover, the RC-ladder networks in both electrodes have been rearranged

An expression for the overpotential due to diffusion of Li^+ ions in the electrolyte can be obtained from (4.92):

$$\eta_{Li^+}^{d,elyt} = \frac{\mu_{1,Li^+}}{n_{Li}F} - \frac{\mu_{m,Li^+}}{n_{Li}F} = \frac{RT}{n_{Li}F} \ln\left(\frac{a_{1,Li^+}}{a_{m,Li^+}}\right) \qquad (4.93)$$

where a_{1,Li^+} and a_{m,Li^+} are the activities of the Li^+ ions in the spatial elements 1 and m of the electrolyte, respectively.

Until now the focus has been on the construction of electronic-network models for rechargeable batteries in general (section 4.1) and NiCd (section 4.2) and Li-ion batteries (this section) in particular. The parameters in these models have to be given proper values before simulations can be performed. This will first be discussed in the case of the NiCd model in the next section, in which simulation results will be compared to measurement results. More simulation results with the NiCd model will be described in section 4.5.1. The parameters of the Li-ion model described in this section will be given in section 4.5.2, followed by a comparison between the results of simulations and measurements.

4.4 Parameterization of the NiCd battery model

4.4.1 Introduction

The simulation results obtained with the battery models described in the previous sections must of course agree as well as possible with results obtained in practice. So the results of the simulations must be quantitatively compared with the results of measurements performed under the same conditions. This section describes the first results of research performed to obtain optimum quantitative agreement between the results of simulations using the NiCd model and measurements [49].

The NiCd model described in section 4.2 is characterized by a chemical and mathematical description of the processes involved, the structure of the simulation model and a set of parameter values. In order to obtain close agreement between simulated and measured battery behaviour, both the design of the model itself and the values of the parameters should be considered. The iterative process for optimizing the battery model to obtain close agreement between the results of simulations and measurements is schematically outlined in Figure 4.24.

The first three parts on the left-hand side of Figure 4.24 represent the steps in the model development outlined in previous sections. In the last two parts, the results of simulations are obtained and compared with the results of measurements. The quantitative agreement between the results of simulations and measurements can be improved in several ways on the basis of the results of the comparison. This is represented by the feedback loops shown on the right-hand side of Figure 4.24.

First of all, the chemical description can be modified, for example by adding to the model the description of a process that was not considered before. This is represented by the feedback loop that points at the first step in Figure 4.24. Secondly, the mathematical description of the considered processes can be modified, as represented by the second feedback loop. Both modifications can be said to modify the model itself. Thirdly, the values of the parameters in the model can be optimized, leaving the original structure of the model intact. This is represented by the third feedback loop.

Three approaches can be used to find parameter values for the battery model, as indicated in Figure 4.24. In the first approach, the parameter values are obtained from the literature, such as the number of electrons that take part in the reactions, standard redox potentials, *etc.* In addition, some design-related parameters may be obtained from battery manufacturers or data books. However, the values of some parameters cannot be found in the literature or have been obtained under conditions that do not hold for the batteries considered here. In that case, the second approach may be useful.

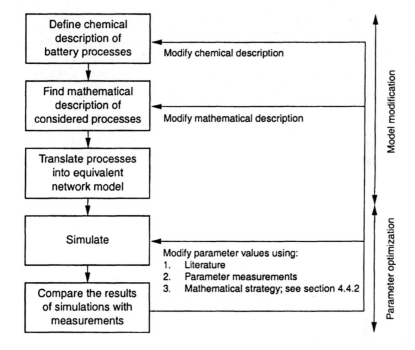

Figure 4.24: Set-up of the modelling strategy and iterative process of parameter optimization and model modification aimed at improving the quantitative agreement between the results of simulations and measurements

The parameter values are found by means of measurements in the second approach. It should be noted that most measurements of battery parameters, such as exchange currents and diffusion coefficients, take a considerable amount of time, because the parameters have to be obtained from a series of voltage and current measurements, which have to be performed on the separate electrodes. Some parameter values will be difficult to obtain even by means of measurements.

A third approach will often have to be used, because not all battery model parameters can be found via the first two approaches. This third approach will be the main focus of this section. It involves the application of a mathematical strategy, based on the comparison of battery curves obtained from simulations and measurements. In principle, the entire set of battery model parameters can be obtained using this strategy. However, the complexity of the approach increases with the number of parameters to be obtained, so this number should not be too high. The strategy will be explained in more detail in section 4.4.2.

Most parameter values in a battery will show a certain spread in practice. This means that whichever of the three above-mentioned approaches is applied, a set of parameter values obtained for a certain battery will not necessarily hold for all batteries of that type. An optimized parameter set will be derived for a single battery in this section.

The results obtained from applying the strategy of section 4.4.2 will be described in section 4.4.3. The results of the simulations obtained with the optimized battery model will be compared with results obtained in measurements in a charging experiment using a single charge current, after which the optimized parameter values will be listed and discussed. The applied mathematical optimization strategy will also be discussed. The quantitative agreement between the results of simulations using the optimized battery model and the results of measurements at various charge currents will be discussed in section 4.4.4. A change in the model based on these results will be suggested in the form of a modified mathematical description, as indicated in Figure 4.24. Finally, in section 4.4.5, the strategy of section 4.4.2 will again be applied at various charge currents using a modified model.

4.4.2   Mathematical parameter optimization

The battery model can be regarded as a black box in the mathematical strategy described in this section. This interpretation of the battery model is shown in Figure 4.25.

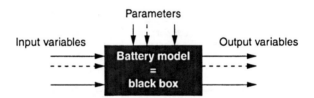

Figure 4.25: Black box interpretation of the battery model

The black box processes a number of input variables and yields a number of output variables. Parameters determine the relationship between the input and output variables. The strategy comprises three main parts, notably (i) defining an admissible parameter space, (ii) identifying a suitable sub-region within this admissible parameter space and (iii) finding the optimum parameter set within the resulting reduced parameter space. Although the battery model is primarily regarded as a black box, knowledge of battery behaviour will be used in the strategy. Examples can be found in the explanation of the three main parts of the strategy below.

The strategy was used in simulated and measured charging experiments involving the NiCd model and a Panasonic P60AA NiCd rechargeable battery, respectively. The design of the NiCd model was based on that of the type of NiCd battery used in the measurements, as described in section 4.2. In the charging experiment, the battery model and the real battery were charged at a 1 C charge rate for two hours at 25°C; see chapter 3 for the definition of C-rate. The simulations were performed using the electronic-circuit simulator PSTAR©, developed at Royal

Philips Electronics N.V. The pressure inside the battery was measured during the experiments with a pressure transducer (Transamerica Instruments, type no. 4702-10) mounted in a hole drilled in the bottom of the battery casing. The battery temperature was measured with a Pt-100 thermocouple placed on the bare metal on the long side of the AA-size casing. More information on the three parts of the strategy is given below.

(i). Defining an admissible parameter space

The input and output variables and the parameters to be taken into account are identified in the first part of the strategy. The input variables include the current (I_{charge}) that flows into the battery and the ambient temperature (T_{amb}). The output variables include the battery voltage (V), temperature (T) and internal oxygen gas pressure (P), and the derivatives of these variables. The derivatives are included because the V, P and T profiles show clear curvatures during overcharging. The inclusion of derivatives in the output variables makes it easier to compare the results of simulations with those of measurements. Another reason for including the derivatives in the output variables is to minimize the number of extreme results obtained in the simulations in the next two steps of the strategy.

As mentioned above, the number of parameters considered should not be too large. Therefore, the parameters with fixed values, such as the number of electrons that take part in a reaction, or with only a limited influence on the output variables in the considered experiment are to be kept constant. They do not participate in the optimization process. The influence of the parameters on the output variables was determined in a sensitivity analysis. In addition, knowledge of battery behaviour was used and, for example, parameters with no or only a negligible influence on the charging behaviour were omitted. A certain range is defined for each parameter considered on the basis of the assumed possible spread in each parameter's value. 34 parameters were selected as variables in the present optimization process, yielding a 34-dimensional parameter space. Within this parameter space, 340 parameter sets were generated by randomly but uniformly spreading the values for each of the 34 parameters throughout the specified range. This was done using the Latin Hypercube method [50], which was improved by further spreading the points in the parameter space with the aid of a simple dedicated software tool to minimize differences in the distances between the neighbouring points.

The admissible parameter space was defined in a charging simulation for each parameter set, as described above. The simulated V, P and T curves were visually inspected and those parameter sets that yielded highly unrealistic results were discarded. Roughly 10% of the original number of parameter sets were discarded because of this reason. This is another example of using battery knowledge in the mathematical process. The model behaviour thus obtained for parameter sets within the admissible parameter space was already in fair agreement with the actual battery's behaviour.

(ii). Identifying a suitable sub-region / reducing the parameter space

In the second step, the simulated output variables of the battery model were compared with the corresponding measured variables of the real battery. This was done for each parameter set within the admissible parameter space defined in the first step. The basic principle of the second step is illustrated in Figure 4.26.

Figure 4.26 illustrates that each simulated and corresponding measured output variable, Var_i, was compared using a cost function, CF_i, given by:

$$CF_i = \sum_{j=1}^{N} W_{i,j} \left(Var_{i,sim,j} - Var_{i,meas,j} \right)^2 \qquad (4.94)$$

where i ranges from 1 to 6, because six output variables are being considered. The factor $W_{i,j}$ denotes the normalizing and weighing factor for the output variable Var_i.

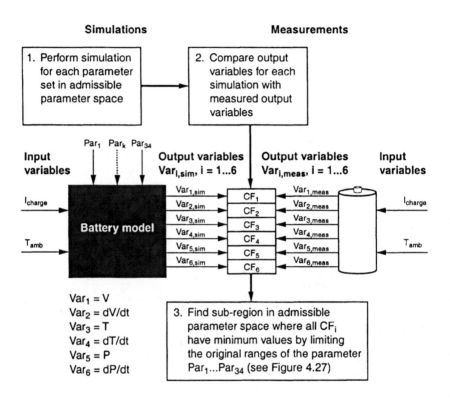

Figure 4.26: Schematic representation of the second step in the optimization process

As can be seen in (4.94), the cost function is a simple Least-Square Error (LSE) function that is weighed and normalized for each of the six output variables. The index j increases from 1 at the beginning of the charging curves to N at the end of the charging curve. N was chosen to be 7200 and the points at which the comparisons were made were each spaced 1 second apart. The use of a quadratic cost function is important, because positive and negative differences are weighed equally.

The output variables need to be normalized because of the large differences in the absolute values of the six output variables. The cost functions need to be of the same order of magnitude when the cost functions CF_i are combined into a single cost function in step (iii) below. The variables were normalized by dividing the squared difference between the simulated and the measured values, shown in (4.94), by the squared difference between the simulated value and the average measured value. The average measured value is a single value that represents the average of all the values measured along the entire curve.

The values were weighed so as to be able to concentrate on certain parts of the curves. In the case of battery charging curves, the most pronounced behaviour in terms of V, P and T occurs after one hour of charging. Therefore, we focussed our attention on the points in the overcharging region in the case of each output variable Var_i by increasing the weight factor towards the end of the curve for each output variable. This is yet another example of the use of battery knowledge in the mathematical strategy. A consequence of this, however, is that the battery behaviour at the beginning of the charging will not be optimized to the same extent as the behaviour during overcharging.

The ranges of the 34 parameters $Par_1..Par_{34}$ were further reduced on the basis of the comparison of the results of simulations and measurements for all parameter sets in the admissible parameter space. This was done to reduce the size of the original parameter space. The new ranges were chosen so as to minimize the cost functions CF_i in the new sub-region. The results of the simulations thus obtained for the parameter sets in the new sub-region were consequently already quite close to the measurements. This close agreement was achieved for all six defined output variables, as all cost functions CF_i already have minimum values in this reduced parameter space.

The basic principle of the reduction of the range of parameters Par_k, $k=1..34$, is schematically represented in Figure 4.27. Six cost functions, CF_i, were calculated for each parameter set. Supposing that there are M parameter sets, then in the end M values of CF_1 are available, M values of CF_2, etc. The values of CF_1 can then be ranked, as can the values of CF_2, *etc.* For each cost function, the parameter sets that yield the lowest values for CF_i can then be identified. This information can be used for the actual reduction of the ranges of the 34 parameters. In the situation illustrated, the five best parameter sets were selected for each cost function. The values of parameter Par_k in these parameter sets are shown along the horizontal lines at the bottom of Figure 4.27. Note that the five best parameter sets of different cost functions are not necessarily the same. This can be inferred from the fact that there is no vertical line with six dots on it. The new range is defined as the region in which the greatest number of parameter values for Par_k are to be found, as shown in the figure. Parameter values within the new range will now yield minimum values for all six cost functions, CF_i. The new range was obtained by considering the uniform distribution of Par_k within its original range.

(iii). Finding the optimum parameter set within the reduced parameter space
In the third step of the strategy, the optimum parameter set is found within the reduced parameter space defined in the previous step. This is done via a standard minimization of a cost function by means of the commonly used simplex method. The cost function comprises the cost functions for the individual output variables as follows:

$$Cost\ function = \sum_{i=1}^{6} CF_i \qquad (4.95)$$

In our experiments, we performed the third step with the Optimize© optimization software developed by Royal Philips Electronics N.V. Starting from a parameter set in the reduced parameter space obtained in the previous step, the software minimizes the cost function in (4.95) by performing many simulations involving different parameter sets in an iterative process. The software stops when the cost function changes remain within a certain preset window. Note that all six output variables are considered simultaneously in this step.

Results of comparisons of the output variables of
simulations and measurements for each parameter set
in admissable parameter space (see Figure 4.26):

Par. set	Cost functions for each parameter set:					
1	$CF_{1,1}$	$CF_{2,1}$	$CF_{3,1}$	$CF_{4,1}$	$CF_{5,1}$	$CF_{6,1}$
2	$CF_{1,2}$	$CF_{2,2}$	$CF_{3,2}$	$CF_{4,2}$	$CF_{5,2}$	$CF_{6,2}$
...	...	...	...	...	...	...
M	$CF_{1,M}$	$CF_{2,M}$	$CF_{3,M}$	$CF_{4,M}$	$CF_{5,M}$	$CF_{6,M}$

↓

Ranking cost functions for each output variable separately:

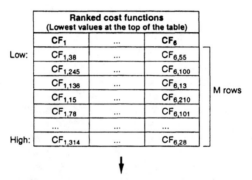

	Ranked cost functions (Lowest values at the top of the table)		
	CF_1	...	CF_6
Low:	$CF_{1,38}$	...	$CF_{6,55}$
	$CF_{1,245}$	...	$CF_{6,100}$
	$CF_{1,136}$	...	$CF_{6,13}$
	$CF_{1,15}$	...	$CF_{6,210}$
	$CF_{1,78}$	...	$CF_{6,101}$
	...	...	...
High:	$CF_{1,314}$	...	$CF_{6,28}$

M rows

↓

Values of Par_k in the five parameter sets yielding the five lowest values of the cost functions CF_i

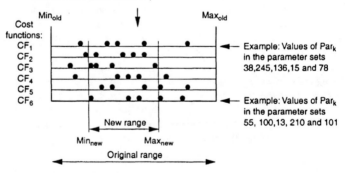

Figure 4.27: Simplified schematic representation of the definition of a new range for parameter Par_k. In the last step shown at the bottom, the values of Par_k in five parameter sets that yield the five lowest values of cost function CF_i are shown on each horizontal line

4.4.3   Results and discussion

Comparison of the results of simulations with the results of measurements
The final results obtained with the mathematical strategy described in the previous section are shown for the simulated and measured *V, P* and *T* curves in Figure 4.28, Figure 4.29, and Figure 4.30, respectively. These figures all show the measured curves, labelled *meas*, plus the curves obtained in the simulation in which the optimum parameter set obtained with the strategy described in the previous section was used. The latter curves have been labelled *opt*. The curves that were obtained in the simulation using the initial parameter set are also shown. These curves have been labelled *start*. These parameter values can also be found in [26], [27] and [28].

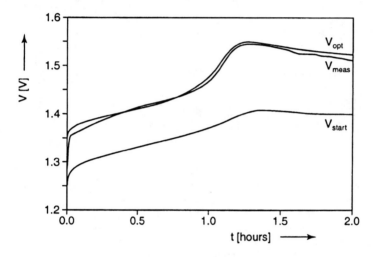

Figure 4.28: Simulated (V_{start} [V] and V_{opt} [V]) and measured (V_{meas} [V]) battery voltage versus time (*t* [hours]) for charging a NiCd P60AA Panasonic battery at a 1 C-rate for 2 hours at 25°C ambient temperature. The battery model developed in section 4.2 was used in the simulations. In addition to the quantitatively optimized simulated voltage (V_{opt}), the voltage curve at the time when the mathematical strategy was first applied using the parameter set found in [26]-[28] is also shown (V_{start})

Figure 4.28 shows that the results obtained in mathematical optimization represent a significant improvement over the V_{start} curve. For charging times larger than roughly one minute, the maximum difference between the optimized simulated voltage curve V_{opt} and the measured voltage curve V_{meas} is only 18 mV, which amounts to 1.4% in the case of a nominal battery voltage of 1.25 V. The difference between the results of simulations and those of measurements seems reasonable, because the measured curve, V_{meas}, will show a certain spread in practice. There are several reasons for this. First of all, the measurement accuracy is finite. In this case, allowance must be made for a voltage measurement resolution of 1 mV. This means that the difference observable in Figure 4.28 is greater than can be explained by measurement accuracy only. Secondly, the battery voltage measured during the charging of several NiCd batteries of the same type will display a certain spread. Unfortunately, no reliable statistical data is available on this spread. Therefore, it is at present not known whether the difference between V_{meas} and V_{opt} observable in Figure 4.28 is within the spread of battery behaviour in practice.

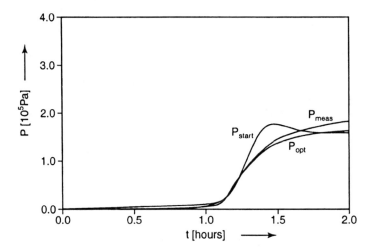

Figure 4.29: Simulated (P_{start} [10^5 Pa] and P_{opt} [10^5 Pa]) and measured (P_{meas} [10^5 Pa]) internal oxygen gas pressure versus time (t [hours]). See Figure 4.28 for conditions

The simulated and measured battery voltage curves V_{opt} and V_{meas} are more or less identical in shape, except for the parts representing the first 20 minutes. First of all, it should be noted that weighing has been applied, which may have led to greater discrepancies between the results of simulations and measurements at the beginning of charging. This can be seen especially in the first few minutes. This difference may also be attributable to processes occurring in the measured battery for which no allowance has been made in the model. A possible example of such a process is hysteresis of the battery's equilibrium potential. Measurements have shown that the equilibrium potential of a NiCd battery depends on whether it is determined after the battery has been charged or discharged [30]. In the latter case, the equilibrium potential is lower than in the former case. As the battery was charged after a discharging step in the measurements, its equilibrium potential will have changed from a low value during the discharging to a higher value during the charging. This could explain the differences between the simulated and measured voltage curves at the beginning of the charging, because the battery model does not allow for any occurrence of hysteresis yet. When we take a closer look at this difference we see that the optimized curve obtained with the model is indeed higher than the measured curve. No appropriate mathematical description of the hysteresis of the equilibrium potential is as yet available. Further research will have to be carried out to check whether the inclusion of hysteresis in the model will improve the quantitative agreement between the results of the simulations and the measurements. Note that this is an example of model modification, as shown in Figure 4.24.

   Figure 4.29 shows that the P_{opt} curve represents a considerable improvement over the P_{start} curve. The difference between the end values of the simulated optimized curve P_{opt} and the measured curve P_{meas} is 18 kPa or 0.18 bar. This difference amounts to 9% of the measured end value of 200 kPa. As with the voltage curves, the two potential causes of spread in the measured curve have to be taken into account. A resolution of 1 kPa holds for the measurement accuracy, so the difference observable in Figure 4.29 cannot be explained by measurement accuracy only.

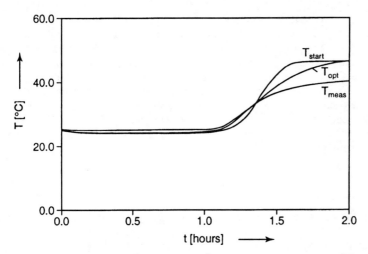

Figure 4.30: Simulated (T_{start} [°C] and T_{opt} [°C]) and measured (T_{meas} [°C]) battery temperature versus time (t [hours]). See Figure 4.28 for conditions

The difference between the optimized and measured pressure curves P_{opt} and P_{meas} during overcharging suggests that the equilibrium between oxygen production and recombination was reached at different oxygen pressures in the simulations and the measurements. Many parameters have to be considered in calculating a battery's behaviour in terms of oxygen production and recombination. It therefore seems reasonable to assume that the observed difference is partly attributable to incorrect parameter values. The P_{meas} and P_{opt} curves are identical in shape and the point at which the oxygen pressure starts to rise is the same in both curves. Moreover, the shape of the simulated P_{opt} curve is much better than that of the P_{start} curve during overcharging. Nevertheless, the discrepancy observable between the results of the simulations and the measurements prompted further research into the kinetics of the oxygen reactions taking place inside the battery [30].

Another improvement is observable between the simulated T_{start} and T_{opt} curves shown in Figure 4.30. However, the difference between the T_{opt} and T_{meas} curves still amounted to 7°C at the end of the charging. This difference cannot be explained by the accuracy of the measurements which is within 0.1°C. The present one-dimensional model for the thermal battery behaviour is probably far too simple. The battery temperature is assumed to be homogeneously distributed throughout the entire battery in the present model. The thermal model involves only one heat capacitance and one heat resistance to the outside world; see Figures 4.15 and 4.19. Taking the parameters for heat capacitance and heat resistance into account in the optimization process did not lead to further improvements. Further research is needed to investigate the influence of using distributed networks on the quantitative agreement between simulated and measured temperatures. For such a network, the battery must be divided into several parts and each part must be able to have its own temperature and will have to be modelled separately. It will then be possible to consider temperature gradients inside the battery, too. The difference between the simulated and measured battery temperatures will also be partly responsible for the observed differences between V_{opt} and V_{meas} and between P_{opt} and P_{meas}, because V and P are temperature-dependent both in the model and in practice.

Discussion of the parameter values
The parameter values of the NiCd model are summarized in Tables 4.3 and 4.4. Some of the battery parameters shown in these tables were inferred from orthogonal model parameters. Parameters are orthogonal when they can be varied independently, which means that a variation in one parameter does not lead to a change in another parameter. The battery parameters have been listed so that they can be understood and compared with measured values more easily. An example is the double-layer capacitance of the two electrodes given in Table 4.4. The values of these double-layer capacitances have actually been calculated using the following simple equation from the model:

$$C^{dl} = C_o^{dl} \cdot A \qquad (4.96)$$

where C_o^{dl} denotes the capacitance per area in [F/m^2] and A the area of the electrode in [m^2].

Table 4.3 lists the parameters that were not varied in the optimization process. The values were obtained from the literature and from the supplier of the P60AA NiCd battery. They include design-related and electrochemical parameters. As a reference to earlier sections, the fourth column lists the equations or figures in which the parameters were used. There where parameters have been used in more than one equation, the first equation in which they were used is referred to in the tables. The appropriate section number is given there where a parameter was only mentioned in the text. The parameter values given in Table 4.3 remained unchanged throughout the simulations and therefore hold for both the *start* and the *opt* curves shown in Figures 4.28, 4.29 and 4.30.

Table 4.4 shows the parameters considered in the application of the strategy described in the previous section. For example, the double-layer capacitances C^{dl} have been inferred from (4.96). In the optimization process, the value of C_o^{dl} remained fixed at 0.2 F/m^2 for both electrodes [27], whereas the areas A were varied separately for each electrode. Consequently, the ratios of the starting values and the optimized values are the same for A and C^{dl} for each of the two electrodes. All the parameters were orthogonal in the actual parameter sets used in the application of the strategy.

The parameters given in Table 4.4 have again been split up into design-related and electrochemical parameters. The second column shows the starting values. These values are in accordance with those presented in previous publications on the NiCd model [26]-[28] and were used for all the curves labelled *start* in Figures 4.28, 4.29 and 4.30. The third column shows the optimized parameter values used for all the curves labelled *opt* in Figures 4.28, 4.29 and 4.30. For validation purposes, the available measured values have been included in the fourth column. These measurements were obtained with an 80AAS NiCd battery, which differs from the type of battery considered in this book, but is based on the same battery system [30]. However, the indicated values serve merely to place the optimized parameter values in a realistic perspective. Finally, the sixth column lists the equations or figures in which the parameters have been used, using the same procedure as for Table 4.3. It should be noted that all temperature-dependent parameters, which are all D_i^o, E_i^o, and I_i^o parameters, are specified at 25°C. The temperature dependence of these parameters is determined by the corresponding value of the activation energy E^a. All I_i^o values are specified at 50% SoC, while all $I_{O_2}^o$ values are specified at the reference pressure of 10^5 Pa=1 bar.

Table 4.3: Fixed parameter values for the NiCd model described in section 4.2 which were not considered in the parameter optimization process

Parameter	Value	Unit	Reference
Design-related parameters			
A_{bat}	$2.7 \cdot 10^{-3}$	m^2	(4.34)
C_{th}^{bat}	14	J/K	Figure 4.15
$Q_{Cd, Max}$	5000	C	Figure 4.16
$Q_{Cd(OH)_2, Ni}$	1000	C	Figure 4.16
$Q_{Cd, Cd}$	800	C	Figure 4.16
R_{th}^{bat}	14	K/W	(4.34)
R_{Ni}	2	mΩ	Figure 4.19
R_{Cd}	2	mΩ	Figure 4.19
Electrochemical parameters			
$(I^o_{O_2})_{Cd}^O$	$1.0 \cdot 10^{-3}$	A	Section 4.2.4
$(I^o_{O_2})_{Cd}^R$	$1.0 \cdot 10^{-3}$	A	(4.66)
M_{Cd}	112.4	g/mol	(4.54)
n_{Ni}	1	-	(4.45)
n_{Cd}	2	-	(4.52)
n_{O_2}	4	-	(4.61)
ρ_{Cd}	$8.6 \cdot 10^6$	g/m^3	(4.54)

Different exchange currents have been used in Tables 4.3 and 4.4 for oxygen evolution and recombination occurring at the two electrodes, $(I^o_{O_2})_{Cd}^R$ and $(I^o_{O_2})_{Cd}^O$ in Table 4.3 and $(I^o_{O_2})_{Ni}^R$ and $(I^o_{O_2})_{Ni}^O$ in Table 4.4; see Figure 4.16. This was explained in section 4.2.4. In Table 4.3, the two exchange currents have been given the same value, which remained constant in the simulations [27], whereas they have different values in the starting and optimized parameter set given in Table 4.4 [27].

Not wishing to draw final conclusions from the optimized parameter values, we would nevertheless like to make some general remarks. First of all, good agreement with the measured parameter values is observable in the case of parameters E^a_{Ni}, $E^a_{DO_2}$, E^o_{Cd}, E^o_{Ni}, I^o_{Ni}, ΔS_{Cd} and ΔS_{Ni}, in other words, seven of the eleven compared parameter values. The difference of a factor of two observed for an exchange current such as I^o_{Ni} is not that great, especially when the range of two orders of magnitude used for this parameter in the optimization strategy is considered. The optimized values of parameters E^a_{Ni}, $E^a_{DO_2}$ and I^o_{Ni} are particularly close to the measured values given in Table 4.4, whereas the starting values differ substantially from the optimized and measured values. This inspires great confidence in the quality of the model and the approach that was used to find the optimized parameter values.

Table 4.4: The parameters of the NiCd model described in section 4.2 which were considered in the parameter optimization process. Values presented in previous publications [26]-[28] and the values obtained using a 80AAS NiCd battery [30] have been included to enable comparison

Parameter	Value in [26], [27] and [28]	Value in optimized parameter set	Measured value	Unit	Reference
		Design-related parameters			
A_{Ni}	11.8	10.9		m^2	(4.46)
A_{Cd}	23.3	32		m^2	(4.53)
C_{Ni}^{dl}	2.4	2.2		F	Figure 4.19
C_{Cd}^{dl}	4.7	6.4		F	Figure 4.19
$Q_{Ni, Max}$	3060	3724		C	Figure 4.16
R_e	10	9.1		$m\Omega$	Figure 4.19
l_{Ni}	$5.5 \cdot 10^{-6}$	$58.3 \cdot 10^{-6}$		m	Figure 4.17
V_g	$1.5 \cdot 10^{-6}$	$1.7 \cdot 10^{-6}$		m^3	(4.70)
		Electrochemical parameters			
$a^b_{OH^-}$	$8.0 \cdot 10^3$	$6.8 \cdot 10^3$		mol/m^3	(4.46)
d_{O_2}	$3.0 \cdot 10^{-7}$	$5.4 \cdot 10^{-7}$		m	(4.67)
d_{OH^-}	$1.0 \cdot 10^{-6}$	$1.3 \cdot 10^{-6}$		m	(4.58)
$D^{o,1}_{H^+}$	$5.0 \cdot 10^{-14}$	$169.9 \cdot 10^{-14}$		m^2/s	Section 4.2.6
$D^{o,2}_{H^+}$, see note 1	-	$80.8 \cdot 10^{-14}$		m^2/s	Section 4.2.6
$D^o_{O_2}$	$5.0 \cdot 10^{-10}$	$9.6 \cdot 10^{-10}$		m^2/s	(4.67)
$D^o_{OH^-}$	$5.0 \cdot 10^{-10}$	$4.0 \cdot 10^{-10}$		m^2/s	(4.58)
E^a_{Cd}	$40.0 \cdot 10^3$	$50.9 \cdot 10^3$	$36.2 \cdot 10^3$	J/mol	(4.72)
E^a_{Ni}	$40.0 \cdot 10^3$	$24.8 \cdot 10^3$	$24.3 \cdot 10^3$	J/mol	(4.72)
$E^a_{O_2}$	$40.0 \cdot 10^3$	$76.8 \cdot 10^3$		J/mol	(4.72)
$E^a_{D_{H^+}}$	$9.6 \cdot 10^3$	$17.8 \cdot 10^3$		J/mol	(4.72)
$E^a_{D_{O_2}}$	$14.0 \cdot 10^3$	$9.2 \cdot 10^3$	$9.4 \cdot 10^3$	J/mol	(4.72)
$E^a_{D_{OH^-}}$	$14.0 \cdot 10^3$	$8.7 \cdot 10^3$		J/mol	(4.72)
E^o_{Cd}	-0.81	-0.83	-0.81	V vs SHE	(4.52)
E^o_{Ni}	0.52	0.53	0.55	V vs SHE	(4.45)
$E^o_{O_2}$	0.40	0.40		V vs SHE	(4.61)
I^o_{Cd}	$2.3 \cdot 10^3$	$27.2 \cdot 10^3$	0.52	A	(4.53)
I^o_{Ni}	36.0	0.82	0.33	A	(4.46)
$(I^o_{O_2})_{Ni}^O$	$117.6 \cdot 10^{-9}$	$1.8 \cdot 10^{-6}$	$2.2 \cdot 10^{-2}$, see note 2	A	(4.65)
$(I^o_{O_2})_{Ni}^R$	$0.5 \cdot 10^{-3}$	$18.8 \cdot 10^{-3}$		A	Section 4.2.4
α_{Cd}	0.5	0.55		-	(4.53)
α_{Ni}	0.5	0.39		-	(4.46)
α_{O_2}	0.5	0.23	0.06, see note 2	-	(4.65)
ΔS_{Cd}, see note 3	-204	-161.9	-191	J/(mol.K)	(4.73)
ΔS_{Ni}, see note 4	130.3	115.9	127.4	J/(mol.K)	(4.73)
ΔS_{O_2}, see note 4	649.0	702.7		J/(mol.K)	(4.73)

Note 1: During the curve-fitting process, the diffusion coefficient $D^o_{H^+}$ was made dependent on the local mol fraction of H^+ ions (x_{H^+}) in the nickel electrode. The local value of x_{H^+} inferred from capacitance C_i

was used for each spatial element i in the RC-ladder network. The following dependence of $D^o_{H^+}$ on x_{H^+} was used [35]:

$$D^a_{H^+} = D^{a,1}_{H^+} \cdot \left(x_{H^+} + \sqrt{\frac{D^{a,2}_{H^+}}{D^{a,1}_{H^+}}} \cdot \left(1 - x_{H^+}\right) \right)^2$$

Note 2: These values hold for sufficiently large current and overpotential values
Note 3: The sign holds for a reduction reaction, so during charging for the cadmium reaction at the cadmium electrode [31]
Note 4: The sign holds for an oxidation reaction, so during charging for the nickel reaction and the oxygen-production reaction at the nickel electrode [31]

Secondly, the parameters I^o_{Cd} and E^a_{Cd} differ significantly from the measured values. A possible reason for this is the fact that in the first step the range of I^o_{Cd} was chosen such as to ensure that the overpotential of the cadmium reaction would be relatively small, or in the order of 1 mV. This means that it was assumed that the kinetics of the cadmium reaction are much faster than those of the nickel reaction. This small cadmium reaction overpotential, and hence the parameters affecting its value, had of course little influence on the overall battery voltage. This made it more difficult to find correct values for these parameters, as they do not influence the battery voltage curve to any great extent. It seems reasonable to assume that this is the reason why the optimization program failed to find an exact value for E^a_{Cd}. The measured value of I^o_{Cd} shows that the assumption that the kinetics at the cadmium electrode are much faster than those at the nickel electrode is not correct in the case of the 80AAS NiCd battery. The measured value of the parameter I^o_{Cd} is not even within the range chosen for I^o_{Cd} in the simulations. Therefore, further research will have to be done into the parameter values of the cadmium reaction of the P60AA type of battery considered here.

Thirdly, a major difference between the optimized and measured values was observed in the case of parameters $(I^o_{O_2})_{Ni}^O$ and α_{O_2}. As can be seen in Table 4.4, the optimized value obtained for $(I^o_{O_2})_{Ni}^O$ is lower than the measured value, whereas the optimized value of α_{O_2} is higher than the measured value. These two parameter values determine the amount of oxygen that is produced at the nickel electrode during overcharging. We used the anodic branch of the Butler-Volmer equation to express the production of oxygen at the nickel electrode during charging, which is repeated here for reference:

$$I_{O_2} = I^o_{O_2} \exp\left\{ \frac{\alpha_{O_2} n F \eta_{O_2}}{RT} \right\} \tag{4.97}$$

As can be inferred from (4.97), a high value of $I^o_{O_2}$ in combination with a low value of α_{O_2} will mathematically yield the same value for I_{O_2} as a low value of $I^o_{O_2}$ in combination with a high value of α_{O_2} at the same overpotential η_{O_2} and at a constant temperature T. Using the values given in Table 4.4, the combination of the optimized values of $I^o_{O_2}$ and α_{O_2} and the combination of measured values yield values for η_{O_2} of 378 mV and 377 mV, respectively, at the considered charge rate of 1 C and at a temperature of 318 K, which is the average temperature in the overcharge region. Although the discrepancy between the measured and optimized parameters for the oxygen evolution at the nickel electrode can be explained by basic mathematics, this discrepancy prompted a more detailed study of the kinetics of the evolution of oxygen at the nickel electrode. More information will be given in the next section.

Discussion of the mathematical optimization strategy
Some remarks can be made on the applied mathematical strategy. First of all, the low weight of the cost function at the start of the charging may have led to the observed discrepancy between the simulation and measurement results. Further research into the possibility of improving the quantitative agreement at the start of the charging should shed more light on this matter. Including the occurrence of hysteresis in the description of the equilibrium potential, as described above, could also help if it should prove impossible to improve the agreement by means of mathematical optimization only.

The mathematical parameter improvement method proved to be a good means for focusing on certain parts of the model in which improvement is still needed. This has sparked research in certain specific directions, such as the investigation of the kinetics of the oxygen evolution reaction at the nickel electrode described above. So all in all, the method has not only led to a significant improvement in agreement between the results of simulations and measurements, but has also prompted further research aimed at improving the model.

Although the model can primarily be regarded as a black box for mathematical parameter improvement, as shown in Figure 4.25, (electro)chemical and battery knowledge was used in several steps in the process. Using this knowledge is crucial to prevent unrealistic results as much as possible. Moreover, several iterations were performed to obtain the final results presented in this section. This means that some guidance of the process is still needed. However, the mathematical parameter optimization method can be readily applied to battery systems other than NiCd systems.

The results presented in this section were obtained in a charging experiment using one charge current only. In the next section, the quantitative agreement between the results of simulations and measurements will first be investigated in the case of the optimized parameter set given in Table 4.4 under various charging conditions.

4.4.4   Quality of the parameter set presented in section 4.4.3 under different charging conditions

Charging simulations have been performed at different currents in order to investigate the quality of the model simulations using the optimized parameter set presented in the previous section. The simulation results were compared with measurements obtained at the same charge current. Figures 4.31 and 4.32 show the simulated and measured V, P and T curves obtained after charging at a 0.1 C charge rate at a 25°C ambient temperature. Figures 4.33 and 4.34 show similar curves obtained at a 2 C charge rate at 25°C ambient temperature. In both cases, a 100% overcharge was applied, as in the case of the 1 C curves presented in the previous section. As can be inferred from the figures showing the curves obtained at these charge currents, the quantitative fit between the simulated and measured curves is worse than that obtained for the charge rate of 1 C applied in the optimization process. Some remarks will be made on both cases below, concentrating on the most prominent differences between the measurements and the simulation results.

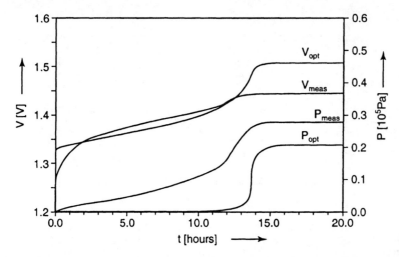

Figure 4.31: Simulated (V_{opt} [V]) and measured (V_{meas} [V]) battery voltage plotted along the left y-axis and simulated (P_{opt} [10^5 Pa]) and measured (P_{meas} [10^5 Pa]) internal oxygen pressure plotted along the right y-axis versus time (t [hours]) obtained in charging a NiCd battery at 0.1 C at 25°C ambient temperature using the NiCd model described in section 4.2 in the simulations using the optimized parameter values given in the third column of Table 4.4

The simulated battery voltage during overcharging is higher than the measured voltage in Figure 4.31. The measured oxygen pressure started to rise early in the charging process already. In the simulation however, the oxygen pressure increased quite abruptly after around 13.5 hours. The pressure levelled off to a steady-state value of 21 kPa after approximately 15.5 hours. Figure 4.32 shows that the shapes of the simulated and measured temperature curves are similar. However, the simulated temperature at the end of the charging is somewhat higher than the measured temperature, which was also the case in Figure 4.30.

The shape of the simulated battery voltage curve differs significantly from that of the measured battery voltage curve obtained in the 2 C charging experiment in Figures 4.33 and 4.34. The main difference concerns the position of the peak, which occurred earlier in the charging process in the simulation. As will be explained in detail in section 4.5, the peak in the battery voltage curve marks the beginning of the production of oxygen at the nickel electrode and the reduction of oxygen at the cadmium electrode. This can also be seen in Figures 4.33 and 4.34, in which the internal oxygen gas pressure and the battery temperature started to rise earlier in the charging process in the simulation. The value of the simulated battery temperature at the end of the charging was again higher in the simulation.

An explanation for the differences between the results of the simulations and the measurements described above can be found by studying the measurement results given in Table 4.4 more closely. For this purpose, a Tafel plot of the current-versus-overpotential measurements of the oxygen-evolution reaction occurring at the nickel electrode is shown in Figure 4.35 [30]. A Tafel plot is a convenient way of plotting current-voltage relationships of electrochemical systems according to

(4.97), because the value of α can be inferred from the slope of the plot, while the value of I^o can be found by extrapolating the curve until the overpotential η^k is zero; see chapter 3.

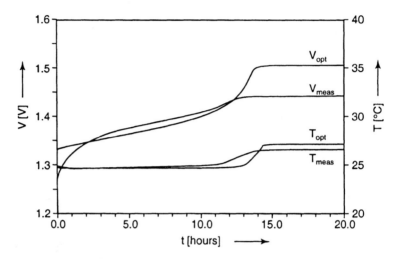

Figure 4.32: Simulated (V_{opt} [V]) and measured (V_{meas} [V]) battery voltage plotted along the left y-axis and simulated (T_{opt} [°C]) and measured (T_{meas} [°C]) battery temperature plotted along the right y-axis versus time (t [hours]). See Figure 4.31 for conditions

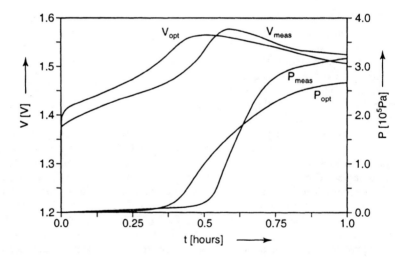

Figure 4.33: See Figure 4.31; the charge rate was 2 C in this case

Figure 4.35 shows that the kinetics of the oxygen production reaction can be described as a two-step process, with the values of α and I^o differing for curves (a) and (b). In fact, as mentioned in footnote 2 of Table 4.4, the measured values of $(I^o_{O_2})_{Ni}^{\ O}$ and α_{O_2} in the table were derived at high current and overpotential values, i.e. from curve (b) in Figure 4.35 [30]. A value of 0.26 for α_{O_2} and a value of $3.1 \cdot 10^{-6}$ [A] for $(I^o_{O_2})_{Ni}{}^O$ can be obtained from curve (a). Curve (c) was obtained from curves (a) and (b) in a calculation similar to that used for a mixed kinetic-diffusion controlled reaction current; see (4.59) and (4.69). This means that:

$$I_{(c)} = \frac{I_{(a)}I_{(b)}}{I_{(a)} + I_{(b)}} \tag{4.98}$$

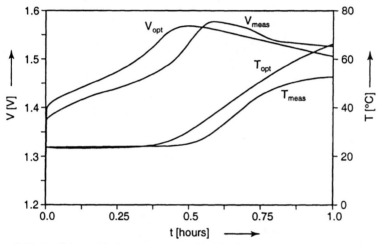

Figure 4.34: See Figure 4.32; the charge rate was 2 C in this case

As can be seen in Figure 4.35, the calculated curve (c) agrees very well with the curve based on measurements at moderate currents. Note that the battery voltage has been plotted along the horizontal axis. The overpotential for the oxygen evolution reaction at the nickel electrode has to be used to infer I^o and α from the Tafel plots. This overpotential can be found by subtracting the oxygen equilibrium potential from the nickel electrode potential, i.e. $\eta_{O_2}=E_{Ni}-E^{eq}_{O_2}$. The oxygen equilibrium potential equals the standard redox potential $E^o_{O_2}$ at the reference pressure of 1 bar, hence $E^{eq}_{O_2}=E^o_{O_2}=0.40V$; see section 4.2.4 and (4.64). The nickel electrode potential can be inferred from the battery voltage by the addition of the cadmium electrode potential, i.e. $E_{Ni}=E_{NiCd}+E_{Cd}$. Neglecting the relatively small overpotential at the cadmium electrode, the cadmium electrode potential is the same as the equilibrium potential $E_{Cd}=E^{eq}_{Cd}=E^o_{Cd}=-0.81V$; see section 4.2.3 and (4.52). In summary, this means that the oxygen overpotential can be inferred from the battery voltage plotted along the horizontal axis by $\eta_{O_2}=E_{NiCd}-0.81-0.40= E_{NiCd}-1.21$ V.

Figure 4.36 shows the same measured Tafel plots as shown in Figure 4.35 along with the Tafel plot calculated for the anodic branch of the evolution of oxygen in the NiCd model; see (4.97). The optimized parameter values given for $(I^o_{O_2})_{Ni}{}^O$ and α_{O_2} in Table 4.4 were used in the calculation of this curve.

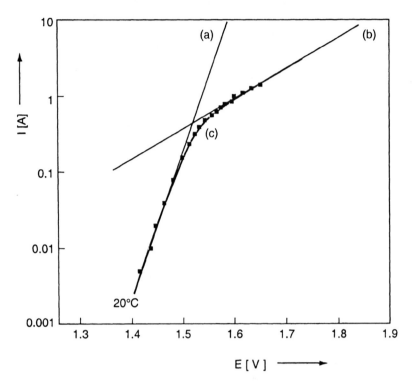

Figure 4.35: Measured Tafel plots representing the oxygen production reaction occurring at the nickel electrode during overcharging at 20°C [30]. The voltage of the NiCd battery used has been plotted along the horizontal axis, while the battery current I has been plotted on a logarithmic scale along the vertical axis. The marks were obtained in measurements. Curve (a) is the linear approximation at low currents, curve (b) at high currents and curve (c) is the calculated transition from curve (a) to (b) for moderate currents, see (4.98)

It can be inferred from Figure 4.36 that the measured (b) and calculated (d) Tafel plots intersect at a current of approximately 600 mA. This is in fact the visualization of the mathematical explanation based on (4.97) that was given for the difference between optimized and measured parameter values for $(I^o_{O_2})_{Ni}{}^O$ and α_{O_2} in section 4.4.3. Although the measurements in Figures 4.35 and 4.36 were performed using a 80 AAS NiCd battery, it is reasonable to assume that a similar two-step approach holds for the P60AA NiCd battery, too. Figure 4.36 will now be used to explain the main differences between the simulation and measurement results described earlier in this section.

For the low current of 0.06 A (0.1 C-rate), curve (d) predicts a larger overpotential for the oxygen evolution reaction in the simulations than the value predicted by curve (a) for the measurements. This means that in the simulations, the rate of the oxygen evolution reaction was lower than in the measurements, resulting in a later start of the oxygen evolution. This is indeed observable in Figure 4.31. As the overpotential of the oxygen production was higher in the simulations, the battery voltage was also higher during overcharging. This is also observable in Figure 4.31.

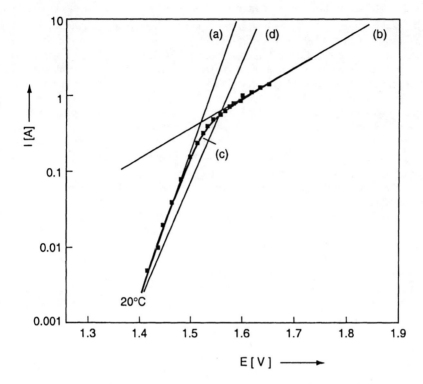

Figure 4.36: Same measured Tafel plots (curves (a)-(c)) as shown in Figure 4.35. Also included is the Tafel plot calculated for the evolution of oxygen at the nickel electrode according to (4.97) (d) as was included in the NiCd model presented in section 4.2, using optimized parameter values for $(I^o_{O_2})_{Ni}^o$ and α_{O_2}; see Table 4.4

At the high charge current of 1.2 A (2 C-rate), the measured battery behaved as predicted by curve (b) in Figure 4.36, in which the region of interest lies to the right of the intersection of curves (b) and (d), because of the high current value. The overpotential for the evolution of oxygen was larger in the measurements than in the simulations. The reverse situation as for 0.1 C occurred. The rate of oxygen evolution was higher in the simulations than in the measurements. This led to an earlier rise in the internal oxygen gas pressure and a lower battery voltage during overcharging in the simulations than in the measurements. This is exactly what is shown in Figure 4.33.

It can be concluded that the main differences between the simulation and measurement results in the presented figures can be explained by the fact that the evolution of oxygen in a NiCd battery seems to proceed in two steps, whereas a one-step process was included in the NiCd model described so far. A model modification to include the two-step process and subsequent parameter value optimization will be considered in the next section.

4.4.5   Results obtained with a modified NiCd battery model and discussion

The modification of the mathematical description of the oxygen evolution reaction that occurs at the nickel electrode suggested in the previous section is in fact an example of the second feedback loop in Figure 4.24. It was shown in the previous section that the anticipated improvement should occur at charge currents ranging from small to large values. The mathematical strategy presented in section 4.4.2 was applied in section 4.4.3 for one charge current only. In this section, step (iii) of the strategy will be repeated using a modified NiCd model, taking into account the two-step behaviour postulated for the production of oxygen at the nickel electrode, for various charge currents simultaneously. In addition to using two different values for $(\Gamma^o_{O_2})_{Ni}^{\ O}$ and α_{O_2}, we also made both α_{O_2} parameters temperature-dependent in accordance with the experimental results [30].

The considered charge rates in the simulations were 0.1 C, 1 C and 2 C. (4.95) was used as cost function and the contributions of the different charge currents were simply added. The different contributions of the different charge currents were not weighed in any way. Step (iii) of the strategy was simply repeated for different charge currents to obtain a first impression of the improvement that can be realized with the modified NiCd model relative to the results presented in the previous section. The ambient temperature was again 25°C. The same measured curves were used as presented in section 4.4.4. The optimized parameter set of Table 4.4 was used as a starting point for all the parameters, except for those describing the oxygen evolution at the nickel electrode. The same ranges as determined in section 4.4.3 were used. The measured values inferred from Figure 4.36 were used for the newly added parameters describing the two-step oxygen-production process [30]. The ranges of these parameters were based educated guesses. The results obtained for a charge rate of 0.1 C are shown in Figures 4.37 and 4.38.

Comparison of the simulated pressure curve shown in Figure 4.37 and the simulated curve shown in Figure 4.31 shows that the repetition of step (iii) using the modified model caused the simulated pressure curve to shift to the left. This is the desired result, because the oxygen production started too late in the simulations in Figure 4.31. So, the kinetics of the oxygen-production reaction at the nickel electrode indeed improved in the simulations. The shape of the simulated curve did not change significantly and the gradual increase in oxygen pressure indicated by the measured curve was not observed. However, the moment at which the pressure increased significantly in the simulation now coincided with the moment at which the pressure increased more in the measurements. The improved kinetics of oxygen production at the nickel electrode did not lead to a significant change in the simulated voltage curve. Comparison of the simulated temperature curve in Figure 4.38 with the simulated temperature curve in Figure 4.32 shows that the repetition of step (iii) using the modified model led to better agreement of the simulated and measured end values.

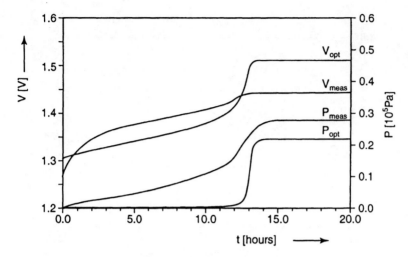

Figure 4.37: Simulated (V_{opt} [V]) and measured (V_{meas} [V]) battery voltage plotted along the left y-axis and simulated (P_{opt} [10^5 Pa]) and measured (P_{meas} [10^5 Pa]) internal oxygen pressure plotted along the right y-axis versus time (t [hours]) for charging a NiCd P60AA battery at 0.1 C at 25°C ambient temperature using the NiCd model described in section 4.2 with the modified mathematical description for the oxygen-production reaction at the nickel electrode in the simulations. The parameter values for the simulated curves were obtained by repeating step (iii) of the strategy presented in section 4.4.2 for charge rates of 0.1 C, 1 C and 2 C simultaneously

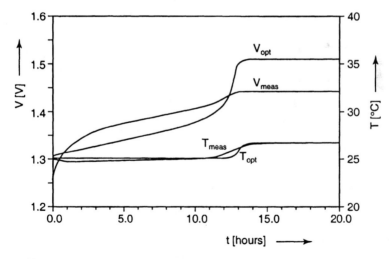

Figure 4.38: Simulated (V_{opt} [V]) and measured (V_{meas} [V]) battery voltage plotted along the left y-axis and simulated (T_{opt} [°C]) and measured (T_{meas} [°C]) battery temperature plotted along the right y-axis versus time (t [hours]). See Figure 4.37 for conditions

The results obtained for a charge rate of 1 C are shown in Figures 4.39 and 4.40.

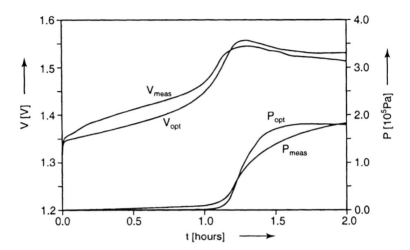

Figure 4.39: See Figure 4.37; the charge rate was 1 C in this case; compare to Figures 4.28 and 4.29

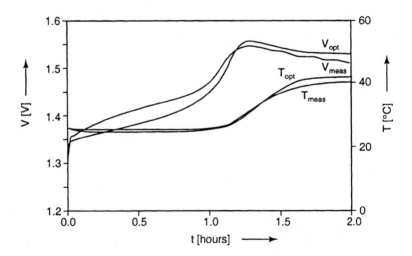

Figure 4.40: See Figure 4.38; the charge rate was 1 C in this case; compare to Figure 4.30

Comparison of the results presented in Figures 4.39 and 4.40 with those given in section 4.4.3 shows that the agreement between the simulation and measurement results changed. This is quite understandable, because the optimization considered in this section was performed for three currents simultaneously. The difference between the simulated and measured battery voltages has increased to around 30 mV in the mid-region in Figure 4.39, which is still acceptable for a nominal battery voltage of 1.25 V.

The difference between the final simulated and measured pressures has become 2.6 kPa instead of 22 kPa in Figure 4.29. This is a significant improvement.

However, the pressure-versus-time curve is steeper in Figure 4.39 than in Figure 4.29, where the curve shown in Figure 4.29 is in better agreement with the measurement. Apparently the simulated oxygen-production process started more abruptly as a result of the better kinetics of the oxygen-production reaction at the nickel electrode.

Comparison of the quantitative agreement between the simulated and measured battery temperatures in Figures 4.40 and 4.30 shows that the simulated end value has improved significantly. The difference is 1.3°C in Figure 4.40, which is smaller than the difference of 7°C in Figure 4.30. The results obtained for a charge rate of 2 C are shown in Figures 4.41 and 4.42.

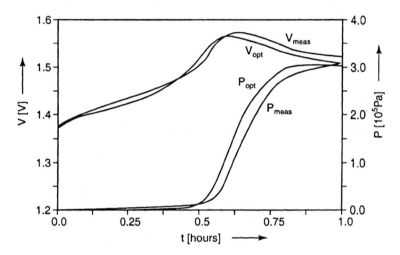

Figure 4.41: See Figure 4.37; the charge rate was 2 C in this case

Comparison of Figures 4.41 and 4.42 on the one hand with Figures 4.33 and 4.34 on the other hand shows that in the case of V, P and T the quantitative agreement between the simulation and measurement results has improved significantly. The agreement is now good for the voltage curve during practically the entire charging period. The maximum difference in value between the simulation and measurement results amounts to 15 mV, or 1.2%, and occurred at the end of the charging. This is comparable with the results presented in section 4.4.2 for a charge rate of 1 C.

The difference between the simulation and measurement results in the final value of the internal oxygen gas pressure is 7 kPa in Figure 4.41, which is a great improvement over the value shown in Figure 4.33. The simulation and measurement results obtained at the moment at which the oxygen production starts still differ slightly. In the simulation, the oxygen production started a little earlier than in the measurement. Nevertheless, the shape of the pressure-versus-time curve is in close agreement with that representing the results of the measurement.

The simulated battery temperature shown in Figure 4.42 is a lot closer in shape to the measured curve than the curve illustrated in Figure 4.34. The difference in the final values between the simulation and measurement results is 4.6°C, which is better than the agreement between the simulation and measurement results obtained for the charge rate of 1 C; see Figure 4.30.

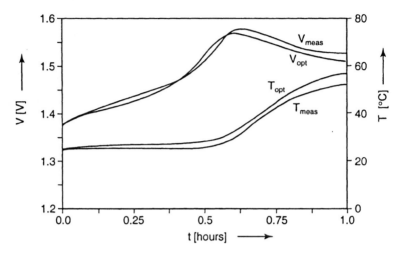

Figure 4.42: See Figure 4.38; the charge rate was 2 C in this case

In summary it can be stated that for the low charge rate of 0.1 C, the model modification has only partly resulted in the desired effect. The kinetics of the oxygen-production reaction at the nickel electrode improved as expected and the oxygen reaction began earlier in the charge process. However, the simulated pressure curve is still too steep and the battery voltage during overcharging remains too high. For a 1 C-rate, the differences between the simulated and measured battery voltages are a little higher than those presented in section 4.4.3, but they are still acceptably small. The end values of the internal oxygen pressure and especially the temperature have improved. The improvement with respect to the results presented in the previous section is evident for the 2 C charge rate. The quantitative agreement between the simulation and measurement results has become even better for this current than for the 1 C charge rate considered in section 4.4.2.

A possible explanation for the marginal improvement in the results at lower charge currents could be the fact that the kinetics of the evolution of oxygen at the nickel electrode are different in a P60AA battery than as shown in Figure 4.35 for a 80AAS battery. Measurement of the Tafel plots shown in Figure 4.35 for the P60AA battery should shed some light on this matter. Step (iii) was moreover performed without considering possibilities of weighing the contributions of the various charge currents in the cost function of (4.95). Another reason could be the fact that the optimum parameter set given in Table 4.4 was taken as a starting point. The strategy presented in section 4.4.2 could be extended for various charge currents and the process could be repeated from scratch. More investigations will have to be carried out to find out whether applying weighing to the cost function or starting the process described in section 4.4.2 again are useful ways of improving the results for lower charge currents.

The optimized parameter set of Table 4.4 will be used in the simulations discussed in the next section, because the experiments described in this section merely served to give a first impression of possible further improvement of the simulations. Therefore, the proposed two-step oxygen-production reaction at the nickel electrode will not be considered in the next section.

4.5   Simulation examples

4.5.1   Simulations using the NiCd model presented in section 4.2

In this section some simulation results will be presented that were obtained using the NiCd battery model presented in section 4.2. The parameter values of the optimized parameter set given in Table 4.4 were used. First, the results of the charging simulation using a 1 C charge rate at 25°C ambient temperature, presented in Figures 4.28, 4.29 and 4.30 will be described in greater detail. For reference, the simulated V, P and T curves have been plotted in a single figure below.

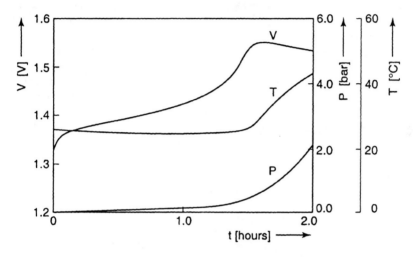

Figure 4.43: Simulated battery voltage (V [V]), internal oxygen gas pressure (P [bar]) and temperature (T [°C]) as a function of time (t [hours]) during charging at 25°C ambient temperature at a 1 C-rate. The optimized parameter values given in Table 4.4 were used

The phenomena observable in the V, P and T curves shown in Figure 4.43 can be understood by considering the simulation model of the NiCd battery shown in Figure 4.19. The current will flow from the left (positive terminal) to the right (negative terminal) during charging. The charge current flows through the nickel reaction path at the nickel electrode (D_{Ni} and electrical port of the transformer), until the moment when the oxygen pressure starts to rise. The molar amounts increase in the chemical capacitances C_{ch}^{NiOOH} and decrease in the chemical capacitances $C_{ch}^{Ni(OH)2}$ in the RC-ladder network. The current flows through the cadmium reaction path (D_{Cd} and the electrical port of the transformer) in the cadmium electrode. The molar amount increases in capacitance C_{ch}^{Cd} and it decreases in capacitance $C_{ch}^{Cd(OH)2}$.

The overpotential across diodes D_{O_2} for the oxygen reaction at the nickel electrode is higher than that for the nickel reaction across diodes D_{Ni}, because the value of E^o_{Ni} is higher than the value of $E^o_{O_2}$; see Table 4.4. However, although the higher overpotential for the oxygen reaction may predict a higher reaction current than for the nickel reaction, according to the Butler-Volmer equation, a negligible amount of oxygen is formed at the nickel electrode at the beginning of the charging.

This is due to the relatively low reaction rate of the oxygen evolution reaction, expressed in a low value for $(I^o_{O_2})_{Ni}{}^o$ in (4.65). No current flows through the cadmium reaction path at the nickel electrode, because under normal charging conditions only $Cd(OH)_2$ is present in the nickel electrode and no metallic Cd, which could be oxidized into $Cd(OH)_2$. No current flows through the oxygen path at the cadmium electrode either, because only a negligible amount of oxygen gas is present during the first stages of charging.

When the charging is continued, the decrease in the molar amount of the $Ni(OH)_2$ species in the nickel electrode will lead to a lower value of I^o_{Ni} through (4.46). As a result, a larger overpotential η^k_{Ni} across diodes D_{Ni} occurs at the same value of the current, as can be inferred from the Butler-Volmer equation given in chapter 3. At the same time, the apparent equilibrium potential $E^{eq^*}_{Ni}$ rises according to (4.49). The increase in both η^k_{Ni} and $E^{eq^*}_{Ni}$ results in a substantial increase in the battery voltage. Consequently, the overpotential of the oxygen reaction across diodes D_{O_2} in the nickel electrode model also rises, as can be understood by applying Kirchhoff's voltage law to the parallel combination of the nickel and oxygen reaction paths. The equilibrium potential of the oxygen reaction remains unchanged until the oxygen pressure starts to rise; see (4.64). An increase in the overpotential of the oxygen reaction eventually allows some oxygen to be formed. This in turn leads to an increase in $(I^o_{O_2})_{Ni}{}^o$ according to (4.65), because the activity of oxygen $a^b_{O_2}$ increases, and the oxygen reaction will gradually take over from the nickel reaction. Overcharging now commences. The increase in $a^b_{O_2}$ leads to an increase in oxygen pressure according to (4.63).

Owing to the available oxygen, current may now also start to flow through the oxygen reaction path at the cadmium electrode, which represents the reduction of oxygen. As a result, the oxygen reduction flow in the chemical domain will have the opposite direction of the oxygen evolution flow induced at the nickel electrode, according to the signs associated with the transformer ports in Figure 4.19. This prevents an unlimited increase in the internal oxygen pressure.

Table 4.4 shows that there is a substantial difference between the standard electrode potentials E^o_{Cd} and $E^o_{O_2}$. As a result, there is a substantial difference between the equilibrium potentials of the cadmium and oxygen reactions. Application of Kirchhoff's voltage law to the parallel connection of the cadmium and oxygen reaction paths at the cadmium electrode reveals that there will be a large overpotential for the oxygen reaction across diodes D_{O_2} at the cadmium electrode. A reaction with a large overpotential will generate a lot of heat according to (4.32). Hence, the heat flow source $J_{th}{}^{in}$ in Figure 4.19b representing the heat generated by the oxygen reduction reaction at the cadmium electrode will attain a large value. As a result, the battery temperature will start to rise more in this overcharge region. Both the overpotentials η and the standard electrode potentials E^o are temperature-dependent through (4.72) and (4.73), respectively. Therefore, the battery voltage decreases, because of the rising temperature. This is due mainly to decreasing overpotentials, as observed in simulations. This can be understood from the fact that the exchange currents I^o increase with increasing temperatures, because the reaction rate constants k_i increase according to (4.72).

The distribution of current between the nickel reaction path and the oxygen reaction path is very important in determining a battery's charging efficiency. Only current that flows through the nickel reaction path leads to effective accumulation of charge inside a battery. Current that flows through the oxygen reaction path leads to pressure build-up and oxygen recombination. Eventually, it leads to energy loss in the form of heat dissipation. The charging efficiency is strongly influenced by both the charge current and the battery temperature. More information on charging efficiency will be given in chapters 5 and 6.

The diffusion of H^+ ions becomes a limiting factor of increasing importance for the rate of the nickel reaction when higher charge currents are applied. This means that the apparent equilibrium potential $E^{eq^*}_{Ni}$ starts to rise earlier in the charging process, because of a higher value of the diffusion overpotential η^d_{Ni}; see (4.49). This in turn leads to an earlier start of the production of oxygen in the charging process, and hence to a lower charging efficiency.

The ratio of the rates of the nickel and oxygen reaction decreases as the battery temperature increases. This is due to the difference in the temperature dependence of the two reactions, which is revealed by the difference in activation energies; see Table 4.4. Figure 4.44 shows the simulated partial currents that flow through the nickel and oxygen reaction paths in the nickel electrode during charging at a constant rate of 0.1 C at ambient temperatures of 0 °C and 60°C. The sum of the two partial currents equals the charge current. The oxygen reaction takes over a lot earlier in the charging process at the higher ambient temperature of 60°C, which leads to a poorer charging efficiency at higher temperatures. This has indeed been confirmed in experiments.

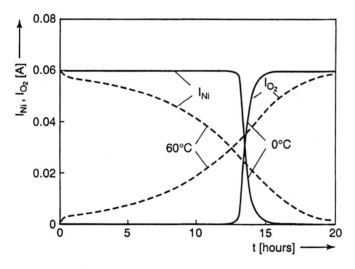

Figure 4.44: Simulated partial nickel (I_{Ni} [A]) and oxygen (I_{O_2} [A]) currents as a function of time (t [hours]) in the nickel electrode during charging at a rate of 0.1 C at 0 °C and 60°C ambient temperature. The optimized parameter values given in Table 4.4 were used

When batteries are not used for some time, they will lose stored energy as a result of self-discharging, especially at higher temperatures. This phenomenon can be easily understood by considering the simulation model of a NiCd battery in Figure 4.19 when no load is applied. The value of the apparent equilibrium potential $E^{eq^*}_{Ni}$ is higher than that of the equilibrium potential $E^{eq}_{O_2}$, because E^o_{Ni} is higher than $E^o_{O_2}$. Therefore, a loop current will flow counterclockwise through the parallel combination of the nickel and oxygen reaction paths in the nickel electrode. This leads to a decrease in stored energy at the nickel electrode and an increase in the molar amount of oxygen and the internal oxygen gas pressure. The oxygen produced is reduced at the cadmium electrode, which leads to a decrease in stored energy in the cadmium electrode through a similar loop current. Figure 4.45 shows the simulated SoC at different ambient temperatures. At 25°C, about 70% of the charged capacity is lost after 100 days. This corresponds well to the reported self-discharge of 60% of the charged capacity of a typical sealed NiCd battery after three months of storage at 25°C [18]. At higher temperatures, the rates of all the electrochemical reactions become higher through (4.72) and consequently the loop currents in the nickel and cadmium electrode models increase. As a result, the self-discharge rate increases at higher temperatures, as can be inferred from Figure 4.45. A higher self-discharge rate at higher ambient temperatures is commonly known from practice and the literature. For example, a 100% loss of charged capacity through self-discharge has been reported at 40°C for a typical sealed NiCd battery in a period of a little over 2 months [18]. This corresponds quite well to the curve simulated at a 45°C ambient temperature in Figure 4.45. The opposite situation occurs at lower temperatures. A 20% loss of charged capacity after 3 months is reported in the literature for a typical sealed NiCd battery at an ambient temperature of 0°C [18]. Again, this corresponds perfectly to the simulated curve shown in Figure 4.45.

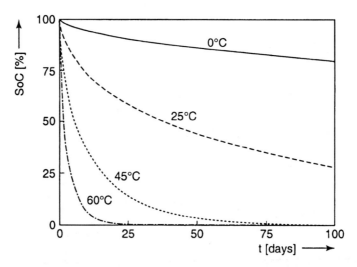

Figure 4.45: Simulated SoC as a function of time (t [days]) at different ambient temperatures. The optimized parameter values given in Table 4.4 were used

Figure 4.46 shows the simulated V, P and T curves during discharging and overdischarging at an ambient temperature of 25°C at a 1 C discharge rate. A 0 V region is clearly recognizable in the voltage curve after approximately 1.5 hours, followed by a region with a reversed battery voltage. In the latter region, both the temperature and the pressure increase and level off to a steady-state value. This is commonly encountered in practice when overdischarging NiCd batteries [18]. What happens during overdischarging can be understood by considering the simulation model of a NiCd battery shown in Figure 4.19.

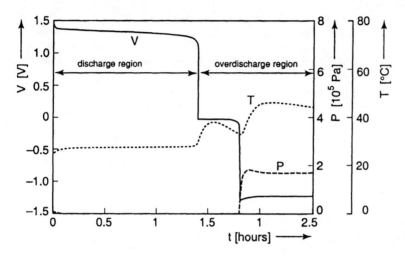

Figure 4.46: Simulated battery voltage (V [V]), internal oxygen gas pressure (P [10^5 Pa]) and battery temperature (T [°C]) curves as a function of time (t [hours]) during discharging and overdischarging at an ambient temperature of 25°C at a 1 C discharge rate. The optimized parameter values given in Table 4.4 were used

The current direction in Figure 4.19a is from the right (negative terminal) to the left (positive terminal) during discharging. During normal discharging, when the battery voltage remains at a positive, relatively constant level, current flows through the cadmium reaction path at the cadmium electrode and the nickel reaction path at the nickel electrode. The battery is effectively discharged in this case. The molar amount of Cd decreases in the cadmium electrode and the molar amount of Cd(OH)$_2$ increases. The molar amount of NiOOH decreases in the nickel electrode, while that of Ni(OH)$_2$ increases. The nickel electrode will run out of NiOOH before all the Cd has been consumed at the cadmium electrode because of the excess amount of Cd at the cadmium electrode, as explained in section 4.2.1. Consequently, the exchange current $I^o{}_{Ni}$ of the nickel reaction will decrease at the end of the discharging. This will lead to a larger overpotential $\eta^k{}_{Ni}$. Moreover, the value of the apparent equilibrium potential $E^{eq*}{}_{Ni}$ will decrease, mainly because diffusion limitation of H$^+$ ions leads to a high value of the diffusion overpotential $\eta^d{}_{Ni}$. As a result, less current flows through the nickel reaction path and a competing reaction will gradually take over. No current can flow through the oxygen reaction path at the nickel electrode because no oxygen is yet present to be reduced. Therefore, the current will start to flow through the cadmium reaction path at the nickel electrode. The resulting battery voltage will approach 0 V because identical reactions now occur at both electrodes. This is the first stage of overdischarging. A lot of heat is generated in this reaction, because the current through the nickel reaction path does not become zero instantaneously and the overpotential of the nickel reaction will have become very large because the electrode potential has reversed. This is illustrated by the temperature rise in Figure 4.46 during the period when the battery voltage is 0 V. As soon as all current has started to flow through the cadmium reaction path at the nickel electrode the temperature will decrease again, because the overpotential of this reaction will be substantially smaller than the nickel overpotential during overdischarging.

When the amount of excess Cd in the cadmium electrode decreases to zero, the current will start to flow through the oxygen reaction path in the cadmium electrode. As a result, the molar amount of oxygen, and hence the oxygen gas pressure, will increase. The potential of the cadmium electrode will now reverse because of the difference in the signs of E^o_{Cd} and $E^o_{O_2}$; see Table 4.4. The oxygen-reduction reaction will start to compete with the cadmium-reduction reaction at the nickel electrode. This is the second stage of overdischarging. The reverse situation as described for the charging process will occur. The cadmium electrode will act as a positive, oxygen-forming electrode, whereas the nickel electrode will act as a negative, oxygen-reducing electrode. As a result, the oxygen pressure will level off, as can indeed be seen in Figure 4.46. Again, this reaction in particular generates a lot of heat because of the high overpotential for the oxygen reduction reaction at the nickel electrode, which leads to a significant increase in temperature.

Voltage reversal may occur when several batteries are connected in series. It is more correct to use the term cells in this case, as explained in chapter 3. Series or parallel connections of cells in battery packs are often used in practice in order to obtain a higher voltage at the same battery capacity in [Ah] or a higher battery capacity in [Ah] at the same battery voltage, respectively. Figure 4.47 shows the simulated battery pack voltage and the internal oxygen gas pressure of one of its cells when six NiCd cell models are connected in series. The battery pack was discharged in a simulation using a 1 C discharge rate at 25°C ambient temperature. Two cases were simulated. In case (a), all six cells inside the pack were given the same capacity as indicated in Table 4.4. In case (b), only one cell was given the same capacity as in Table 4.4 and the others were given capacities decreasing in steps of 10% of the capacity indicated in Table 4.4 to reflect the spread in cell capacity. This spread depends on the type, manufacturer, usage history and age of the pack. The heat resistances between the six cells and between the cells and the environment were taken to be zero for simplicity. In more advanced simulations, thermal models of the heat flows between the cells in the battery pack or from individual cells to the environment could also be considered.

Curves (a) in Figure 4.47 have the same shape as the V and P curves in Figure 4.46. All the cells will become empty at exactly the same moment because all the cells have exactly the same capacity, and the battery pack voltage $V(a)$ is simply the sum of six identical voltage curves. The pressure curve $P(a)$ is shown for only one cell, but it is the same for all six cells. The oxygen pressure rises when all the cells have reversed their voltages, which can likewise be observed for one cell in Figure 4.46. No pressure build-up will take place in practice in the case of identical cell capacities because when the battery pack voltage decreases and approaches 0 V, the electronic equipment will no longer operate and the discharge of the battery pack will stop.

Curves (b) in Figure 4.47 clearly show the result of the different capacities inside the battery pack. Curve $V(b)$ shows the total pack voltage and curve $P(b)$ shows the pressure build-up inside the cell with the lowest maximum capacity. This cell will be the first fully discharged cell and its voltage will approach 0 V. This can be seen in curve $V(b)$ as the first drop in the battery pack voltage. The battery pack voltage is still high enough to power the load, and discharging will continue because the other cells still have a voltage of around 1.3 V. Subsequently, the cell with the second lowest capacity reaches a voltage of approximately 0 V, which is represented by the second drop in the battery pack voltage. Then the cell with the third lowest capacity reaches a voltage of 0 V and a third drop in the battery pack voltage is observed. The fourth drop in the battery pack voltage is due to voltage reversal of the cell with the lowest maximum capacity. As a consequence, the oxygen pressure inside this cell will increase, as can be inferred from curve $P(b)$. A vertical line has been added to the figure at that moment for convenience. The battery pack voltage at this moment is still around 2.5 V, which may allow discharging to continue even further. High pressure-build-up inside the cells may eventually lead to venting and loss of electrolyte, which degrades the cell's performance. The simulation clearly shows that the battery pack is as strong as its weakest cell. Therefore, the capacities of cells inside a battery pack should be as similar as possible and the load should be shut off at an appropriate voltage level. The choice of this voltage level will be a compromise between the run time of the load and the lifetime of the battery pack. The simulation clearly shows the need for a Battery Management System as already described in chapter 2.

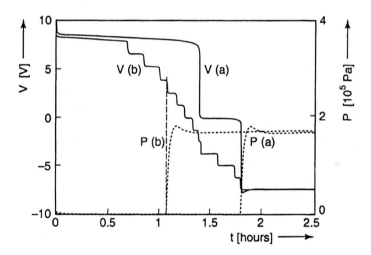

Figure 4.47: Simulated voltage (V [V]) and internal oxygen gas pressure (P [10^5 Pa]) as a function of time (t [hours]) of a battery pack with six NiCd cells connected in series during discharging at a 1 C-rate and at an ambient temperature of 25°C. Curves (a) represent the situation when all six cells have the same capacity; see Table 4.4. Curves (b) represent the situation when the six cells decrease in steps of 10% of the capacity indicated in Table 4.4

4.5.2   Simulations using the Li-ion model presented in section 4.3

Some simulation results obtained using the Li-ion battery model presented in section 4.3 will be presented in this section. The values of the parameters used in this model were found in a simpler way than for the NiCd battery model. The number of parameters used in the Li-ion model is considerably smaller than that used in the NiCd model, mainly because of the absence of modelled side-reactions. The curves simulated with the Li-ion model were fitted to measured curves obtained with the

CGR17500 Li-ion battery from Panasonic [40]. As a starting point, many parameter values relating to the battery's design were obtained from the battery manufacturer and other parameter values were, where possible, obtained from the literature. In addition, the parameters used in the equations for the equilibrium potentials for the positive electrode in (4.79) and (4.80) and for the negative electrode in (4.88) and (4.89) were found by curve fitting, using curves measured for the equilibrium potentials of the separate electrodes.

The parameter values for the exchange currents $I^o_{LiCoO_2}$ and $I^o_{LiC_6}$ were found in impedance measurements performed with the CGR17500 battery [40]. Polar plots of the real and imaginary parts of the impedance of a battery can be used to derive parameters relating to the kinetic and diffusion processes occurring inside the battery, as will be described in more detail in chapter 6. However, it is not possible to distinguish between the contributions of the two individual electrodes to these processes. Therefore, it has in this book been assumed that the kinetic parameters inferred from the measured battery impedance can be assigned exclusively to the $LiCoO_2$ electrode. This means that it is assumed that the kinetics of the LiC_6 electrode are relatively fast in comparison with those of the $LiCoO_2$ electrode. The value of the double-layer capacitance $C^{dl}_{LiCoO_2}$ was also inferred from the impedance plot under the assumption mentioned previously. The double-layer capacitance value of $C^{dl}_{LiC_6}$ was chosen identical to $C^{dl}_{LiCoO_2}$.

The parameter values used in the simulations using the Li-ion model described in this section are summarized in Table 4.5. As in Table 4.4, a distinction has been made between design-related parameters and electrochemical parameters. The fourth column of this table lists the equations or figures in which the parameters have been used. Temperature dependence of the model equations has not been implemented yet, as mentioned in section 4.3. Therefore, temperature-related parameter values have not been included in Table 4.5 and all the parameter values are valid at 25°C. The I^o_i values are specified at 50% SoC. A PTC has been included in series with the battery for safety reasons (see chapter 3), which is also considered in the model by means of a series resistance. Its value R_{PTC} can be found in Table 4.5.

Some examples of simulations performed using the Li-ion model with the parameter values of Table 4.5 will be described in the remainder of this section. First, some charging simulations will be discussed and the simulation results will be compared with measurements obtained with a CGR17500 Li-ion battery. A Li-ion battery is charged in two modes, a constant current (CC) mode followed by a constant voltage (CV) mode. More information will be given in chapter 5. Figure 4.48 shows the simulated and measured voltage and current curves for charging a CGR17500 Li-ion battery at a constant rate of 1 C in CC mode and a constant voltage of 4.1 V in CV mode. The charge time is two hours.

The CC and CV modes can be easily distinguished in Figure 4.48. The current is constant in CC mode, while the battery voltage rises gradually to the value it attains in CV mode. In the latter region, the battery voltage remains constant while the battery current drops. As can be seen in the figure, the quantitative agreement between the simulation results and the measurements is quite good. The largest discrepancy between the simulation results and measurements is observable in the voltage curves. As in the NiCd model, the simulated charging voltage is higher than that measured at the beginning of the charging. A possible reason for this, apart from incorrect parameter values, could again be the occurrence of processes in the measured battery which have not been taken into account in the Li-ion model. The change from CC mode to CV mode occurs somewhat later in the charging process in the simulations because the simulated voltage is lower than the measured voltage at the moment when the battery approaches the maximum voltage level in CV mode.

Table 4.5: Most important parameter values of the Li-ion battery model described in section 4.3

Parameter	Value	Unit	Reference
Design-related parameters			
$C^{dl}_{LiC_6}$	0.4	F	Figure 4.21
$C^{dl}_{LiCoO_2}$	0.4	F	Figure 4.21
$Q^{max}_{LiCoO_2}$	5184	C	See note 1
$Q^{max}_{LiC_6}$	3060	C	-
l_{pos}	$10 \cdot 10^{-6}$	m	Figure 4.21
l_{neg}	$10 \cdot 10^{-6}$	m	Figure 4.21
l_{elyt}	$2.8 \cdot 10^{-4}$	m	Figure 4.21
R_{leak}	$80 \cdot 10^{3}$	Ω	Section 4.3.4
R_{PTC}	$50 \cdot 10^{-3}$	Ω	-
ε_r	10	-	Figure 4.11
Electrochemical parameters			
A_{elyt}	$20 \cdot 10^{-3}$	m^2	See note 2
$D_{Li^+}^{pos}$	$1.5 \cdot 10^{-13}$	m^2/s	Section 4.3.6
$D_{Li^+}^{neg}$	$1.5 \cdot 10^{-13}$	m^2/s	Section 4.3.6
$D_{Li^+}^{elyt}$	$2 \cdot 10^{-11}$	m^2/s	Section 4.3.6
$D_{PF_6}^{elyt}$	$3 \cdot 10^{-11}$	m^2/s	Section 4.3.6
$E^o_{LiCoO_2}$	3.89	V vs Li	(4.78)
$E^o_{LiC_6}$	0.04	V vs Li	(4.87)
$I^o_{LiCoO_2}$	$251 \cdot 10^{-3}$	A	(4.82)
$I^o_{LiC_6}$	15.3	A	(4.90)
n_{Li}	1	-	(4.78)
$U_{pos,1}$	-3	-	(4.79)
$U_{pos,2}$	20	-	(4.80)
$U_{neg,1}$	-10	-	(4.88)
$U_{neg,2}$	-2	-	(4.89)
$x^{pos}_{phasetransition}$	0.75	-	(4.81)
$x^{neg}_{phasetransition}$	0.25	-	Section 4.3.3
z_{Li^+}	+1	-	Section 4.3.4
z_{PF_6}	-1	-	Section 4.3.4
α_{LiCoO_2}	0.4	-	(4.82)
α_{LiC_6}	0.15	-	(4.90)
$\zeta_{pos,1}$	0	-	(4.79)
$\zeta_{pos,2}$	17.25	-	(4.80)
$\zeta_{neg,1}$	0	-	(4.88)
$\zeta_{neg,2}$	2	-	(4.89)

Note 1: The rated capacity of the $LiCoO_2$ electrode, i.e. $0.72Ah \times 3600(sec/hour) = 2592$ C, is only roughly half of the full capacity, because x_{Li} is only cycled between 0.5 and 0.95, as described in section 4.3.2. Hence, the value of the full capacity is a factor of two higher than the rated capacity and this value has been used in the equations describing the behaviour of the $LiCoO_2$ electrode

Note 2: Surface area electrolyte, which is the area of spatial element in the electrolyte; see Figure 4.5

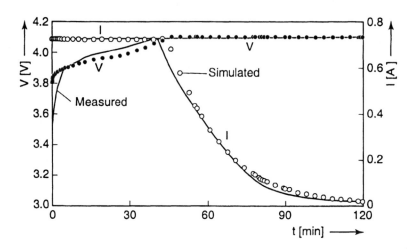

Figure 4.48: Simulated (marked lines) and measured (solid lines) voltage (V [V]) and current (I [A]) curves as functions of time (t [min]) for charging a Li-ion battery at a 1 C charge rate in CC mode and a 4.1 V charge voltage in CV mode for two hours in total at a battery temperature of 25°C

Figures 4.49 and 4.50 show the simulated and measured battery voltage and current curves, respectively, for charging a Li-ion battery at different charge rates, ranging from 0.1 C to 1 C. With each charge current, CC/CV charging is stopped when the battery current drops below 10 mA in CV mode. This ensures that almost equal capacities in [Ah] are reached for each charge current. This can be inferred from the figures. The quantitative agreement between the measurements and the simulation results is better for lower charge currents for both the voltage and current curves. The voltage simulated during the initial stages of charging is higher than the measured voltage, as observable in Figure 4.48, whereas this situation is reversed later in CC mode. Therefore, the CV mode later takes over from the CC mode in the simulations.

Figure 4.51 shows the charge time necessary to charge the battery to a capacity of 770 mAh as a function of the current in CC mode when the charge voltage in the CV region is 4.1 V. Charging is stopped when the battery current in CV mode reaches the value of 10 mA.

The simulated and measured charge times in Figure 4.51 show an almost perfect fit, as opposed to the simulated curves in Figures 4.49 and 4.50, which show some deviations from the measured curves. Figure 4.51 indicates that the charge time initially decreases strongly as a function of the charge current. However, the charge time hardly decreases when the charge rate becomes larger than 1 C. This is because the charge time is almost completely determined by the time at which the battery is charged in CV mode at higher charge currents. In this mode, the charge time is independent of the charge current. The overpotentials in the battery are much higher than at smaller charge currents when higher charge currents are applied in CC mode. As a result, the CV mode is reached much quicker in the charging process. More information on the influence of the parameters I_{max} and V_{max} on the charge time and other factors will be given in chapter 5.

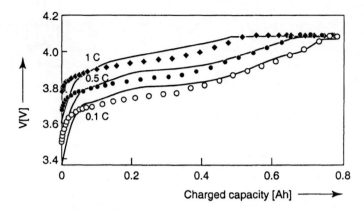

Figure 4.49: Simulated (marked lines) and experimental (solid lines) battery voltage (V [V]) curves as functions of the charged capacity [Ah] for charging a Li-ion battery with various charge currents in the CC mode and a 4.1 V charging voltage in the CV mode at a battery temperature of 25°C. The charge rate increases from bottom to top: (○) 0.1 C, (●) 0.5 C, (♦) 1 C

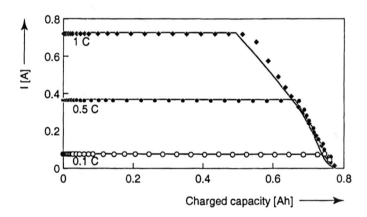

Figure 4.50: Simulated (marked lines) and experimental (solid lines) battery current (I [A]) curves as a function of the charged capacity [Ah]. See Figure 4.49 for conditions

Figure 4.52 shows simulated and measured battery voltage curves as functions of the discharged capacity during discharging at various discharge currents. The figure shows that the quantitative agreement between the results of the simulations and the measurements is quite good for discharge rates lower than 1 C. However, a discrepancy is observable in the region from 0 to 0.3 Ah of discharged capacity for the discharge rates of 1 C and 1.5 C. In this region, the simulated battery voltage is higher than the measured battery voltage. This difference at higher discharge currents could not be reduced in further optimization of the model parameters. This suggests the absence in the model of a process which does occur in real life. Alternatively, the mathematical description of a process included in the model might be incorrect.

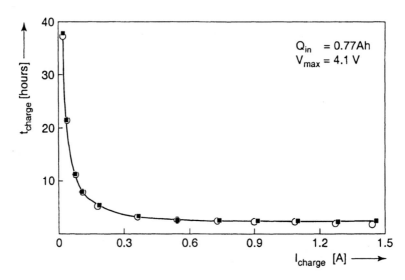

Figure 4.51: Simulated (o) and experimental (+) charge time necessary to charge a Li-ion battery to a capacity of 770 mAh (t_{charge} [hours]) as a function of the charge current (I_{charge} [A]) during CC/CV charging with $V_{max} = 4.1$ V in CV mode at a battery temperature of 25°C

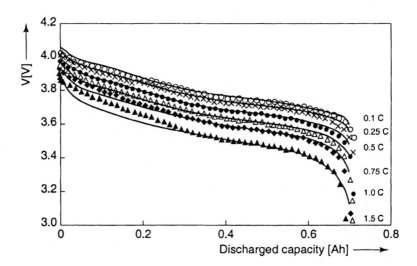

Figure 4.52: Simulated (marked lines) and experimental (solid lines) battery voltage curves (V [V]) as functions of the discharged capacity [Ah] during discharging a Li-ion battery with various discharge currents at a battery temperature of 25°C. The discharge rate increases from top to bottom: (o) 0.1 C (✖) 0.25 C (•) 0.5 C (Δ) 0.75 C (♦) 1 C (▲) 1.5 C

The kinetics of the LiCoO$_2$ electrode were assumed to be much slower than the kinetics of the LiC$_6$ electrode, as discussed previously. Therefore, the kinetics of the LiCoO$_2$ electrode have far more influence on the battery voltage than the kinetics of the LiC_6 electrode in the simulations. A possible explanation for the difference between the simulation results and the measurements at higher discharge currents in the region from 0 to 0.3 Ah discharged capacity is that in the model, the kinetics of the $LiCoO_2$ electrode have been assumed to be independent of the phase formation. As discussed in section 4.3.2, a phase transition occurs in the $LiCoO_2$ electrode at $x_{Li}=0.75$. This means that, starting from a fully charged battery, or $x_{Li}=0.5$ in the $LiCoO_2$ electrode, the phase transition at $x_{Li}=0.75$ occurs at approximately 0.36 Ah discharged capacity. This value of 0.36 Ah agrees more or less with the value at which the discrepancy between the simulation results and the measurements occurs in Figure 4.52. It seems reasonable to assume that the model predicts the battery voltage during discharging quite well for $x_{Li}>0.75$ for all discharge currents, because the kinetics modelled for the $LiCoO_2$ electrode apparently apply to this region. However, a deviation occurs at higher discharge currents for $x_{Li}<0.75$, which could be explained by a deviation in kinetics occurring in reality which has not been taken into account in the model. It is also possible that the kinetics of the LiC_6 electrode play a more important role in determining the voltage in reality.

Another remarkable feature that can be seen in Figure 4.52 is that the total discharged capacity is more or less the same for all discharge currents. This behaviour differs from the behaviour of, for example, NiCd batteries, where the discharged capacity varies more with the discharge current [18]. An explanation for this difference is that the diffusion of H^+ ions in the nickel electrode of a NiCd battery is limited due to the relatively large diffusion length in comparison with that of a $LiCoO_2$ or LiC_6 electrode in a Li-ion battery. For example, SEM photographs of a nickel electrode of a 80AAS NiCd battery have shown that the size of the grains is 50 μm [40]. This value corresponds well to the value of 58.3 μm given for the thickness of the nickel electrode l_{Ni} in the optimized parameter set of Table 4.4. This value is about five times higher than the grain size in a $LiCoO_2$ or LiC_6 electrode; see l_{pos} and l_{neg} in Table 4.5. Therefore, the influence of the diffusion of H^+ ions in the nickel electrode of a NiCd battery is greater than the influence of the diffusion of Li^+ ions in the electrodes of a Li-ion battery. Diffusion limitation within the electrode causes part of the electrode material to become inaccessible during discharging. This effect becomes stronger with an increasing discharge current, which may explain the dependency of the discharged capacity on the discharge current in NiCd batteries.

Figure 4.53 shows the simulated overpotentials that occur in the Li-ion model during discharging at a rate of 1.5 C. It can be seen in Figure 4.53 that, except at the end of the discharging process, the diffusion of Li^+ ions in the electrodes hardly contributes to the total overpotential. This means that the phenomenon that part of the electrode material becomes inaccessible during discharging is not dominant in a CGR17500 Li-ion battery. This explains the observed independence of the discharged capacity with respect to the discharge rate between 0.1 C and 1.5 C. The simulations show that the total overpotential is determined mainly by the reaction overpotential of both electrodes (η^k_{total}) in the considered Li-ion battery and the overpotential across the electrolyte due to diffusion of Li^+ ions ($\eta^d_{Li^+}$). The overpotential due to the diffusion of Li^+ ions in the electrolyte is obtained from (4.93).

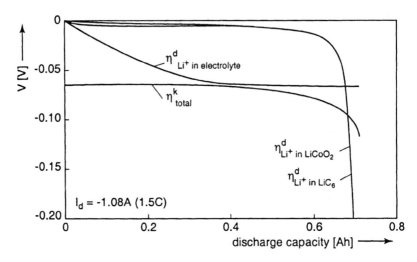

Figure 4.53: Simulated overpotentials for the Li-ion model during discharging at a 1.5 C-rate at a battery temperature of 25°C

4.6   Conclusions

In this chapter it has been shown that the general approach to battery modelling based on physical system dynamics can be used to model basic building blocks that can be used to model various battery system. The general approach has been applied to two battery systems in this chapter, NiCd batteries and Li-ion batteries. The application of the general approach to NiCd batteries leads to a battery model with which the battery voltage (V), current (I), temperature (T) and internal oxygen gas pressure (P) can be simultaneously and coherently simulated under a wide variety of conditions. The battery voltage and the behaviour of the electrolyte of Li-ion batteries have been simulated. The chosen approach allows other processes to be added to the models, as long as a mathematical description is available. Moreover, it can also be applied to other types of batteries and battery systems. As long as the basic structure of the battery in terms of, for example, electrode capacities and volumetric characteristics, the types of reactions taking place, *etc.*, as well as the mathematical description of all the processes are known, the basic building blocks described in this chapter can be used to develop a network model of the battery. An advantage of the applied approach is that the parameters used in the model all have a physical or chemical meaning.

To enhance the predictive nature of the models, a good quantitative agreement between the results of simulations and measurements is important. Such agreement depends on the design of the model, including such aspects as the considered processes, the correctness of the mathematical descriptions of the considered processes and the values of the parameters in the model. The emphasis in section 4.4 has been on parameter estimation. A method has been used that should increase the chances of finding the optimum parameter set as much as possible, because the estimation of many parameter values of a complex non-linear model is a complicated process. The first results presented in section 4.4.3 show that it is possible to find a good quantitative agreement between simulation and measurement results when the model is optimized for one operating condition. This good quantitative fit holds for V, P and T simultaneously. Comparison of optimized and measured parameter values, where available, shows that the chosen approach yields parameter values close to the measured values.

The quantitative agreement between the simulation results and the measurements decreases when other operating conditions are considered. A modification in the mathematical description of one of the modelled processes has been considered in a first attempt to improve the quantitative agreement for more than one charging condition. Part of the iterative process has been repeated with this modified model. The results are promising and the improvements realized so far conform to our expectations. However, especially the oxygen pressure curves observed in the simulations and the measurements still differ considerably in shape for low charge currents. Hence, more iterations in the iterative optimization process are needed to further improve the quantitative agreement. Even with simple manual tuning of parameters quite good quantitative agreement can already be obtained for the Li-ion model, mainly because it involves fewer parameters.

The simulation examples given in section 4.5 show that the models can be used to gain insight into the processes that take place inside the batteries. The models act as a 'transparent battery', in which, for example, the development of the overpotentials of the individual reactions can be studied. The user can determine which externally observable characteristics correspond to which process, as has been illustrated in various examples. Besides for gaining insight into what happens inside a battery, the models can also be used to generate V, I, P and T quickly and repeatably. This offers the designer of a BMS a quick means for predicting battery behaviour early in the design process. An example given in this chapter has shown the effect of connecting cells with unequal capacities in series in a battery pack. A possible mechanism for breakdown of the pack and ways of preventing it have been described. More examples of using the model in various forms will be given in the following chapters.

The quantitative agreement is not optimum in all situations, as has been illustrated in some examples. Besides incorrect parameter values another reason could be the omission of certain processes from the models. The lack of proper physical or chemical and associated mathematical descriptions of some of these processes is the reason for the omission. These processes can be readily included in the model once the corresponding mathematical descriptions have been found.

As a first example of these omissions, aging effects, which lead to the gradual loss of battery capacity when a battery ages, have not been included, because a good physical or chemical description of this process is still lacking. One of the possible reasons for capacity loss could be irreversible reactions taking place during the life of a battery. These reactions consume active material and form non-active species that do not take part in energy storage. Such an irreversible reaction could be modelled as a parallel path to the main reactions.

A second example of an omission from the current battery models is the effect of electrode porosity on the battery behaviour. A real electrode consists of a large number of grains of active material that are connected to the current collector. There are pores between the grains, which should be filled by the electrolyte solution to enable the transfer of ions to and from the interface between the electrode grains and the electrolyte solution. However, depletion of the electrolyte solution from the pores has been proven to have a significant influence on the behaviour of a battery [51]. A possible way of modelling the effect of porosity is to extend the models from the one-dimensional approach that has been described in this chapter to a two-dimensional approach. The electrodes can be modelled as many small identical electrodes connected in parallel, with each small electrode denoting a particle of active material [40]. Each grain is modelled in the same way as in the electrode models described for the NiCd and Li-ion battery models. The grains are on one side all connected to one node that denotes the current collector. On the other side, each grain contacts the model of the electrolyte at a different spot. In the case of a Li-ion battery model, the electrolyte model should be a two-dimensional extension of the model described in this chapter. For the NiCd model, this means that the approach of modelling the electrolyte by a simple resistance must be abandoned. The electrolyte should be modelled by a two-dimensional RC-ladder network that describes the diffusion and migration of OH^- and K^+ ions in the electrolyte.

A third example of an omitted process is the hysteresis that occurs in the equilibrium potential. It has been found in measurements with both NiCd and Li-ion batteries that the equilibrium potential of a battery during charging is higher than the equilibrium potential during discharging. No theoretical explanation has been reported in the literature so far. Therefore, this process cannot be included in the models at this moment. However, it is believed that the occurrence of this process might provide an explanation for the differences between the simulation results and the measurements in both battery models at the initial stages of charging.

As a fourth example, the memory effect has not been included in the NiCd model. This effect causes a reversible capacity loss when the battery is repeatedly partially charged and discharged. Although, as mentioned in chapter 3, it is commonly believed that the memory effect is associated with crystallization processes that occur at the cadmium electrode, no mathematical description of this process is known at this moment.

A fifth and final example is the omission of side-reactions from the Li-ion model. One of the side-reactions that may occur is the formation of metallic lithium from Li^+ ions. This reaction will occur at potentials that are negative with respect to a lithium reference electrode. Such potentials can occur at the negative electrode under certain conditions. Metallic lithium is extremely reactive, and any deposited Li atoms will readily react with the electrolyte to form a passivation layer or SEI; see section 4.3.1. This will lead to a loss of Li^+ ions in the system, and hence to a capacity loss because the formation of a passivation layer is an irreversible process. Hence, modelling this side-reaction is also a way of modelling capacity loss. The lithium deposition reaction can be modelled as a path parallel to the main reaction at the negative electrode [40]. Both the main and the side-reactions can induce currents through the capacitance that contains the molar amount of Li^+ ions at the negative electrode surface. The SEI layer can be modelled by a resistance connected in series with the electrode model. The value of this resistance can be made to depend on the amount of material that is converted in the side-reaction in order to model an increasing resistance as a function of cycle life.

Apart from the addition of processes to the models, the temperature-dependence of the various modelled processes also deserves some attention. It has been shown in section 4.4 that, although reasonable quantitative agreement can be obtained between simulated and measured temperature curves, the present temperature model is probably too simple. An extension has been proposed, in which the battery temperature is no longer homogeneously distributed through the battery (pack). This will require further attention in future research. No temperature dependence of model parameters has yet been considered in the present Li-ion model. This can readily be included in the future.

4.7 References

[1] R.C. Rosenberg, D.C. Karnopp, *Introduction to Physical System Dynamics*, McGraw-Hill Series in Mechanical Engineering, New York, 1983

[2] J. Bouet, F. Richard, Ph. Blanchard, "A Discharge Model for the Ni Hydroxide Positive Electrode", in *Nickel Hydroxide Electrodes*, D.A. Corrigan, A.H. Zimmerman (Eds.), PV 90-4, The Electrochemical Society Proceedings Series, Pennington N.J., pp. 260-280, 1990

[3] J.W. Weidner, P. Timmerman, "Effect of Proton Diffusion, Electron Conductivity and Charge-Transfer Resistance on Nickel Hydroxide Discharge Curves", *J. Electrochem. Soc.*, vol. 141, no. 2, pp. 346-351, February 1994

[4] E. Karden, P. Mauracher, F. Schöpe, "Electrochemical Modelling of Lead/Acid Batteries under Operating Conditions of Electric Vehicles", *J. Power Sources*, vol. 64, no. 1, pp. 175-180, January 1997

[5] D. Fan, R.E. White, "A Mathematical Model of a Sealed Nickel-Cadmium Battery", *J. Electrochem. Soc.*, vol. 138, no. 1, pp. 17-25, January 1991

[6] D. Fan, R.E. White, "Mathematical Modelling of a Nickel-Cadmium Battery- Effects of Intercalation and Oxygen Reactions", *J. Electrochem. Soc.*, vol. 138, no. 10, pp. 2952-2960, October 1991

[7] T.F. Fuller, C.M. Doyle, J. Newman, "Simulation and Optimization of the Dual Lithium Ion Insertion Cell", *J. Electrochem. Soc.*, vol. 141, no. 1, pp. 1-10, January 1994

[8] C.M. Doyle, T.F. Fuller, J. Newman, "Modelling of Galvanostatic Charge and Discharge of the Lithium/Polymer/Insertion Cell", *J. Electrochem. Soc.*, vol. 140, no. 6, pp. 1526-1533, June 1993

[9] S.A. Ilangovan, S. Sathyanarayana, "Impedance Parameters of Individual Electrodes and Internal Resistance of Sealed Batteries by a New Non-destructive Technique", *J. Applied Electrochem.*, vol. 22, no. 5, pp. 456-463, May 1992

[10] S.C. Hageman, "Simple PSpice Models Let You Simulate Common Battery Types", *EDN*, pp. 117-132, October 28, 1993

[11] S.C. Hageman, "PSpice Models Nickel-Metal-Hydride Cells", *EDN*, p. 99, February 2, 1995

[12] S. Waaben, I. Moskowitz, J. Frederico, C.K. Dyer, "Computer Modelling of Batteries from Non-linear Circuit Elements", *J. Electrochem. Soc.*, vol. 132, no. 6, pp. 1356-1362, June 1985

[13] J.J.L.M. van Vlerken, P.G. Blanken, "Lumped Modelling of Rotary Transformers, Heads and Electronics for Helical-Scan Recording", *IEEE Trans. Magn.*, vol. 31, no. 2, pp. 1050-1055, March 1995

[14] W.S. Kruijt, H.J. Bergveld, P.H.L. Notten, *Principles of Electronic-Network Modelling of Chemical Systems*, Nat.Lab. Technical Note 261/98, Philips Internal Report, October 1998

[15] A.J. Bard, L.R. Faulkner, *Electrochemical Methods: Fundamentals and Applications*, John Wiley&Sons, New York, 1980

[16] K.J. Vetter, *Electrochemical Kinetics: Theoretical and Experimental Aspects*, Academic Press, New York, 1967

[17] M. Barak (Ed.), T. Dickinson, U. Falk, J.L. Sudworth, H.R. Thirsk, F.L. Tye, *Electrochemical Power Sources: Primary & Secondary Batteries*, IEE Energy Series 1, A. Weaton&Co., Exeter, 1980

[18] D. Linden, *Handbook of Batteries*, Second Edition, McGraw-Hill, New York, 1995

[19] J.R. MacDonald (Ed.), *Impedance Spectroscopy*, John Wiley&Sons, New York, p. 59, 1987

[20] J. Horno, M.T. Garcia-Hernandez, "Digital Simulation of Electrochemical Processes by the Network Approach", *J. Electroanal. Chem.*, vol. 352, no. 1&2, pp. 83-97, June 1993

[21] G.M. Barrow, *Physical Chemistry*, Fourth Edition, McGraw-Hill, Tokyo, Japan, 1979

[22] C.H. Hamann, W. Vielstich, *Electrochemie II: Elektrodenprozesse, Angewandte Elektrochemie*, Taschentext 42, Verlag Chemie, Weinheim, 1981

[23] P.H.L. Notten, W.S. Kruijt, H.J. Bergveld, "Electronic-Network Modelling of Rechargeable NiCd Batteries", Ext. Abstract I-5-05, *46th ISE Meeting*, Xiamen, China, 1995

[24] W.S. Kruijt, H.J. Bergveld, P.H.L. Notten, *Electronic-Network Modelling of Rechargeable NiCd batteries*, Nat.Lab. Technical Note 211/96, Philips Internal Report, 1996

[25] P.H.L. Notten, W.S. Kruijt, H.J. Bergveld, "Electronic-Network Modelling of Rechargeable NiCd Batteries", Ext. Abstract 96, *191st ECS Meeting*, Montreal, Canada, 1997

[26] W.S. Kruijt, H.J. Bergveld, P.H.L. Notten, "Electronic-Network Modelling of Rechargeable Batteries: Part I: The Nickel and Cadmium Electrodes", *J. Electrochem. Soc.*, vol. 145, no. 11, pp. 3764-3773, November 1998

[27] P.H.L. Notten, W.S. Kruijt, H.J. Bergveld, "Electronic-Network Modelling of Rechargeable Batteries: Part II: The NiCd System", *J. Electrochem. Soc.*, vol. 145, no. 11, pp. 3774-3783, November 1998

[28] H.J. Bergveld, W.S. Kruijt, P.H.L. Notten, "Electronic-Network Modelling of Rechargeable NiCd Cells and its Application to the Design of Battery Management Systems", *J. Power Sources*, vol. 77, no. 2, pp. 143-158, February 1999

[29] W. Randszus, H.A. Kiehne (Eds.), *Gasdichte Nickel-Cadmium Akkumulatoren*, Varta Batterie AG, Hannover, Germany, 1978

[30] P.H.L. Notten, A.A. Jansen van Rosendaal, W.S. Kruijt, H.J. Bergveld, "The oxygen evolution and recombination kinetics in sealed rechargeable NiCd batteries", in preparation, 2002

[31] S.G. Bratsch, "Standard Electrode Potentials and Temperature Coefficients in Water at 298.15K", *J. Phys. Ref. Data*, vol. 18, no. 1, pp. 1-21, 1989

[32] *Gmelins Handbuch der Anorganischen Chemie*, Verlag Chemie GmbH, Leipzig-Berlin, 8th Ed., vol. 33, p. 74, 1925

[33] P.H.L. Notten, *Chronopotentiometric Measurements at Ni/Cd Batteries and Separate Electrodes*, Nat.Lab. Technical Note 268/92, Philips Internal Report, 1992

[34] M. Avrami, "Kinetics of Phase Change: II. Transformation-Time Relations for Random Distribution of Nuclei", *J. Chem. Phys.*, vol. 8, no. 2, pp. 212-224, February 1940

[35] S. Motupally, C.C. Streinz, J.W. Weidner, "Proton Diffusion in Nickel Hydroxide Films: Measurement of the Diffusion Coefficient as a Function of State-of-Charge", *J. Electrochem. Soc.*, vol. 142, no. 5, pp. 1401-1408, May 1995

[36] Y. Okinaka, "On the Oxidation-Reduction Mechanism of the Cadmium Metal-Cadmium Hydroxide Electrode", *J. Electrochem. Soc.*, vol. 117, no. 3, pp. 289-295, March 1970

[37] J. O'M. Bockris, D.M. Drazic, *Electro-Chemical Science*, Taylor&Francis Ltd., London, 1972

[38] S.U. Falk, A.J. Salkind, *Alkaline Storage Batteries*, John Wiley&Sons, New York, 1969

[39] R.E. Davis, G.L. Horvarth, C.W. Tobias, "The Solubility and Diffusion Coefficient of Oxygen in Potassium Hydroxide Solutions", *Electrochim. Acta*, vol. 12, no. 3, pp. 287-297, March 1967

[40] W.S. Kruijt, H.J. Bergveld, P.H.L. Notten, *Electronic-Network Model of a Rechargeable Li-ion Battery*, Nat.Lab. Technical Note 265/98, Philips Internal Report, October 1998

[41] W.S. Kruijt, E. de Beer, P.H.L. Notten, H.J. Bergveld, "Electronic-Network Model of a Rechargeable Li-ion Battery", *The Electrochemical Society Meeting Abstracts*, vol. 97-1, Abstract 104, p. 114, Paris, France, August 31- September 5, 1997

[42] R. Fong, U. von Sacken, J.R. Dahn, "Studies of Intercalation into Carbons Using Nonaqueous Electrochemical Cells", *J. Electrochem. Soc.*, vol. 137, no. 7, pp. 2009-2013, July 1990

[43] K. Brandt, "Historical Development of Secondary Lithium Batteries", *Solid State Ionics*, vol. 69, no. 3&4, pp. 173-183, August 1994

[44] W.R. McKinnon, R.R. Haering, "Physical Mechanisms of Intercalation", in *Modern Aspects of Electrochemistry*, vol. 15, B. Conway, R.E. White, J. O'M. Bockris (Eds.), Plenum Press, New York, pp. 235-304, 1983

[45] T. Ohzuku, A. Ueda, "Phenomenological Expression of Solid-State Redox Potentials of $LiCoO_2$, $LiCo_{1/2}Ni_{1/2}O_2$, and $LiNiO_2$ Insertion Electrodes", *J. Electrochem. Soc.*, vol. 144, no. 8, pp. 2780-2785, August 1997

[46] J.N. Reimers, J.R. Dahn, "Electrochemical and In Situ X-Ray Diffraction Studies of Lithium Intercalation in Li_xCoO_2", *J. Electrochem. Soc.*, vol. 39, no. 8, pp. 2091-2097, August 1992

[47] T. Ohzuku, Y. Iwakoshi, K. Sawai, "Formation of Lithium-Graphite Intercalation Compounds in Nonaqueous Electrolytes and Their Application as a Negative Electrode for a Lithium Ion (Shuttlecock) Cell", *J. Electrochem. Soc.*, vol. 140, no. 9, pp. 2490-2498, September 1993

[48] P.A. Derosa, P.B. Balbuena, "A Lattice-Gas Model Study of Lithium Intercalation in Graphite', *J. Electrochem. Soc.*, vol. 146, no. 10, pp. 3630-3638, October 1999

[49] H.J. Bergveld, W.J.J. Rey, R. du Cloux, W.S. Kruijt, P.H.L. Notten, "Quantitative Optimization of Electronic-Network Models for Rechargeable Batteries", to be submitted, 2002

[50] M.D. McKay, R.J. Beckman, W.J. Conover, "A Comparison of Three Methods for Selecting Values of Input Variables in the Analysis of Output from a Computer Code", *Technometrics*, vol. 21, no. 2, pp. 239-245, February 1979

[51] C.M. Doyle, J. Newman, A.S. Gozdz, C.N. Schmutz, J.M. Tarascon, "Comparison of Modelling Predictions with Experimental Data from Plastic Lithium Ion Cells", *J. Electrochem. Soc.*, vol. 143, no. 6, pp. 1890-1903, June 1996

Chapter 5
Battery charging algorithms

One of the links in the energy chain is the charger, as explained in chapter 1. In the charger, electrical energy from the mains is transferred via magnetic energy into electrical energy suitable for charging the battery. Two types of control can be distinguished in the charger or power module, notably control of the energy conversion process and control of the charging process; see chapter 2. The latter control process is implemented in the form of a charging algorithm. A charging algorithm is an example of a battery management function with which the battery itself is monitored and the energy conversion process in the charger is controlled in order to charge the battery in an efficient way.

Batteries with different chemistries require different charging algorithms. Examples of charging algorithms for NiMH and Li-ion batteries have been described in section 2.4. The charging algorithms for these battery chemistries will be discussed in greater detail in this chapter. The charging algorithms for NiCd and NiMH batteries will be dealt with in section 5.1, the charging algorithm for Li-ion batteries in section 5.2, in which the influence of the current and voltage parameters on the charging process will also be discussed.

An example of the use of the NiCd battery model for comparing two different charging methods will be given in section 5.1.3. Simulation results will be compared with measurement results. The results of simulations using the Li-ion model will be used to demonstrate the difference between normal and fast charging in section 5.2.4. A possible mechanism for the loss of cycle life encountered in fast charging in practice will be derived from the results of simulations. Finally, conclusions will be drawn in section 5.3.

5.1   Charging algorithms for NiCd and NiMH batteries

5.1.1   Charging modes, end-of-charge triggers and charger features

NiCd and NiMH batteries are usually both charged with a constant current. When fast charging is applied, this current is switched to a lower value when a certain end-of-charge trigger occurs. In practice, this trigger is based on time, voltage or temperature. More than one trigger is used in most chargers. The first trigger occurring during the charge process leads to a change in the charge current. By switching to a lower current, the battery is protected from excessive internal pressure build-up and temperature rise.

As a general rule, the higher the charge current, the more care should be taken to avoid overcharging by switching off the high current in time. Therefore, the triggers that are used depend first of all on the applied charge current. Moreover, some triggers are more suitable for NiCd and some for NiMH batteries. This will be discussed in section 5.1.2. Finally, the choice of a certain end-of-charge trigger depends on the battery's acceptable cycle life. The greater the amount of overcharging is allowed by applying a certain end-of-charge trigger, the lower the battery's cycle life will be, because overcharging has a negative influence on the cycle life. In general, the charging algorithm has to be more sophisticated when the amount of overcharging has to be limited, because deriving triggers from measurements is more complicated in that case. Charge control, and hence the battery management system, consequently become more complicated when a battery with a longer cycle life is required. This was already described in section 2.1.

Charging modes
The charging modes are categorized by the value of the charge current as indicated below [1]. The battery will be charged with more than 100% of its rated capacity in most cases to obtain a fully charged battery. This is needed because some of the applied energy is lost in side-reactions. Moreover, several charging modes will sometimes be combined. An example is the application of the sequence fast charging, top-off charging and trickle charging. See chapter 3 for the definition of C-rate for charge currents.

1. Low-rate charging
In the case of low-rate charging the charge rate is around 0.1 C and the charging is stopped by a timer. The charge time is 14-16 hours in the case of a fully discharged battery and a 150% applied charge. Low-rate charging is used mainly in cheap overnight chargers, in which the charging 'algorithm' is very simple, because only a timer is used. In the simplest case, the battery is charged by a DC voltage source with a series resistance. The charge current decreases, because the battery voltage gradually rises during the charging. The term 'semi-constant current charging' is then used. In practice, examples can be found of continuous charging with a low rate.

2. Quick charging
A 0.3 C charge rate is applied in quick charging. The charge time for a fully discharged battery when the total charge applied to the battery ranges from 120% to 150% is 4 to 5 hours. Quick charging is usually ended with a timer.

3. Fast charging
In the case of fast charging, the charge rate is higher than or equal to 1 C in practice. This means that the maximum charge time is just over one hour. The actual charge time will depend on the battery's SoC before charging and on the amount of overcharging resulting from the chosen end-of-charge trigger. For example, a 120% charge with a 1 C charge rate implies a charge time of 1 hour and 12 minutes when the charging is started with a completely empty battery. Ultra-fast chargers with charge rates over 4 C, hence with charge times less than 15 minutes, have been reported for NiCd batteries [2],[3]. It should be noted that the charged capacity will be lower than 100% in this case. Proper end-of-charge triggers based on voltage and temperature measurements have to be implemented for fast charging of both NiCd and NiMH batteries to prevent the risk of serious damage to the batteries.

4. Top-off or equalization charging
Top-off charging at a rate of around 0.1 C is applied directly after fast charging. It is more often applied to NiMH batteries than to NiCd batteries. Battery manufacturers sometimes recommend it. As the term indeed implies, top-off charging is used to charge a battery to a full 100% charge after the fast-charge current has been switched off. Another reason to apply top-off charging occurs when several cells are charged in series, as in a battery pack. Top-off charging is then applied to ensure that all the cells are fully charged after an end-of-charge trigger has been generated by the combination of cells. Therefore, the term equalization charging is sometimes

encountered in the literature [2], although the process does not lead to equal cell capacities. In most cases, the top-off charging period is terminated by a timer, for example after half an hour or after one hour.

5. Trickle or maintenance charging
A relatively small trickle or maintenance charge rate of 0.03 C to 0.05 C is applied to the battery to compensate for self-discharge. Trickle charging serves to maintain a 100% charged battery when it is left in the charger for some time. Trickle charging begins directly after fast charging has ended when no top-off charging is applied. Alternatively, it can also start directly after quick or top-off charging has ended, when applicable. There is generally no time limit for trickle charging, so it can continue for an indefinite amount of time.

6. Reflex or 'burp' charging
This method of charging is not often encountered in practice. It involves charging a battery with current pulses followed by short rest periods with zero current, a short period of discharge and another short rest period. The inventors of this charging scheme claim that a battery can be charged more efficiently with Reflex charging. They argue that the discharge pulses or 'burps' help to maintain a low internal gas pressure and battery temperature. However, no hard proof of this is given. An example can be found in [3].

End-of-charge triggers
Depending on the charging mode, several end-of-charge triggers or combinations of these triggers can be used [1]-[4]. The end-of-charge triggers most frequently encountered are listed below. After a short description, these triggers will be compared in terms of the amount of overcharging they cause. This is performed in Figure 5.1, which compares most of the end-of-charge triggers mentioned below.

1. Timed end-of-charge trigger
A timer is used in this method to switch off the charge current. It can be used as a stand-alone end-of-charge trigger to terminate low-rate or quick charging. Moreover, a timed end-of-charge trigger is often used to terminate a period of top-off charging. However, detrimental overcharging may occur in the case of higher charge currents when only a timer is used to end the charging of a battery which was not fully discharged when the charging began. The timed end-of-charge trigger is therefore used only as a fail-safe limit in such cases. This means that a timer will eventually force a switch-over of the charge current to a trickle-charge current when all other end-of-charge triggers fail.

2. Maximum temperature end-of-charge trigger
As described in chapter 4, the temperature of NiCd batteries will rise quite steeply at the end of charging. The same holds for NiMH batteries. This rise in temperature can be used to derive end-of-charge triggers. For example, a battery can be charged until a maximum battery temperature has been reached. Obviously, a disadvantage of this method is the fact that a battery will virtually not be charged at high ambient temperatures, whereas it may be substantially overcharged at low ambient temperatures. Therefore, a maximum battery temperature is usually used as a fail-safe limit for fast charging.

3. ΔT end-of-charge trigger

With a ΔT end-of-charge trigger, the ambient temperature is also taken into account. Two temperature sensors have to be used, one of which measures the ambient temperature and the other the battery temperature. Charging is stopped when the difference between battery and ambient temperatures reaches a certain predetermined value. This value depends on several factors, including the battery size, configuration, heat capacity and, where applicable, the number of cells. It must therefore be determined separately for each type of battery before implementation. Care should be taken that the battery temperature does not influence the ambient temperature sensor. The ambient temperature sensor is usually attached to a dummy thermal mass which is thermally isolated from the battery. This end-of-charge trigger is usually used to terminate fast charging.

4. ΔT/Δt end-of-charge trigger

In order to set off a ΔT/Δt end-of-charge trigger, the change in time of the battery temperature has to be greater than a predetermined value. In this case, the influence of the ambient temperature is virtually eliminated. A reasonable predetermined value is 1°C/min. This end-of-charge trigger is usually used to terminate fast charging.

5. Maximum voltage end-of-charge trigger

The charge current is in this case switched to a trickle-charge current when the battery voltage reaches a certain predetermined value. This is a highly unreliable method, because the battery voltage depends on many factors, such as the battery's series resistance, temperature and age and the type of battery concerned. Currentless measurement of the battery voltage can eliminate the influence of the series resistance. This is applied in most fast chargers that use end-of-charge triggers based on battery voltage.

The problem that arises when a maximum voltage is used to switch the charge current to a trickle-charge current is twofold. When the maximum voltage level is chosen too low, the battery will not be fully charged. When it is taken too high, the danger exists that the battery voltage will never reach this level. This is especially true because the battery heats up at the end of charging which leads to a drop in battery voltage. A maximum battery voltage is used only as a fail-safe limit for fast and quick charging on account of the problems mentioned above.

6. -ΔV end-of-charge trigger

The voltage of a NiCd battery is a function of temperature and therefore it drops slightly during overcharging; see chapter 4. The charging of NiCd batteries is often stopped when the battery voltage or the average cell voltage in a battery pack has dropped 10 to 20 mV. -ΔV end-of-charge triggers are less suitable for NiMH batteries because the drop in voltage at the end of the charging is significantly lower. A voltage drop of 10 mV per cell is recommended to set off the trigger in the case of NiMH batteries. The allowed drop in voltage should not be too small, as this would make the trigger sensitive to noise. On the other hand, a too large value will lead to detrimental overcharging. This is especially true at lower charge currents and elevated ambient temperatures, because then the battery voltage drop will be lower and it will take longer for the trigger to be set off. This end-of-charge trigger is not recommended for charge rates below 0.5 C, because the -ΔV value is then hardly detectable. This end-of-charge trigger is usually used to terminate fast charging.

7. Voltage plateau or dV/dt=0 end-of-charge trigger

This trigger is based on the shape of the battery voltage curves during the charging of NiCd and NiMH batteries. As described above, before the voltage drop occurs both types of battery show a voltage plateau. Here, the first derivative of the battery voltage is zero, in other words $dV/dt=0$. Care should be taken to ensure that the circuitry with which this end-of-charge trigger is implemented is sufficiently sensitive, because the battery voltage may also be fairly flat in the mid-region of charging. Of course, this should not trigger the circuitry. The battery voltage samples are usually filtered by calculating the running average value over several samples before calculating the derivative. This is done to reduce the influence of noise disturbances on the calculation of the derivative. The trigger is set off when the derivative is zero. The voltage plateau end-of-charge trigger is usually used to terminate fast charging.

8. Inflection-point or d²V/dt²=0 end-of-charge trigger [2]

Like the previous two methods, this method is based on the general shape of the battery voltage curve during charging. Before the battery voltage reaches a plateau, an inflection point occurs. At this point, the first derivative of the battery voltage is at its maximum and hence the second derivative of the battery voltage is zero, or $d^2V/dt^2=0$. Similar remarks as given above for the dV/dt end-of-charge trigger hold with respect to sensitivity and filtering. The trigger is set off when the calculated second derivative is zero. This trigger is not very often encountered in commercial fast chargers because the number of calculations that have to be performed is rather large.

9. Pressure-based end-of-charge trigger

In NiCd batteries, the internal gas pressure rises at the end of charging as illustrated in chapter 4. This also holds for NiMH batteries. This phenomenon could theoretically be used to derive a trigger. The pressure could be measured by a pressure sensor that is in contact with the battery's interior through a hole drilled in the battery casing. This is however not very practical. Alternatively, the pressure could be measured by strain gauges set around the battery casing. However, no commercial application of strain gauges has so far been introduced.

Comparison of the end-of-charge triggers

The end-of-charge triggers based on voltage and temperature measurements are summarized in Figure 5.1. The use of a maximum time (t_{max}), temperature (T_{max}) and voltage (V_{max}) as a fail-safe limit is also shown. The depicted voltage and temperature curves hold for a NiCd battery charged at a rate of 1 C. Although similar, these curves will look somewhat different for NiMH batteries, as will be explained in the next section.

Figure 5.1 shows that use of a $-\Delta V$ end-of-charge trigger leads to significant overcharging. In this figure, the $-\Delta V$ threshold Par_1 has been chosen to be 20 mV as an example. More sophistication should be included in the system to ensure less overcharging with a voltage-based end-of-charge trigger. Figure 5.1 clearly illustrates that overcharging is virtually prevented with the inflection-point or $d^2V/dt^2=0$ end-of-charge trigger. As a consequence, the temperature and internal gas pressure hardly rise, as explained in chapter 4. However, calculating the second derivative of the battery voltage involves considerably more signal-processing effort than determining a $-\Delta V$ voltage drop.

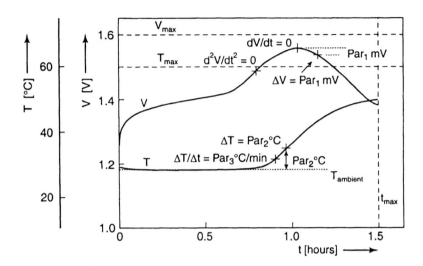

Figure 5.1: Overview of end-of-charge triggers based on voltage and temperature measurements. Typical shape of battery voltage (V [V]) and temperature (T [°C]) for charging a NiCd battery at a 1 C charge rate at 28°C ambient temperature. These curves have been drawn by hand as an illustration of their typical shape

Both temperature-related triggers result in less overcharging than the $-\Delta V$ end-of-charge trigger, as can be seen in Figure 5.1. Obviously, the values of parameters Par_2, the threshold for the ΔT trigger, and Par_3, the threshold for the $\Delta T/\Delta t$ trigger, should be chosen low enough, because otherwise the battery temperature rises too high. Less overcharging and overheating means less loss of cycle life. Figure 5.2 shows the advantage in terms of cycle life of using a $\Delta T/\Delta t$ trigger instead of a $-\Delta V$ trigger in a NiMH battery [1]. The cycling conditions were charging at a 1 C-rate, resting for 0.5 hour, discharging to a 1 V battery voltage and resting again for 2 hours.

A battery's cycle life is usually defined as the number of charge/discharge cycles that a battery can experience until 80% of the initial capacity remains. Figure 5.2 shows that, according to this definition, the cycle life is only 500 cycles when a $-\Delta V$ trigger is used, whereas the cycle life is more than 700 cycles when a $\Delta T/\Delta t$ trigger is used. Hence, although the implementation of the latter trigger requires more effort as it implies including a temperature sensor and calculating the derivative of the measured temperature versus time, it yields a better cycle life.

Charger features
Apart from one or more end-of-charge triggers and fail-safe methods, such as maximum time or temperature, various battery parameters are also checked before charging is started. First of all, most fast chargers check for any faults, such as a short-circuit or an open-circuit. Such faults could indicate a damaged battery. In addition, most fast chargers check the initial battery voltage and temperature. The battery voltage and temperature both have to be within a certain range to enable fast charging. The battery is in an overdischarged state when the initial battery voltage is below a minimum level, for example below 0.9 V. In this case, it has to be charged with a relatively small current until the battery voltage has risen above this

minimum level. This is done to prevent battery damage. Afterwards, fast charging can proceed with a higher current. The battery can only be charged with a relatively small current when the initial battery temperature is out of range, for example outside the range from 0°C to 40°C. Fast charging with a higher current may proceed when the battery temperature re-enters the required range.

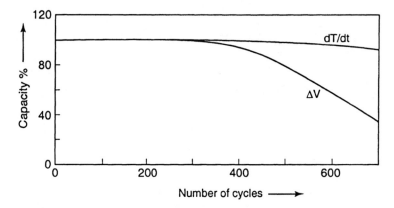

Figure 5.2: Battery capacity as a function of the number of charge/discharge cycles of a NiMH battery using either a $-\Delta V$ or a $\Delta T/\Delta t$ end-of-charge trigger to terminate fast charging at a 1 C charge rate (measured curves redrawn from [1] with permission)

Premature set-off of any of the applied end-of-charge triggers may occur at the beginning of fast charging, for example on account of battery voltage peaks. Therefore, the use of some of the triggers is inhibited in the initial stages of fast charging in most commercial fast chargers.

Especially NiCd batteries may suffer from the memory effect; see chapter 3. A method for restoring full battery capacity involves forcing a complete discharge until the battery voltage has become 0.9 V. In order to accomplish this, some commercial fast chargers offer an optional discharge feature. The user can start the discharging by pushing a button. The user is recommended to fully discharge the battery after a predetermined number of charge/discharge cycles.

5.1.2   Differences between charging algorithms for NiCd and NiMH batteries

Although the charging characteristics of NiCd and NiMH batteries are quite similar, each type of battery requires its own charging algorithm, mostly because NiMH batteries are more sensitive to overcharging than NiCd batteries. In the case of these batteries, the end-of-charge triggers should therefore be chosen more carefully, because of their impact on the battery's cycle life. In general, the triggers that are recommended for NiMH batteries are more sophisticated and lead to less overcharging than the triggers used for NiCd batteries. It is not recommended to charge NiMH batteries in less than one hour, because of their sensitivity to overcharging.

Figure 5.3 gives a schematic representation of the main differences in charging characteristics between NiCd and NiMH batteries [1]. The voltage curves in this figure look similar. However, the shape of the voltage curve of the NiMH battery is less pronounced than that of the NiCd battery. This difference becomes increasingly evident at lower charge rates and higher temperatures, although the voltage characteristic of a NiCd battery also becomes less pronounced under these conditions. Because of this, $-\Delta V$ triggers are less suitable for NiMH batteries, as already discussed in the previous section. Another reason for this is that such triggers lead to considerable overcharging, to which NiMH batteries are more sensitive than NiCd batteries.

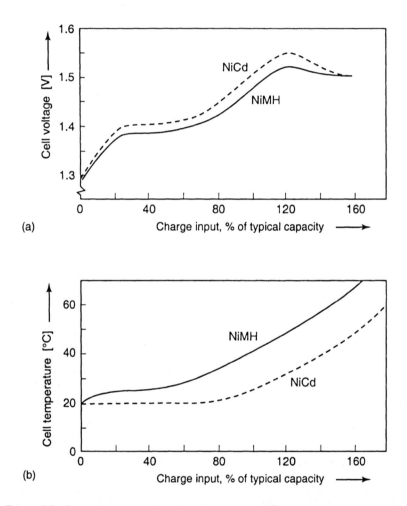

(a)

(b)

Figure 5.3: Comparison of typical characteristics of NiCd (dashed curves) and NiMH batteries (solid curves) during charging at a moderate constant-current charge rate: (a) voltage characteristic (b) corresponding temperature development (measured curves redrawn from [1] with permission)

As can be seen in Figure 5.3, the temperature of NiCd batteries rises only at the end of charging. The moment at which the temperature starts to rise more or less coincides with the steeper rise in battery voltage. This is due to the side-reactions that take over from the main storage reactions; see chapter 4. The net charge reaction, which takes place roughly before 80% charge input has been reached in Figure 5.3, is endothermic in a NiCd battery. As a result, the temperature remains relatively constant in this region. In a NiMH battery however, the net charge reaction is exothermic. This means that the temperature rises even far before 80% of the end of the charge process is reached. The exothermic nature of the side-reaction leads to an even greater increase in battery temperature when the battery approaches full charge. This is similar to the increase in temperature in NiCd batteries at the end of charging.

Another difference in temperature behaviour is observable during overcharging. At moderate charge rates, the battery temperature will level off to a steady-state value in NiCd batteries. This was illustrated in several charge curves in chapter 4. In NiMH batteries, however, the battery temperature continues to rise during overcharging [1]. This and the higher battery temperature near the end of charging in NiMH batteries makes it clear why overcharging should be prevented as much as possible in NiMH batteries. A high battery temperature is a disadvantage with respect to cycle life, and high temperatures occur already relatively early during overcharging.

The type of end-of-charge trigger most frequently used in NiCd batteries is the $-\Delta V$ trigger, because of its simplicity and the fact that NiCd batteries are relatively robust with respect to overcharging. An alternative has to be used when the charge rate is lower than, for example, 0.5 C, because the voltage drop then becomes virtually undetectable. In that case, a simple alternative could be timed charging with thermal cut-off protection.

The end-of-charge trigger recommended for NiMH batteries is either the $dV/dt=0$ or one of the temperature-difference triggers ΔT or $\Delta T/\Delta t$. As was shown in Figure 5.1, these end-of-charge triggers lead to less overcharging than $-\Delta V$ triggers. Figure 5.2 showed that $\Delta T/\Delta t$ triggers yield a significant improvement in cycle life over $-\Delta V$ triggers. $dV/dt=0$ triggers may also be used when no clear peak is observable in the voltage characteristic, for example at low charge currents or high temperatures. A combination of a $dV/dt=0$ and a $\Delta T/\Delta t$ trigger was used for a NiMH battery in the example of the cellular phone in section 2.4. A ΔT trigger was used in the example of the low-end shaver.

5.1.3   Simulation example: an alternative charging algorithm for NiCd batteries

The battery and charging electronics are packed close together in a small volume in some portable products, such as shavers. The main problem encountered in practice is a high internal temperature, which has a negative influence on the battery's cycle life. A simulation of charging a NiCd battery inside a shaver will demonstrate this problem in the first part of this section. An alternative charging algorithm has been simulated as a possible solution [5]. Such simulations have to be performed using a thermal model of a shaver, which will be described next.

Thermal shaver model
The charger is a Switched-Mode Power Supply (SMPS) in flyback configuration [6]. It consists of a rectifying mains buffer, a transformer, a High-Voltage IC (HVIC) that regulates the charge current, a rectifying secondary diode and an electrolytic smoothing output capacitor. An example of such a charger inside a shaver was shown in Figures 2.9 and 2.10 in chapter 2.

A simplified thermal model of the electronics in the shaver that are relevant during charging can be made on the basis of the results of temperature-difference and relaxation measurements. Three heat-producing components can be distinguished in the shaver:

- The NiCd battery.
- The HVIC that controls the charge current.
- The remaining supply parts (transformer, rectifying mains buffer, diode).

The thermal model of the electronics in the shaver can be coupled to the thermal model of the battery shown in Figure 4.19(b) using the principles of physical system dynamics [7]. The thermal shaver model takes into account the thermal resistance R_{th} from the battery to its surroundings and the thermal properties of the battery's surroundings. This thermal model therefore replaces R_{th} and C_{th}^{amb} in Figure 4.19(b) and is connected at node T_{bat}. This is illustrated in Figure 5.4, in which the remaining supply parts are denoted as *supply* [5].

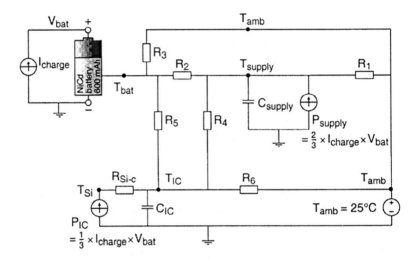

Figure 5.4: Thermal and electrical simulation model of a shaver with an integrated NiCd battery and charger (see also Table 5.1)

The complete network connected to node T_{bat} is represented in the thermal domain. Each heat-producing component is represented by a heat source P_i, which models the dissipation, and a heat capacitance C_i. This combination is coupled to the constant ambient temperature, modelled by a constant temperature source T_{amb}, and to the two other heat-producing components by means of heat resistances R_i. The ambient temperature remains constant because the shaver itself will not heat up its environment. However, the heat-producing components inside the shaver will significantly influence each other's temperature. An additional heat resistance R_{Si-c} from the silicon to the outside of the package of the HVIC is taken into account. The actual charging process is modelled in the electrical domain by means of a current source I_{charge}.

A 50% efficiency has been found for the power supply in practice. This means that the dissipated power equals the electrical power $I_{charge}V_{bat}$ supplied to the battery. Again, it has been found in practice that approximately 1/3 is dissipated in the IC and 2/3 in the remaining supply parts. The ambient temperature and the temperatures of the remaining supply parts, HVIC package and battery are represented at the nodes T_{amb}, T_{supply}, T_{IC} and T_{bat}, respectively. The values of the components in the thermal model are summarized in Table 5.1. These values were derived from measurements.

Table 5.1: Values of the components of the thermal network shown in Figure 5.4

Component	Description	Value	Unit
R_1	Heat resistance from remaining supply parts to ambient	47	K/W
R_2	Heat resistance from battery to remaining supply parts	93	K/W
R_3	Heat resistance from battery to ambient	41	K/W
R_4	Heat resistance from remaining supply parts to HVIC	50	K/W
R_5	Heat resistance from battery to HVIC	166	K/W
R_6	Heat resistance from HVIC to ambient	232	K/W
R_{Si-c}	Heat resistance from internal silicon to outside package of HVIC	15	K/W
C_{IC}	Heat capacitance of HVIC	2	J/K
C_{supply}	Heat capacitance of remaining supply parts	8	J/K
C_{bat}	Heat capacitance of battery (see Figure 4.19(b))	36	J/K

Simulation of a galvanostatic charging algorithm inside a shaver

The result of the simulation of the use of a galvanostatic charging algorithm is shown in Figure 5.5. Galvanostatic charging means that the current remains constant until it is switched to a trickle-charge current. This is applied in shavers nowadays. The simulation model shown in Figure 5.4 was used. The parameter set of Tables 4.3 and 4.4 was used in the NiCd battery model and the values specified in Table 5.1 were used for the heat resistance and capacitance of the battery. The charging efficiency is defined as the ratio of the effectively stored electrical charge and the total applied electrical charge. A constant charge current I_{charge} of 1.2 A was applied for 1 hour.

Figure 5.5 shows a rise in battery temperature to 62°C at the end of charging. The charging efficiency clearly drops at the end of the charging, which is partly due to the elevated battery temperature. A lower charging efficiency at a higher battery temperature was demonstrated earlier in Figure 4.44. Moreover, the self-discharge rate is higher at elevated temperatures. This was shown earlier in Figure 4.45. The battery's State-of-Charge (SoC) drops as a result of self-discharge after the charging has stopped, because the high battery temperature slowly drops to the ambient temperature; see chapter 6 for a definition of SoC. After 2 hours, the battery's SoC is 94%, while the charging efficiency at that point is 80%.

Simulation of a thermostatic charging algorithm inside a shaver

The result of simulating an alternative charging algorithm is shown in Figure 5.6. The same simulation conditions as in Figure 5.5 were chosen, except that the charge current I_{charge} was made dependent on the battery temperature in such a way that the battery temperature never exceeded 35°C. This may be called *thermostatic* charging, i.e. charging with a maximum temperature. A total charge time of 1 hour was used, as in the galvanostatic simulation illustrated in Figure 5.5.

Figure 5.6 illustrates the maximum battery temperature of 35°C and the decrease in charge current during charging. The drop in charging efficiency is virtually zero due to the lower battery temperature. Even after two hours, the charging efficiency was still 99.5%, as opposed to the 80% observable in Figure 5.5.

Moreover, the drop in SoC was virtually zero after the charge current was interrupted. The SoC was 80% after one hour of charging, which is 14% lower than the value in Figure 5.5. However, the applied electrical energy was considerably less than in the case of galvanostatic charging and this led to less dissipation. An advantage of thermostatic charging is that a considerably longer cycle life may be anticipated, due to a reduction in maximum battery temperature. The results of the simulated galvanostatic and thermostatic charging experiments inside a shaver are summarized in Table 5.2.

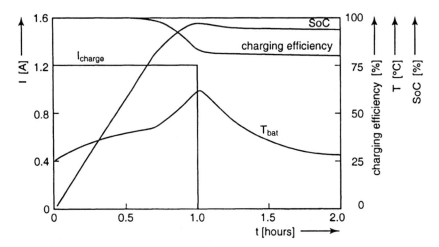

Figure 5.5: Simulated charge current (I_{charge} [A]), State-of-Charge (*SoC* [%]), charging efficiency [%] and battery temperature (T_{bat} [°C]) as functions of time (t [hours]) for charging for one hour with a current of 1.2 A at 25°C ambient temperature. A NiCd battery model with the parameter set of Tables 4.3 and 4.4 was used in the simulation model shown in Figure 5.4

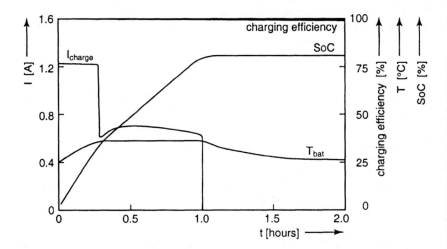

Figure 5.6: Same curves as in Figure 5.5 for the simulation of an alternative charging algorithm. In this case, the battery temperature was not allowed to rise above 35°C

Table 5.2: Comparison of the results of the simulated galvanostatic (Figure 5.5) and thermostatic (Figure 5.6) charging experiments inside a shaver

Method:	t_{charge} [min]	SoC [%]	Charging eff. [%]	T_{max} [°C]
Galvanostatic	60	94	80	62
Thermostatic	60	80	99.5	35

In the above example the same charge times were chosen for galvanostatic and thermostatic charging. Galvanostatic and thermostatic charging can also be compared on the basis of charge times. This will be considered next, in both simulations and measurements.

Simulations and measurements of fast thermostatic charging
Thermostatic charging can also be applied for faster charging at low battery temperatures. This will be demonstrated below by means of simulations and measurements. For simplicity, the simulations and experiments were performed using a NiCd battery in free space. Therefore, the thermal environment was taken to be the same as in Figure 4.19(b) in the simulations and the thermal shaver model was omitted. This means that a thermal resistance R_{th} of 25 K/W was assumed from the battery to the environment and a battery heat capacitance C_{th}^{bat} of 14 J/K was used. These values are in accordance with the values in Table 4.3. The environment itself had a fixed temperature of 25°C, which was forced in the simulation by using a constant temperature source connected to node T_{amb} in the thermal model part of Figure 4.19(b). The battery was charged in either a galvanostatic or a thermostatic mode. A current of 1.2 A was used for the galvanostatic charging. The measurements were performed using a Gates 600 mAh NiCd battery. A charge time of 40 minutes was chosen for the measurements because the capacity of this battery is lower than that of the NiCd model used in the simulations. A charge time of 60 minutes was applied in the galvanostatic simulation.

The charge current started at a value of 4 A in the case of thermostatic charging and was lowered to zero in eleven steps. Ten closely spaced temperature thresholds in a range from 29°C to 31.5°C were used in both the simulations and the measurements. The current was stepped down each time the battery temperature reached the value of a temperature threshold. The current was set to zero and the charge time ended when the battery temperature reached the final threshold of 31.5°C. This implementation enabled us to use the available equipment in a relatively simple set-up. The two charging methods are summarized in Table 5.3.

The simulated results of the galvanostatic charging are shown in Figure 5.7. The figure shows that the battery was overcharged, which led to a battery temperature of 54°C. The attained SoC value was 95% and the charging efficiency was 81% at the moment when the charge current was switched off after 60 minutes.

The simulated result of thermostatic charging is shown in Figure 5.8. The current steps listed in Table 5.3 can be clearly recognized in this figure. A charge time of 35 minutes resulted in this case. The SoC was 89% and the charging efficiency was 94% at the end of the charging. The highest battery temperature, of 31.5°C, occurred at the end of the charging. Hence, although the charge time used in the thermostatic simulation was around a factor of two lower than that used in the galvanostatic simulation, the SoC was about the same. In addition to the benefit of a charge time reduced by a factor of two, the battery temperature remained more than 20°C lower than in the galvanostatic charging simulation. The results obtained with the two charging methods in simulations are summarized in Table 5.4.

Table 5.3: Characteristics of the galvanostatic and thermostatic charging methods used in the simulations and measurements

Temperature T [°C]	Charge current I [A]	
	Galvanostatic	*Thermostatic*
T < 29.25	I = 1.2 A for all temperatures for a	4
29.25 ≤ T < 29.5	fixed time of 60 minutes in the	3.6
29.5 ≤ T < 29.75	simulations and 40 minutes in the	3.2
29.75 ≤ T < 30	measurements	2.8
30 ≤ T < 30.25		2.4
30.25 ≤ T < 30.5		2
30.5 ≤ T < 30.75		1.6
30.75 ≤ T < 31		1.2
31 ≤ T < 31.25		0.8
31.25 ≤ T < 31.5		0.6
T ≥ 31.5		0

Table 5.4: Comparison of the results of simulated galvanostatic and thermostatic charging according to Table 5.3

Method:	t_{charge} [min]	SoC [%]	Charging eff. [%]	T_{max} [°C]
Galvanostatic	60	95	81	54
Thermostatic	35	89	94	31.5

The measured results of the galvanostatic and thermostatic charging experiments are compared in Table 5.5.

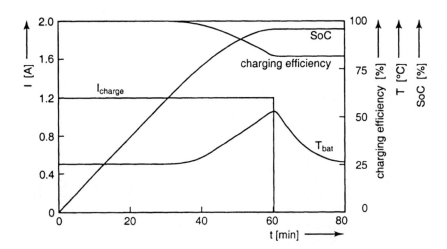

Figure 5.7: Simulated charge current (I_{charge} [A]), State-of-Charge (*SoC* [%]), charging efficiency [%] and battery temperature (T_{bat} [°C]) as functions of time (*t* [min]) for *galvanostatic* charging of a NiCd battery model for one hour using 1.2 A at 25°C ambient temperature. The parameter set of Tables 4.3 and 4.4 was used

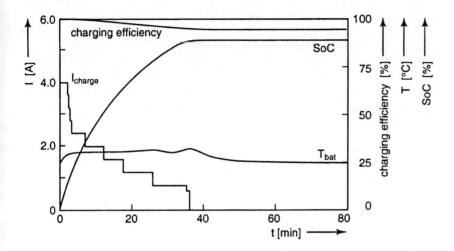

Figure 5.8: Same simulated curves as in Figure 5.7 for *thermostatic* charging according to Table 5.3 at 25°C ambient temperature

Table 5.5: Comparison of the results of measured galvanostatic and thermostatic charging according to Table 5.3

Method:	t_{charge} [min]	SoC [%]	Charging eff. [%]	T_{max} [°C]
Galvanostatic	40	96	85	34
Thermostatic	20	98	94	31.5

Comparison Tables 5.4 and 5.5 illustrates the reduction by a factor of two of the charge time realized with thermostatic charging in both the simulations and the measurements. The SoC for galvanostatic charging is somewhat higher than for thermostatic charging in the simulations. The SoC was practically equal in the case of galvanostatic and thermostatic charging in the measurements. The charging efficiency in thermostatic charging was higher than that in galvanostatic charging in both the simulations and the measurements. The simulated and measured efficiencies agreed very well with each other.

The temperature at the end of the galvanostatic charging was only slightly higher than that at the end of the thermostatic charging in the measurements; see Table 5.5. In the case of the simulations, this difference was significantly higher. This means that the battery was overcharged more severely in the galvanostatic charging simulation. The SoC and the temperature at the end of the charging will decrease when the charge time of 60 minutes is lowered in the simulations. In that case, the simulation result of galvanostatic charging will approach the measurement result shown in Table 5.5. However, the difference in charge time between galvanostatic and thermostatic charging will then be less than a factor of two.

Further measurements indicated that even a charge time of 12 minutes to an SoC of 94% was possible. The charge current was not stepped down further than 2 A in that case and the maximum charge temperature was chosen 33.5°C. This illustrates that the thermostatic charging algorithm can be optimized towards minimum charge time (12 minutes versus 20 minutes) at the cost of a slight increase in battery temperature (33.5°C versus 31.5°C).

Simulation of fast thermostatic charging inside a shaver
Similar simulations of fast thermostatic charging were performed for a NiCd battery inside a shaver using the simulation model illustrated in Figure 5.4. The temperature thresholds were chosen to be a little higher in order to obtain a reasonable SoC because of the higher overall battery temperature. The conditions are listed in Table 5.6 and the simulation results are summarized in Table 5.7.

Table 5.6: Thermostatic charging regime used to simulate the charging of a NiCd battery inside a shaver; see also Figure 5.4 and Table 5.1

Temperature T [°C]	Charge current I [A]
T < 37	4
37 ≤ T < 39	3.6
39 ≤ T < 40	3.2
40 ≤ T < 41	2.8
41 ≤ T < 41.5	2.4
41.5 ≤ T < 42.0	2
42.0 ≤ T < 42.2	1.6
42.2 ≤ T < 42.4	1.2
42.4 ≤ T < 42.6	0.8
42.6 ≤ T < 42.8	0.6
T ≥ 42.8	0

Table 5.7: Result of the simulation in which the thermostatic charging regime of Table 5.6 was used to charge a NiCd battery inside a shaver; see also Figure 5.4 and Table 5.1

Method:	t_{charge} [min]	SoC [%]	Charging eff. [%]	T_{max} [°C]
Thermostatic	56	93	94.5	42.8

Comparison of Tables 5.7 and 5.2 shows that the same SoC can be attained with the thermostatic charging regime of Table 5.6 as in galvanostatic charging in 4 minutes less charging time. This difference in charging time is a lot less impressive than the result obtained in charging a NiCd battery in free space. However, the previous disadvantage of 14% less SoC has now been eliminated. The temperature at the end of the charging was still 19.2°C lower than in the case of galvanostatic charging. Hence, the advantage of an anticipated longer cycle life still holds.

5.2   Charging algorithm for Li-ion batteries

5.2.1   The basic principle
A Li-ion battery has to be charged with a so-called CC/CV regime. This was already described and illustrated in chapter 4; see Figure 4.48. The basic voltage and current characteristics for charging a Li-ion battery are repeated in Figure 5.9.

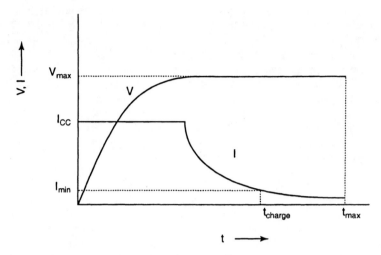

Figure 5.9: Battery voltage (*V*) and current (*I*) characteristics during the application of a CC/CV charging regime to a Li-ion battery

A battery manufacturer specifies both the values of the current I_{CC} and the maximum voltage V_{max}. The charge rate in CC mode ranges from 0.7 C to 1 C, which is related to the allowed current density. The value of V_{max} depends on the type of Li-ion battery. It can be either 4.1V ± 1% or 4.2 V ± 1%. The reason for the tight requirements with respect to the accuracy of this voltage was explained in chapter 2; see Figure 2.5. The value of I_{min} is usually chosen to be between a 0.05 C and a 0.1 C-rate. Alternatively, the total charging time is usually set to 2 hours when t_{max} is used to terminate the charging. As an illustration, the charge time t_{charge} when I_{min} is used to end charging is smaller than t_{max} in Figure 5.9.

There is a difference between using I_{min} or t_{max} to terminate the charge. The advantage of using I_{min} is the fact that the battery is then charged to the same capacity, expressed in [%] of the full battery capacity, at the same battery temperature and internal impedance. In addition, the time for which the battery is in CV mode is minimized, which is advantageous for the battery's cycle life. The time for which the charge regime is in CV mode may be relatively long when t_{max} is used to end the charging, especially when an almost full battery is being charged. The difference between using I_{min} or t_{max} is explained in more detail below.

The current that flows into the battery in CV mode is determined by the difference between the externally applied voltage V_{max} and the battery's equilibrium potential E^{eq}_{bat}. This difference equals the total overpotential η_{bat}^{total}, including the ohmic voltage drop. As charging proceeds in CV mode, E^{eq}_{bat} rises in dependence on the battery's SoC and because V_{max} is fixed, the current will drop accordingly. This happens irrespective of whether I_{min} or t_{max} is used to end charging.

The total overpotential depends on the battery's SoC and impedance and on current and temperature; see chapter 6. When the battery's temperature and internal impedance remain constant, the total overpotential depends on current I and SoC only. When charging is always stopped at the same current I_{min}, this will always occur at the same SoC under the condition of constant temperature and internal impedance. This SoC can be found from the equation $E^{eq}_{bat}(SoC)=V_{max}-\eta_{bat}^{total}(I_{min}, SoC)$, where SoC is the only unknown quantity. This equation expresses that the battery voltage at the terminals during charging, V_{max}, equals the sum of the battery equilibrium potential E^{eq}_{bat} and the total overpotential η_{bat}^{total}; see (Eq. 3.20) and chapter 6. However, when charging is stopped at t_{max} charging will not necessarily always stop at the same current value. For example, when a half-full battery is charged with a charge time t_{max} the current at the end of charging will be different from the situation where an empty battery has been put on charge for the same charge time of t_{max}. Therefore, the SoC at the end of charging will not be the same, because the equation $E^{eq}_{bat}(SoC)=V_{max}-\eta_{bat}^{total}$ (I, SoC) will yield different SoC values for different values of I. Note that a difference in battery temperature or impedance between charge cycles will always lead to differences in SoC at the end of charging, irrespective of the use of either I_{min} or t_{max}.

A Li-ion battery may only be charged within a certain battery voltage and temperature range. Charging is inhibited when either the battery voltage or temperature is outside this range. A charger will check for fault conditions, such as short-circuits or open-circuits, before charging a battery. An electronic safety switch is always connected in series with a Li-ion battery (see chapter 2), which will interrupt the current flow when the battery voltage or temperature drifts beyond the safety ranges. These safety voltage and temperature ranges are of course wider than the ranges within which charging is allowed. Hence, the safety switch will still be able to interrupt the charge current when the charger safety fails. The Positive-Temperature Coefficient resistor (PTC) also acts as an extra safety device (see chapter 2) interrupting the charge current when the safety IC fails.

5.2.2   The influence of charge voltage on the charging process

A higher value of V_{max} leads to a higher capacity, as illustrated in chapter 2, but at the same time the cycle life decreases drastically. The reason for the higher capacity is the wider range of the mol fraction of Li^+ ions (x_{Li}^{pos}) over which the positive electrode can be cycled. It was described in chapter 4 that x_{Li}^{pos} cycles between 0.5 and 0.95 in the positive $LiCoO_2$ electrode under normal conditions, with a mol fraction of 0.5 corresponding to a fully charged battery. When the value of V_{max} is increased, x_{Li}^{pos} is allowed to become lower than 0.5, which means that more Li^+ ions are removed from the positive electrode. This leads to a larger capacity.

A possible reason for the loss of cycle life is the decomposition of the electrolyte at a relatively high voltage. A high voltage is present at the positive electrode when the battery is charged in CV mode. The decomposition reactions also involve Li^+ ions, which irreversibly react with compounds of the electrolyte to form decomposition products. The longer the battery remains at the relatively high CV voltage, the more electrolyte will be decomposed, using up valuable active material, thereby lowering the capacity. So besides high values of V_{max}, the application of V_{max} for long periods of time should also be avoided. This was already mentioned in the previous section.

Another possible reason for the loss of cycle life is the fact that Li metal deposition is allowed to occur at the electrode surface when the potential of the negative LiC_6 electrode becomes lower than 0 V. The Li metal atoms formed at the electrode surface can react with electrolyte compounds and form a passivation layer on the remaining metallic Li. This layer insulates the remaining lithium underneath, which results in a loss of capacity. There will moreover be a risk of dendrite formation, which could lead to an internal short-circuit.

The value of V_{max} is specified with a high degree of accuracy, because battery manufacturers want to guarantee a certain capacity and a certain cycle life. A specified accuracy of $\pm 1\%$ means an accuracy of ± 41 mV for a value of V_{max} of 4.1 V. Figure 2.5 showed that such an accuracy already implies a fair amount of spread in a battery's capacity and cycle life. Therefore, some battery manufacturers demand an accuracy even smaller than $\pm 1\%$ in CV mode.

The value of V_{max} at the battery terminals is extremely important in determining both the charged capacity and cycle life. Additional series resistance in the current path between the battery and the charger should be avoided, because V_{max} has to be accurately controlled at the battery terminals. This can be achieved in two ways.

The first possibility is to measure the battery terminal voltage when no current flows and to control this voltage to V_{max}. This can be referred to as 'resistance-free' measurement. In this case, the charge current is interrupted for a short period during which the voltage is measured at the battery terminals. The current value or duration of the next current pulse can be adapted, depending on the difference between this measurement and the desired value of V_{max}. The second possibility is to use two additional currentless wires to sense the battery voltage while the charge current is still flowing through the power lines. Again, the sensed voltage is to be controlled to V_{max}. Both methods are compared in Figure 5.10. A resistance R_{series} models the total series resistance that is always present outside the battery. This resistance includes the resistance of the wires, the wire contacts, the PTC and the safety switch. It is divided equally between both wires in the figure. This figure is based on a very simple model of the battery that shows only the ohmic internal series resistance R_{ohmic}, the charge-transfer overpotential η^{ct}_{bat}, which includes all kinetic and diffusion overpotentials, and the total equilibrium potential E^{eq}_{bat}. It is assumed that η^{ct}_{bat} hardly changes when the current is interrupted for a short time because of the relatively long RC time.

As can be seen in Figure 5.10a, interruption of the current not only eliminates the influence of R_{series}, but also the influence of R_{ohmic} inside the battery. This explains the term 'resistance-free' measurement. The sum $E^{eq}_{bat} + \eta^{ct}_{bat}$ is controlled to V_{max}. The use of voltage sense wires only eliminates the influence of R_{series}, as can be seen in Figure 5.10b. The reason is that current flows continuously through R_{ohmic} and the voltage IR_{ohmic} cannot be eliminated. A voltage of $E^{eq}_{bat} + \eta^{ct}_{bat} + IR_{ohmic}$ is sensed and controlled to a value of V_{max}. Hence, a later transition from CC to CV mode occurs in the case of Figure 5.10a than when voltage sense wires are used as in Figure 5.10b. This later transition would be beneficial for reducing a battery's charge time, because the current in CC mode is higher than that in CV mode. However, how 'resistance-free' measurement influences a battery's cycle life is not known.

5.2.3 The influence of charge current on the charging process

The use of a higher value of I_{CC} would instinctively lead to a lower charge time. However, the CV mode will be entered earlier during the charging process, because the total battery overpotential and the ohmic voltage drop inside the battery are higher at a higher charge current. The charge time does not depend on the charge current in CV mode, but on the course of the battery equilibrium potential. This effect was illustrated by the results of a simulation and a measurement example in Figure 4.51. The total charge time hardly decreases when I_{CC} is made to be higher than a 1 C-rate.

An increase in I_{CC} does not lead to a decrease in charge time, but it has a negative influence on the cycle life. This is because a battery has a relatively high voltage V_{max} for a relatively long time. As described above, this enables the generation of decomposition products. The increase in both I_{CC} and V_{max} and an additional compensation for the ohmic battery resistance R_{ohmic}, as shown in Figure 5.10a, decrease the charge time. However, this will have an even more dramatic effect on a battery's cycle life.

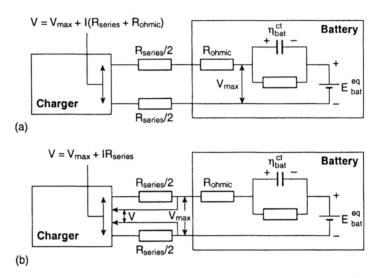

Figure 5.10: Two methods for controlling V_{max} in CV mode: (a) 'resistance-free' measurement by means of current interruption and (b) using additional currentless voltage sense lines. The battery is represented by a simple model

5.2.4   Simulation example: fast charging of a Li-ion battery

The results of simulations using two different charging regimes will be described in this section [8]. The usual charging parameters I_{CC} (0.7 C-rate or 0.5 A) and V_{max} (4.1 V) were used in charging regime A, as recommended by Panasonic. Increased values of I_{CC} (2.5 C-rate or 1.8 A) and V_{max} (4.25V) were used in charging regime B. The parameters of the two charging regimes are summarized in Table 5.8. The simulated charge time needed to reach the nominal capacity of 720 mAh has been included in the fourth column. The build-up of capacity during the simulations is illustrated in Figure 5.11.

Table 5.8: Results of simulations using two different CC/CV charging regimes, the Li-ion model described in chapter 4 and the parameter set of Table 4.5

Charging regime	I_{CC} [A]	V_{max} [V]	t_{charge} to nominal capacity of 720 mAh [min]
A	0.5	4.1	95
B	1.8	4.25	30

Table 5.8 and Figure 5.11 show that the charge time to the nominal capacity is only 30 minutes in the case of charging regime B, whereas it is about 95 minutes in the case of charging regime A. Hence, increasing I_{CC} and V_{max} indeed leads to a decrease in charge time. Moreover, Figure 5.11 shows that charging a battery to a higher

voltage leads to a significant increase in capacity. The reason for this was explained in section 5.2.2. The rest of this section will focus on the possible causes of capacity loss observed when charging regime B is applied.

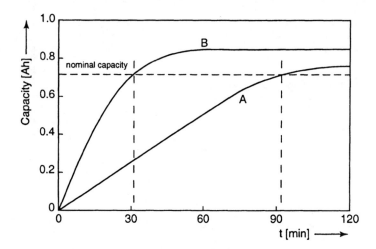

Figure 5.11: Simulated build-up of capacity ([Ah]) versus time (t [min]) for charging regimes A and B; see Table 5.8

There are two possible mechanisms for the reduction in cycle life, as explained in section 5.2.2. In the first place, electrolyte decomposition may occur at the $LiCoO_2$ electrode due to the high electrode potential. In the second place, Li metal deposition may take place at the LiC_6 electrode, which is induced when the electrode potential drops below 0 V. The breakdown mechanisms themselves have not yet been implemented in the Li-ion model. However, the electrode potentials can be easily visualized in the simulations. That way some initial insight into the probability of either of the breakdown mechanisms occurring can be obtained. The development of the $LiCoO_2$ electrode potential is illustrated in Figure 5.12 for charging regimes A and B.

Figure 5.12 shows that the positive electrode potential never exceeds the value of 4.1 V in the case of charging regime A. The maximum electrode potential is about 4.15 V in the case of regime B. The results of electrolyte decomposition experiments using a $LiMn_2O_4$ positive electrode can be found in the literature [9]. The electrolyte concerned consisted of a mixture of ethyl carbonate (EC), dimethyl carbonate (DMC) and lithium hexafluorophosphate ($LiPF_6$), which is slightly different from the electrolyte used in the CGR17500 Li-ion battery. In these experiments, decomposition of the considered electrolyte at a positive $LiMn_2O_4$ electrode started at a voltage of 5 V. Although the positive electrode material in the CGR17500 battery is $LiCoO_2$ instead of $LiMn_2O_4$ and the electrolyte composition is slightly different, it seems reasonable to assume that electrolyte decomposition cannot be the main cause of the capacity reduction observed at the simulated electrode potential values. However, it should be noted that experiments will have to be performed to investigate electrolyte decomposition at a $LiCoO_2$ electrode to confirm this.

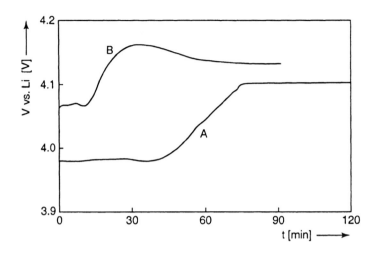

Figure 5.12: Simulated LiCoO$_2$ electrode potential versus a Li metal reference electrode ([V]) as a function of time (t [min]) for charging regimes A and B; see Table 5.8

The negative electrode potential simulated for the two charging regimes is shown in Figure 5.13. The negative electrode potential does not drop below 0 V during the 95 minutes needed to charge the battery to 720 mAh in the case of regime A. However, the electrode potential drops below 0 V in the 30 minutes needed to charge the battery to 720 mAh in the case of charging regime B. The electrode potential drops to −120 mV when charging is continued with regime B.

A possible cause of capacity loss in the case of regime B could be Li metal deposition according to the simulation result shown in Figure 5.13, because the negative electrode potential drops below 0 V. Care should be taken in interpreting of absolute values in these simulations, because the quantitative agreement between the results of the simulations and measurements has yet to be investigated under various conditions. For example, the simulation shows that 30 minutes of charging with regime B results in a negative electrode potential of −12 mV. It still has to be investigated by means of further measurements whether this value is in agreement with measurements, and whether this negative potential is large enough to allow a significant amount of Li deposition.

Simulations to determine modified negative electrode capacity
Simulations using the Li-ion battery model can be used to experiment with different electrode capacities. Such simulations can easily be used to explore, for example, the impact of the capacity of the negative LiC$_6$ electrode on the charging behaviour, because the simulations only take a limited amount of time and involve no safety risks. Figure 5.14 shows the simulated negative electrode potential versus a Li metal reference electrode during charging with regime B at different negative electrode capacities. The negative electrode capacities used in the simulations range from 792 mAh to 1275 mAh. The nominal value is 850 mAh or 3060 C; see also Table 4.5, parameter $Q^{max}_{LiC_6}$.

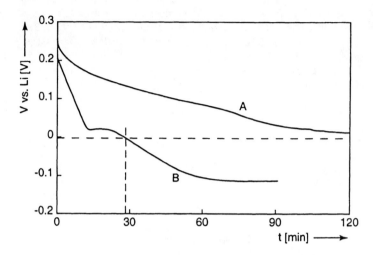

Figure 5.13: Simulated LiC$_6$ electrode potential versus a Li metal reference electrode ([V]) as a function of time (t [min]) for charging regimes A and B. The negative electrode capacity was 850 mAh

Figure 5.14 shows that an increase in the negative electrode capacity leads to a shift in the moment at which the negative electrode potential drops below 0 V. For example, an increase in the negative electrode capacity from the nominal value of 850 mAh to 935 mAh, leads to an increase in time from 27 minutes to 34 minutes. This means that a 10% increase in the nominal negative electrode capacity would already prevent the risk of the negative electrode potential dropping below 0 V during fast charging, because the fast charging time is 30 minutes. This would prevent the assumed possible Li metal deposition during fast charging with regime B. It is assumed that a change in the negative electrode capacity hardly influences the battery voltage and that the charge time consequently remains 30 minutes. Dedicated experiments should be performed to check this.

5.3   Conclusions

In this chapter charging algorithms for NiCd, NiMH and Li-ion batteries have been described. Charging algorithms are an important aspect of the implementation of a Battery Management System, as batteries with different chemistries have to be charged using different dedicated charging algorithms.

It has been shown that simulations using the battery models developed in this book are useful tools in designing such charging algorithms. Although not all processes have been included in the models, such as side-reactions that lead to irreversible capacity loss, simulations serve as quick means for finding potentially better charging algorithms or possible reasons for poor cycle life performance. Additional experiments will have to be performed to verify any significant, interesting phenomena found in the simulations. The results of the simulations help focus the attention in experiments. This leads to a decrease in measurement time. In the case of the NiCd example, measurements with the thermostatic charging algorithm confirmed the phenomena observed in the simulations. In the case of the

Li-ion example, measurements of, for example, the electrode potentials and morphology are still required. Moreover, the processes of electrolyte decomposition at the positive electrode and Li metal deposition at the negative electrode should be studied in greater detail. The results of such studies can then possibly be implemented in the model when a mathematical description of these processes becomes available. This will enable simulation of degradation processes in the future and could lead to redesign of the battery itself.

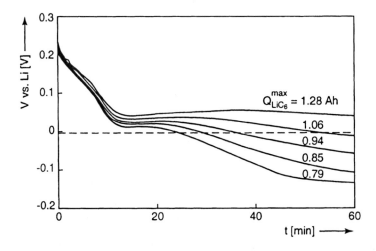

Figure 5.14: Simulated LiC_6 electrode potential versus a Li metal reference electrode ([V]) as a function of time for charging regime B and different negative electrode capacities $Q^{max}_{LiC_6}$

5.4   References

[1]   D. Linden, *Handbook of Batteries*, Second Edition, McGraw-Hill, New York, 1995
[2]   G. Cummings, D. Brotto, J. Goodhart, "Charge Batteries Safely in 15 Minutes by Detecting Voltage Inflection Points", *EDN*, pp. 89-94, September 1, 1994
[3]   F. Goodenough, "Battery-Based Systems Demand Unique ICs", *Electronic Design*, pp. 47-61, July 8, 1993
[4]   R. Cates, R. Richey, "Charge NiCd and NiMH Batteries Properly", *Electronic Design*, pp. 118-122, June 10, 1996
[5]   H.J. Bergveld, W.S. Kruijt, P.H.L. Notten, "Electronic-Network Modelling of Rechargeable NiCd Cells and its Application to the Design of Battery Management Systems", *J. Power Sources*, vol. 77, no. 2, pp. 143-158, February 1999
[6]   K.K. Sum, M.O. Thurston, *Switch Mode Power Conversion; Basic Theory and Design*, Electrical Engineering and Electronics Series, vol. 22, Marcel Dekker, New York, 1984
[7]   R.C. Rosenberg, D.C. Karnopp, *Introduction to Physical System Dynamics*, McGraw-Hill Series in Mechanical Engineering, New York, 1983
[8]   W.S. Kruijt, H.J. Bergveld, P.H.L. Notten, *Electronic-Network Model of a Rechargeable Li-ion Battery*, Nat.Lab. Technical Note 265/98, Philips Internal Report, October 1998
[9]   J.M. Tarascon, D. Guyomard, "New Electrolyte Compositions Stable over the 0 to 5 Voltage Range and Compatible with the $Li_{1+x}Mn_2O_4$/carbon Li-ion Cells", *Solid State Ionics*, vol. 69, no. 3&4, pp. 293-305, August 1994

Chapter 6
Battery State-of-Charge indication

One of the tasks of a Battery Management System is to keep track of a battery's State-of-Charge (SoC), as discussed in chapter 1 of this book. Information on the SoC can be used to control charging and discharging and it can be signalled to the user. An accurate SoC determination method and an understandable and reliable SoC display to the user are important features for a portable product. However, many examples of poor accuracy and reliability can be found in practice. This can be pretty annoying, especially when a portable device suddenly stops functioning whereas sufficient battery capacity is indicated. A poor reliability of the SoC indication system may induce the use of only part of the available battery capacity. For example, the user may be inclined to recharge the battery every day, even when enough battery capacity is indicated on the portable device. This will lead to more frequent recharging than strictly necessary, which in turn leads to an earlier wear-out of the battery. The effect of inaccuracy of SoC indication can be even worse when the SoC value is also used to control charging. The battery is either not fully charged or it is overcharged. In the former case, the battery will be recharged more often than needed, which will lead to an earlier wear-out. In the latter case, frequent overcharging will lead to a lower cycle life. Some methods with which an SoC indication system can be realized will be described in this chapter.

The definition of SoC and related terms will be given first in section 6.1. Existing SoC indication methods will be categorized and some practical examples will be given. Measurement results obtained with a commercially available SoC indication IC from Benchmarq (bq2050) will be described in section 6.2 as an illustration of a system encountered in practice. The main shortcomings of this system will be identified. A simple mathematical model for a Li-ion battery will be derived in sections 6.3 and 6.4 on the basis of simulations using the battery model of chapter 4. On the basis of the results presented in sections 6.3 and 6.4, a set-up for a new SoC system will be proposed in section 6.5. Some measurement results obtained with the proposed set-up will be described in section 6.6 and the results will be compared with those obtained with the bq2050. Conclusions will be drawn in section 6.7.

6.1 Possible State-of-Charge indication methods

6.1.1 Definitions

The term SoC may be confusing. The main reason for this is that a distinction must be made between the charge or energy inside the battery and the portion of this charge or energy that will actually be available under the current discharge conditions. A significant difference may occur between these two, for example when a battery is being discharged at low temperatures. The charge or energy that can then be supplied by the battery is significantly smaller than the charge or energy that is actually present. The question now arises whether SoC expresses the charge or energy that is present or the portion of the charge or energy that will be available under the current conditions. In either case, the term SoC may express: (1) the actual SoC of the battery, which is the quantity we are trying to estimate, (2) the estimated

SoC, which is a calculated variable and (3) the SoC displayed to the user. Type (2) will usually equal type (3). For psychological reasons however, type (3) may temporarily differ from type (2) for the sake of consistency.

Consistency is a complex issue, because batteries show complex behaviour sometimes. This may lead to incomprehension by the user, even when the SoC is correctly estimated and displayed. For example, when a user is in a cold environment, the battery will appear empty earlier than at room temperature. This will lead to incomprehension when this is displayed to a user and he/she moves back to room temperature, because then the displayed SoC increases without recharging. Although correct, this 'inconsistency' could be hidden from the user. The issue of consistency will not be dealt with further in this book, so type 3 will not be further considered.

In an attempt to avoid further confusion, the following definitions will be used in this book:

State-of-Charge (SoC):
The charge (in [Ah]) that is *present inside* the battery. The SoC reflects an estimated value in most cases in this chapter. SoC can also be expressed in [%] of the maximum possible charge, for example on a bar graph in which 100% reflects a full battery and 0% reflects an empty battery.

Remaining capacity (Cap$_{rem}$):
The charge (in [Ah]) that is *available* to the user under the valid discharge conditions. Again, Cap$_{rem}$ reflects an estimated value in most cases. This means that Cap$_{rem}$ is equal to or smaller than SoC, depending on the conditions. Cap$_{rem}$ can also be expressed in [%] of the maximum possible charge, for example on a bar graph in which 100% reflects a full battery and 0% reflects an empty battery.

Remaining time of use (t$_{rem}$):
The estimated time that the battery can supply charge to a portable device under the valid discharge conditions before it will stop functioning when the battery voltage will drop below the End-of-Discharge voltage V_{EoD}.

Charge obtained from the battery (Q$_{out}$):
The charge in [Ah] obtained from a battery during discharging. The value of Q$_{out}$ is zero at the start of each subsequent discharge.

The term SoC will also be used as a collective noun, such as in the title of this chapter. This means that an *SoC indication system* may estimate the battery's SoC and/or Cap$_{rem}$ and/or t$_{rem}$; see the definitions given above. In the explanation of the principles of SoC indication methods in the remainder of this section, the term SoC will be used as a collective noun for simplicity. The distinction between SoC and Cap$_{rem}$ will be made in later sections.

The remaining time of use will be most interesting for a user of a portable device. This time t$_{rem}$ in [s] can be inferred from Cap$_{rem}$ in two ways, depending on the type of load:

$$t_{rem} = \frac{Cap_{rem}}{I} \cdot 3600 = \frac{[Ah]}{[A]} \cdot \frac{[s]}{[h]}$$

(6.1)

for a current-type load (I) and

$$t_{rem} = \frac{\int_{V_{EoD}}^{V_{bat}} Cap_{rem}(V)dV}{P} \cdot 3600 = \frac{[Wh]}{[W]} \cdot \frac{[s]}{[h]}$$

(6.2)

for a power-type load (P) with V_{bat} expressing the battery voltage at the moment t_{rem} is estimated. The integral of Cap_{rem} over the voltage range applicable during the subsequent discharge expresses the energy obtained from the battery. It is impossible to exactly anticipate future load conditions. Therefore, t_{rem} can only be inferred from the past current or power consumption, assuming that it remains constant until V_{EoD} is reached. As an alternative, a worst- and best-case value of t_{rem} can be displayed. The best case applies to the minimum expected load, while the worst case applies to the maximum expected load. For example, a cellular phone may indicate talk time-left and standby time-left. The current will increase for a power-type load when the battery voltage decreases during discharge. This has to be taken into account in the estimation in (6.2). In the remainder of this chapter, only current-type loads will be considered for simplicity. Therefore, Cap_{rem} will be expressed in [Ah] in accordance with the definitions given above.

6.1.2   Direct measurements

The direct measurement method is based on a reproducible and pronounced relation between a measured battery variable and the SoC. This battery variable should be electrically measurable in a practical set-up. Examples of such battery variables are battery voltage (V), battery impedance (Z) and voltage relaxation time (τ) after application of a current step. Most relations between battery variables depend on the temperature (T); see chapters 3 and 4. Therefore, besides the voltage or impedance, the battery temperature should also be measured. The relation f_T^d between the measured battery variable and the SoC, in which d means direct and T the temperature as a parameter, can be stored in the system. The basic principle of an SoC indication system based on direct measurement is shown in Figure 6.1.

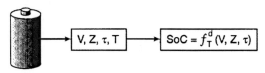

Figure 6.1: Basic principle of an SoC indication system based on direct measurement

The main advantage of a system based on direct measurement is that it does not have to be continuously connected to the battery. The measurements can be performed as soon as the battery has been connected, after which the SoC can be directly inferred from the function f_T^d.

The main problem is determining the function f_T^d, which should describe the relation between the measured battery variable and the SoC under all applicable conditions, including spread in battery behaviour. Conditions include the discharge current, which may vary quite a lot depending on the application, temperatures (for example outdoor use involves a substantially wider temperature range than indoor use), storage times, *etc*. A battery's behaviour depends strongly on these conditions. In general, the greater the amount of variation in conditions in practical use, the less accurate a system based on direct measurement will be. The reason for this is that it is difficult to derive f_T^d for all conceivable conditions. To make things worse, all batteries will wear out during use. This implies a change in battery behaviour during lifetime. Adaptive systems may be able to cope with battery spread and aging. This will be described in more detail in section 6.1.4. Possible battery variables will be discussed below.

Battery voltage (V)
Many existing SoC systems are based on voltage measurement, for example in cellular phones. The battery voltage is then measured and translated into a number of bars shown in a bar graph. The accuracy and reliability of these systems is generally poor, especially when temperature dependence has not been taken into account and in the case of batteries with a limited voltage variation for most of the discharged capacity. For example, the voltage variation from a full to an empty NiCd or NiMH battery is limited to approximately 200 mV. The discharge curve depicted in Figure 6.2 illustrates why SoC systems based on voltage measurement will usually have a poor accuracy. The voltage variation ΔV around voltage V can result from variations in current, resistance or temperature. Due to the relatively flat discharge curve, this voltage variation will lead to a considerable variation ΔSoC around the estimated SoC value.

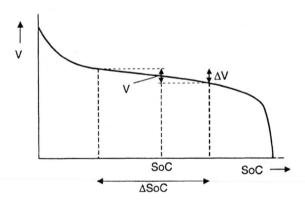

Figure 6.2: Schematic discharge curve illustrating why SoC indication systems based on voltage measurement will usually have a poor accuracy

In order to get more insight in the various components that constitute the battery voltage V_{bat}, consider the simplified battery model shown in Figure 6.3. The EMF (Electro-Motive Force [1],[2]) voltage denotes the sum of the equilibrium potentials

E^{eq} of the two battery electrodes. The η^k voltage is the sum of the kinetic overpotentials of the two electrodes, while η^d expresses the sum of the diffusion overpotentials. Finally, the η^Ω voltage is the ohmic overpotential, which depends on the total series resistance in the battery. The sign and value of all the overpotentials depend on the direction and value of the battery current (I).

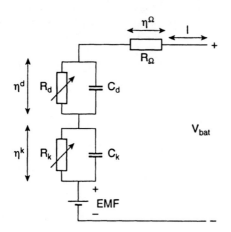

Figure 6.3: Simplified battery model in the electrical domain

Kinetic and diffusion overpotentials will gradually increase or decrease when the battery current is applied or interrupted, respectively. This occurs with a certain time 'constant', which is expressed by two RC combinations, R_kC_k and R_dC_d, in Figure 6.3. The non-linear resistances R_d and R_k depend on the battery's SoC and general condition, such as storage time and age and on current and temperature. This dependence is expressed by arrows in Figure 6.3. The value of R_Ω is variable in a portable device with a detachable battery pack, for example, because of contaminated contacts. In other words, all the overpotentials depend on many factors. All these contributing factors have to be taken into account in the function f_T^d when the battery voltage is measured during current flow, which is virtually impossible. The voltage relaxations have to be taken into account when the current is interrupted. Possible methods for measuring only the EMF of a battery will be described in more detail in section 6.3.

An example of changing battery characteristics in order to improve the accuracy of an SoC system based on voltage measurement is the Infolithium system applied in Sony camcorders [3]. In the batteries used in this system, the applied electrode materials realize a steeper discharge curve, which is beneficial for the accuracy; see Figure 6.2. However, the accuracy of such a system is still limited when the portable device is used under a wide variety of conditions.

Battery impedance (Z)
The battery impedance is a frequency-dependent complex quantity, which describes the relation between battery voltage and current. It can be measured in the frequency domain and derived from measurements in the time domain [4]. In the former case, the AC impedance valid in a certain bias point is measured. This is referred to as impedance spectroscopy [1],[5]. In the latter case, the voltage difference before and

after the application of a current step can be divided by the magnitude of the current step. However, the time at which the voltage samples are taken then strongly influences the value inferred from the measurements. Therefore, it is not correct to use the term impedance for this value.

A simplified polar impedance plot is shown in Figure 6.4a, where $f_1 < f_2$. The negative imaginary axis has been indicated at the top, as is common practice in electrochemistry [1],[5]. This yields capacitive behaviour in the first quadrant and inductive behaviour in the fourth quadrant. An example of measuring voltage differences ΔV_1 and ΔV_2 after the application of a current step ΔI is shown in Figure 6.4b.

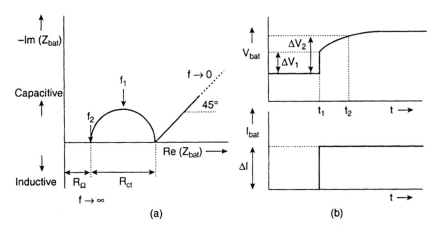

(a)   (b)

Figure 6.4: (a) Measurement of battery impedance in the frequency domain: simplified polar plot showing battery impedance as a function of frequency, (b) Measurement of voltage differences ΔV_1 and ΔV_2 after the application of a current step ΔI in the time domain

At low frequencies, the battery impedance will be represented by a straight line in the polar plot under a 45° angle; see Figure 6.4a. This part of the impedance plot corresponds to the process in which diffusion limitations are dominant. A semicircle will appear at higher frequencies. This semicircle, with diameter R_{ct}, is due to kinetic aspects of the charge transfer reactions. Frequency f_1 equals $1/(2\pi R_{ct}C^{dl})$, in which R_{ct} denotes the charge transfer resistance inferred from $RT/(nFI^0)$ (see chapters 3 and 4), while C^{dl} expresses the double-layer capacitance. In practice, this semicircle corresponds to the electrode reaction with the poorest kinetics. A second semicircle corresponding to the electrode reaction with the fastest kinetics will then be virtually undetectable. R_{ct} will be in the order of 0.2 Ω at 50% SoC in practical Li-ion batteries [6]. The impedance is real and equal to the total ohmic series resistance R_Ω, which may have a value up to 300 mΩ in practical Li-ion batteries (see chapter 2) at the frequency f_2, which is in the order of 10 kHz in practical Li-ion batteries [6]. The impedance of a practical battery will be inductive at frequencies larger than f_2; see Figure 6.4a. The electrode geometry and the internal connections determine the inductance.

Figure 6.4b illustrates that the value of $\Delta V_1/\Delta I$ inferred from measurements at the time t_1 is smaller than the value of $\Delta V_2/\Delta I$ inferred at the time t_2. The value of $\Delta V_1/\Delta I$ corresponds to R_Ω in Figure 6.4a. However, the value of $\Delta V_2/\Delta I$ will also depend on kinetic and possibly even diffusion-related aspects of the battery voltage. It is therefore not correct to use the term impedance for these values, although this is common practice in the electrochemical literature [4].

Measurement of battery impedance as a function of frequency as shown in Figure 6.4a is not practical for SoC indication in a portable device, because a signal with a frequency sweep has to be applied. Some dependence of the impedance on the SoC can be found in a laboratory set-up, but this dependence is usually smaller than the dependence on temperature [4]. For this reason, and the fact that this form of measurement is not very practical, it is not applied in practice. However, impedance spectroscopy is very useful for studying battery behaviour in detail in a laboratory set-up when it is used to characterize separate electrodes.

The measurement of R_Ω at the time t_1 shown in Figure 6.4b is used mainly in portable products as a means of indicating the battery's condition (State-of-Health or SoH) [4],[7],[8]. Battery wear-out can be detected by an increased series resistance. Therefore, assessing the value of R_Ω by simply applying a current step can serve to test whether the battery is of poor quality and should be replaced. This is used mainly in industrial applications like Uninterruptible Power Supplies (UPS), in which a large number of batteries are placed in series and parallel.

Voltage relaxation (τ)

Apart from just taking two voltage samples before and after the application of a current step as shown in Figure 6.4b, the complete battery voltage relaxation curve can also be measured and used as a criterion for the SoC. This technique is referred to as chronopotentiometry, which is the equivalent of impedance spectroscopy in the time domain. However, because no reliable information on the SoC can be obtained in impedance spectroscopy, the same remark holds for chronopotentiometry. Therefore, this technique is not applied in practice.

6.1.3   Book-keeping systems

Book-keeping systems are based on current measurement and integration. This can be denoted as coulomb counting, which literally means 'counting the charge flowing into or out of the battery'. This yields an accurate system when all the charge applied to the battery can be retrieved under any condition, at any time, which means that Cap_{rem} is always the same as the SoC. However, chapters 3 and 4 have shown that things are a lot more complicated in batteries. Therefore, other battery variables such as voltage and temperature are also measured in practice. The basic principle of an SoC indication system based on book-keeping is shown in Figure 6.5.

The function $f^{bk}_{V,T,I}$, where bk means book-keeping and V, I and T are parameters in the function, is based on the content of the coulomb counter, which is the integral of current I over time. Other battery variables (V and T) are measured to compensate the content of the coulomb counter for the battery behaviour. The value of I itself is also used to compensate the counter.

The following battery behaviour can be compensated for in a book-keeping system:

• **Charging efficiency:** Side-reactions occur at the end of charging in, for example, NiCd and NiMH batteries. As a result, not all charge applied to the battery is effectively stored. The charging efficiency depends on the SoC, I and T. This phenomenon was illustrated by means of a simulation in section 4.5.

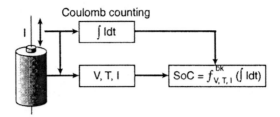

Figure 6.5: Basic principle of an SoC indication system based on book-keeping

- **Discharging 'efficiency':** Depending mainly on the SoC, T and I, only part of the available charge inside a battery can be retrieved. The term 'efficiency' is somewhat misguiding here, because it is possible to retrieve the charge under different conditions. For example, a battery that may seem empty after it has been discharged with a relatively high current can still be discharged further after a rest period and/or with a lower current. In general, less charge can be obtained from a battery at low temperatures and/or large discharge currents. A battery's age also influences the discharging efficiency, for example due to an increased internal resistance of the battery. The compensated SoC value equals Cap_{rem} under the valid discharge conditions.
- **Self-discharge:** Any battery will gradually lose charge, which becomes apparent when the battery is left unused for some time. This cannot be measured by a coulomb counter, as no net current flows through the battery terminals. The self-discharge rate of a battery depends strongly on temperature, as well as on the SoC. A simulated example of self-discharge of a NiCd battery was given in section 4.5.
- **Capacity loss:** The maximum possible battery capacity in [Ah] decreases when a battery ages. Capacity loss depends on many factors. In general, the more the battery is misused, for example overcharged and overdischarged on a regular basis, the larger the loss will be. Some types of batteries have been optimized towards maximum cycle life, which means that their capacity decreases more slowly as a function of the number of charge/discharge cycles [1]. Voltage measurement is used to deal with capacity loss in most commercial book-keeping systems. An example will be given in section 6.2, in which the capacity parameter is updated based on the measurement of a battery discharge curve.
- **Storage effects/memory effect:** Some types of batteries will temporarily show deviating behaviour after long times of storage. The behaviour will depend not only on the storage time, but also on temperature. Some types of batteries, such as NiCd batteries, show the memory effect as described in chapter 3. This will lead to a temporal and reversible capacity loss, as opposed to the irreversible capacity loss due to aging.

A practical example of a book-keeping system is shown in Figure 6.6. Apart from an analogue measuring function, involving A/D conversion and preprocessing, a microcontroller or microprocessor is used to perform the actual book-keeping calculations in the form of function $f^{bk}{}_{V,T,I}$. Two types of memory are needed. Basic battery data, such as the amount of self-discharge as a function of T and the discharging efficiency as a function of I and T, is read from the ROM. The RAM is used to store the history of use, such as the number of charge/discharge cycles, which can be used to update the maximum battery capacity.

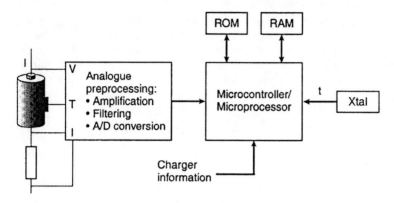

Figure 6.6: Practical set-up of a book-keeping system

An important difference with respect to a direct measurement system is the presence of the variable 'time' in the function $f^{bk}_{V,T,I}$. The time reference t is obtained from a crystal oscillator (Xtal). As a result, a memory at least has to remain connected to the battery at all times. A microcontroller inside the device can be used to calculate the SoC and store regular SoC updates in the memory inside the battery pack with a time stamp when a battery pack is connected to a portable device. When the pack is detached, the microcontroller can read the latest time stamp upon renewed attachment and estimate the time for which the battery has been unused. In extreme cases, the entire system can be integrated within the battery pack. An example will be given in section 6.2.

Errors will accumulate over time in a book-keeping system. Possible error sources include measurement inaccuracy and limited accuracy of the compensations in $f^{bk}_{V,T,I}$. Therefore, the system must be calibrated from time to time. The estimated SoC is reset to an assumed value in a calibration point. For example, the charger can signal to the SoC system when an end-of-charge trigger occurs; see chapter 5. This serves as a 'battery full' calibration point, as indicated in Figure 6.6 by the arrow labelled *Charger information*. The effectiveness of this calibration of course depends on the accuracy of the 'battery full' trigger. Moreover, the system can be calibrated to 'empty' when the battery voltage drops below a certain level, defined as 'battery empty'. However, this is rather tricky, because when a battery is discharged with a large current at a low temperature, the voltage will drop below the 'battery low' level while the battery still contains a considerable amount of charge. The reason for this is the discharging 'efficiency' phenomenon described above.

The compensations in $f^{bk}_{V,T,I}$ are rather empirical in most practical systems [9], [10]; see also section 6.2. However, the behaviour cannot be accurately extrapolated when the battery is exposed to conditions for which the empirical equations have not been derived. In an ideal case, battery behaviour is described in the form of an electrochemical/physical model such as the models described in chapter 4 [11], in which the SoC is calculated on the basis of the external inputs V, I and T. Such a comprehensive description could be too complicated to handle in practical systems.

A combination of an empirical and an electrochemical/physical model will be suggested in section 6.5.

6.1.4   Adaptive systems

The main problem in designing an accurate SoC indication system is the unpredictability of both battery and user behaviour. Battery behaviour depends strongly on conditions, including age, some of which may be unanticipated. Moreover, spread in behaviour of batteries of the same type and batch makes life more difficult. A possible solution is to add adaptivity to a system based on direct measurement, book-keeping or a combination of the two. The basic principle of adding adaptivity to an SoC indication system is depicted in Figure 6.7.

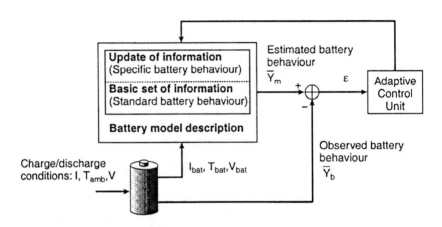

Figure 6.7: Basic principle of an adaptive SoC indication system

The basis in Figure 6.7 is the battery model description. The measured battery variables I_{bat}, T_{bat} and V_{bat} are the inputs of this model, which estimates battery behaviour in the form of output vector $\overline{Y}_m$ on the basis of these inputs. Vector $\overline{Y}_m$ contains at least the SoC, but could also contain additional battery variables, such as an estimated value of the battery series resistance. Another possibility would be to estimate the battery voltage on the basis of the I_{bat} and T_{bat} measurements, and to compare this estimated value with the measured value V_{bat}. The model may contain the function f_T^d of section 6.1.2 or the function $f^{bk}_{V,T,I}$ of section 6.1.3, or a combination of the two. The system starts with a basic set of information, which describes standard battery behaviour for the type of battery concerned.

Adaptivity of the model is based on a comparison of $\overline{Y}_m$ with observed battery behaviour in the form of vector $\overline{Y}_b$. This comparison is made whenever possible. It results in an error signal ε, which is input to an Adaptive Control Unit. The unit updates the information in the model by updating parameter values or even by changing the model description. As a result, the model is adapted on the basis of behaviour specific to the battery to which the system is connected and the error between estimation and observation is minimized.

A first simple example of adaptivity was already given in section 6.1.3. The parameter *maximum battery capacity* is updated from time to time in a system, taking into account capacity loss on the basis of observed differences between estimated and observed SoC values. A second example of an adaptive system can be found in [12]. The model for a lead-acid battery is based on electrochemical and physical theory. Adaptivity is based on the comparison of estimated and measured values of the battery charge/discharge current and battery voltage. The errors between estimations and measurements are used to update the parameters in the battery model by means of a Kalman filter. A similar approach can be found in [13].

A third example can be found in [14], where the relation between the charge removed during a sampling period and the discharge current is updated using estimations of the battery's EMF, which has a direct relation with the SoC. More detailed information on this system will be given in section 6.5.3. A fourth example of adding adaptivity to a system on the basis of direct measurement can be found in [15]. Here, the discharge time remaining until a predefined end-of-discharge voltage has been reached is estimated using a predefined function. This function describes the course of the discharge voltage of a standard battery for various discharge currents. The parameters in this function are updated on the basis of measurement of the battery's voltage. More information will be given in section 6.5.3. A fifth example of adding adaptivity to an SoC system is found in [16]. Here, a neural network is used to model battery behaviour in the form of a discharge curve.

In conclusion, it is fair to say that some kind of adaptivity is needed in an SoC system to take changes in battery behaviour over time into account. Many examples in which adaptivity has been successfully applied can be found in the literature. However, it is premature to assume that by simply adding a learning ability, the battery model will eventually take all relevant battery behaviour into account to enable accurate SoC estimation. The extent to which this is possible depends strongly on the battery itself, and even more on the variety of conditions under which the battery is used.

6.1.5   Some remarks on accuracy and reliability

The accuracy of an SoC indication system is limited by systematic and random errors. Systematic errors result from an incorrect or incomplete inclusion of battery behaviour in a system. For example, a systematic error in the estimation will occur when no dependence on discharge current of the remaining capacity Cap_{rem} has been taken into account in a book-keeping system. Systematic errors may lead to a situation in which a portable device stops operating while the estimated remaining capacity Cap_{rem} is still larger than 0%. Therefore, most SoC systems deliberately give a pessimistic estimation. This is achieved by defining a safety margin or reserve capacity to take into account all uncertainties in the estimation. The reserve capacity is still available for use when the indication shows 0% of remaining capacity. This leads to an even larger value of the systematic error. In general, the less accurate the battery behaviour is described in the system, the greater the systematic error will be. The magnitude of the systematic error depends on the condition of use. Some systems make only a small systematic error under standard conditions. However, under non-standard conditions, such as discharges at low temperatures, the systematic error increases.

Random errors lead to a probability distribution around the estimated SoC and Cap_{rem} values. Random errors are caused by spread in behaviour of batteries of the same type and by measurement inaccuracy. Random errors are taken as the basis for system accuracy definitions in most commercial presentations of SoC indication systems, mainly in the form of measurement accuracy. This can be achieved by defining the system's error for a condition under which the systematic error is small or even negligible. However, the systematic error becomes significantly larger than the random error especially under non-standard conditions. This leads to a much lower accuracy.

A large estimation error under conditions that are frequently encountered in practice leads to a low perceived reliability. Consequently, the user decides to recharge the battery more often than needed, which leads to more charge/discharge cycles in the same time. This may lead to earlier wear-out of the battery. Therefore it is important to define precisely under which conditions the battery is likely to be used. The estimation error under these conditions at least should be acceptable.

6.2   Experimental tests using the bq2050

6.2.1   Operation of the bq2050

The bq2050 IC from Benchmarq is a book-keeping system for Li-ion batteries. It has been designed for integration in a battery pack. The battery voltage, current and temperature are measured. The system is based on coulomb counting, as explained in section 6.1.3. The counter is compensated for charge and discharge current as well as temperature. Self-discharge compensation corrected for temperature is applied. Finally, a means for updating the maximum available battery capacity is included to deal with capacity loss. The estimated SoC and remaining capacity Cap_{rem} can be displayed on a Light-Emitting Diode (LED) display or communicated to a host processor in the system through a one-wire serial interface.

The system uses several registers for storing estimated and measured battery variables, such as Cap_{rem} and temperature, and registers for storing an identification code and programming pin settings. The possible settings include the maximum capacity, the initial SoC when the system is first connected to the battery, and the type of Li-ion battery, which is programmed by selecting the type of negative electrode present in the Li-ion battery: graphite or coke. Moreover, the self-discharge correction can be enabled or disabled and two different values can be selected for V_{EoD}. The first value, $V_{EoD,1}$, is used as an early warning, while the second value, $V_{EoD,2}$, serves as a 'battery empty' reference.

Four registers are important in the calculation of remaining capacity in the bq2050:

- NAC: Nominal Available Charge
- CAC: Compensated Available Charge
- LMD: Last Measured Discharge
- DCR: Discharge Count Register

The operation of the bq2050 can be understood from the diagram in Figure 6.8 [10]. All four registers count in mAh units. The NAC and CAC registers are the actual results of the capacity estimations. Basically, NAC is a coulomb counter, which is only compensated for self-discharge. The CAC register reflects the capacity that is accessible under the current discharge conditions or the charge that is stored under the current charge conditions. It is obtained from the NAC register by multiplying its contents by a factor smaller than or equal to one. Therefore, the value stored in the

CAC register is always equal to or smaller than the contents of the NAC register. The multiplication factors change and CAC acquires a different value as soon as the conditions (I and/or T) change. In terms of the definitions given in section 6.1.1, the NAC register contains the estimated SoC value, while the CAC register contains the estimated Cap_{rem} value.

The maximum battery capacity is stored in the LMD register. The NAC register, and consequently the CAC register, never count higher than the value of the LMD register. The content of the LMD register equals the content of the PFC (Programmed Full Count) register when the system is first connected to the battery. This PFC register can be programmed by the user. The DCR register plays an important role in updating the LMD register to take capacity loss into account. The update mechanism will be described in more detail later in this section.

The NAC register is increased proportional to the integral of charge current over time when the battery is charged. The NAC register is also decreased with the self-discharge rate, because self-discharge is a continuous process. This rate is dependent on the content of the NAC register divided by a temperature-dependent factor $f_{sd}(T)$, which is smaller at higher temperatures, resulting in a higher self-discharge rate at higher SoC and temperature. This was described in chapter 4 and is in agreement with practical observations [1]. For temperatures lower than $10°C$, multiplication of the content of the NAC register by 0.9 yields the content of the CAC register. This is presumably done to take a lower charging efficiency at lower temperatures into account.

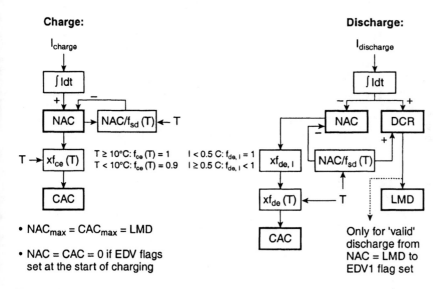

Figure 6.8: Diagram of calculation of remaining capacity in bq2050 (modified from [10])

The NAC and CAC registers are reset to zero before charging starts when both 'battery empty' (EDV) flags have been set during the previous discharge cycle. This is an example of calibration, as discussed earlier. The battery empty flags are two bits that are set when the battery voltage drops below $V_{EoD,1}$ and $V_{EoD,2}$, respectively. This does not necessarily mean that the NAC and CAC registers are zero at that moment, because the remaining charge inside the battery could still be considerable when the battery voltage drops below V_{EoD}, as explained above. The EDV flags may not be set at discharge rates larger than 2 C to prevent the risk of the error due to resetting the NAC and CAC registers to zero becoming too large. In agreement with the general definition in chapter 3, the C-rate is defined on the basis of the battery capacity C, which equals the content of the LMD register in the bq2050.

When the battery is discharged, the content of the NAC register decreases proportional to the integral of the discharge current over time. The content of the CAC register may be smaller than that of the NAC register during current flow to take the discharging efficiency into account. This is achieved by multiplying NAC with the factors $f_{de,I}$ and $f_{de}(T)$. The higher the discharge current and the lower the temperature, the smaller these factors will be. For discharge rates smaller than 0.5 C and temperatures above 10°C, the contents of the NAC and CAC registers are equal, because the correction factors are unity in that case.

The content of the DCR register is increased during discharging. The content of the DCR register is reset to zero when the NAC register reaches its maximum value during charging, which is NAC=LMD. The DCR register is increased on the basis of coulomb counting and self-discharge, as long as NAC>0 during discharging. The situation may occur that NAC is decreased to zero while the battery voltage is still larger than $V_{EoD,I}$. The DCR register is then only increased by coulomb counting and self-discharge is ignored during the subsequent discharge time, because NAC=0 leads to a zero self-discharge rate; see Figure 6.8.

The content of the DCR register is used only to update the LMD register after a 'valid' discharge. This is a discharge that starts at NAC=LMD, which means that discharging starts from a 'full' battery, according to the system. The discharge process may moreover not be interrupted with recharges of more than 256 NAC counts, and self-discharging during the whole discharge process may not exceed more than 18% of the PFC value. This means that leaving the battery on the shelf and discharging it completely by means of self-discharge will not lead to a battery capacity update. Finally, a temperature below 0°C when the $V_{EoD,2}$ level is reached inhibits a 'valid' discharge. Summarizing the demands for 'valid' discharge, one could say that a 'valid' discharge is a discharge from a full counter to a battery voltage below $V_{EoD,2}$ with hardly any interruptions and at a reasonable temperature.

6.2.2   Set-up of the experiments

A CGR17500 Li-ion battery from Panasonic was connected to the Printed-Circuit Board (PCB) of the evaluation kit of the bq2050. The serial port of the IC was read by a computer, which made a log file of the status of the internal registers. The battery was also connected to automatic cycling equipment, which also retained a log file in which voltage, current and the integral of current over time were stored. The measurement set-up is shown in Figure 6.9. The bq2050 was programmed with the following settings:

- One Li-ion cell with a graphite negative electrode
- Maximum capacity: PFC=704 mAh (this value is the closest to the nominal capacity of the CGR17500 battery of 720 mAh)
- V_{EoD} values: $V_{EoD,I}$ =3.04 V (early warning), $V_{EoD,2}$ =2.94 V (battery empty)
- Self-discharge compensation enabled
- NAC=LMD=PFC on initiation

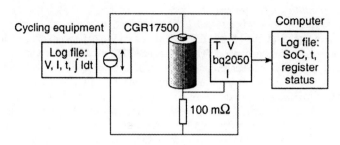

Figure 6.9: Measurement set-up for experiments using the bq2050 IC

Two types of experiments were performed to check the accuracy of the bq2050:

Experiment 1

The battery was successively charged and discharged in the first experiment. Charging was performed with a current of 500 mA (0.7 C-rate) in CC mode and a voltage of 4.1 V in CV mode and a charge-termination current of 10 mA, as explained in chapter 5. Discharging was performed with DC currents with different values until the battery voltage dropped below 2.8 V, which is below $V_{EoD,1}$ and $V_{EoD,2}$. Each discharge cycle was directly followed by a charge cycle. The battery was discharged in successive cycles with 150 mA (0.2 C-rate), 700 mA ($\approx$1 C-rate), 2 A (2.7 C-rate), 700 mA and again with 150 mA. The experiment was performed at 25°C and at –5°C.

Experiment 2

The charging algorithm in the second experiment was identical to that used in experiment 1. Both DC currents with different values and pulse-shaped currents were applied during discharging. Again, discharging proceeded until the battery voltage reached 2.8 V and each discharge cycle was followed by a charge cycle. The battery was first discharged with 150 mA, then with 2 A, then twice with 300 mA. The battery was discharged twice with a pulse-shaped current after this, with the profile illustrated in Figure 6.10. Such pulse-shaped currents can be found in cellular phones based on the Time-Division Multiple Access (TDMA) principle, such as GSM (Global System for Mobile communication) phones. More information on cellular telephony will be given in chapter 7. The value of 300 mA is somewhat lower than the average value of the pulse-shaped current. Experiment 2 was performed only at 25°C.

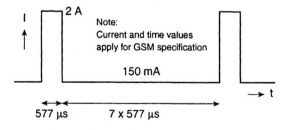

Figure 6.10: Pulse-shaped current used in experiment 2

Initiation in both experiments

The system and battery were initialized before both experiments were started to bring them in a well-defined state. The settings of the bq2050 were chosen such that NAC equalled LMD and hence PFC upon connection of the battery to the system. The battery was first discharged to 2.8 V with a current of 150 mA. This first discharge was a 'valid' discharge at a temperature of 25°C, because it proceeded from NAC=LMD all the way down to the setting of both EDV flags. The new value for LMD then equalled the capacity that was removed from the moment the battery was connected until the moment the EDV flags were set. This is not necessarily the maximum battery capacity. The battery was subsequently charged using the same procedure as described above. The bq2050 updated its LMD value again after a subsequent 'valid' discharge with 150 mA, which corresponded to the full battery capacity because the battery was completely charged. To be on the safe side, the battery was charged again using the same procedure, discharged with 150 mA and charged. The result of this initiation procedure is a full battery and an initialized bq2050.

6.2.3 Results and discussion

Definition of error in empty estimation (E)

The battery user will perceive an empty battery when the battery voltage drops below 2.8 V. Assume that this happens at $t=t_{empty}$. In an ideal case, the CAC register will be zero at exactly $t=t_{empty}$. An error E of the 'empty' estimation will occur in practice when the CAC register is either zero before $t=t_{empty}$ or still has a content larger than zero at $t=t_{empty}$. In the first case, the error E will be defined as the charge Q_{out} that was obtained from the battery from the moment when the CAC register became zero until $t=t_{empty}$ divided by the actual LMD value. The error will then be defined as negative, which means that the estimation is pessimistic. In the latter case, the error E will be defined as positive. The content of CAC at $t=t_{empty}$ will then be divided by the LMD value. This situation will be most annoying to a user.

Accuracy considerations

The error E may either be inferred from $E=(Q_{out}/LMD) \cdot 100\%$, with $E<0$, or from $E=(CAC_{t=t_{empty}}/ LMD) \cdot 100\%$, with $E>0$ as explained above. This means that the accuracy ΔE with which this error E will be determined is given by the following general definition:

$$\Delta E = |E| \cdot \left(\frac{\Delta x}{|x|} + \frac{\Delta LMD}{|LMD|} \right) \quad (6.3)$$

where x can be either Q_{out} or $CAC_{t=t_{empty}}$ depending on the sign of E as discussed above and the error itself can be represented by $E \pm \Delta E$ to take inaccuracy into account. The value of the LMD register is changed in units of 1 mAh, which means that the accuracy of the register content is ±0.5 mAh. ΔLMD will be taken 0.5 mAh, because it is a parameter that is not updated on a regular basis. However, the CAC register is a variable and will be updated all the time when current flows. In order to quantify ΔCAC, we should realize that the bq2050 stored data once every minute in the experiments. On the other hand, the cycling equipment stored battery voltage data every time the voltage changed with 2 mV. The measurement error in the battery voltage made by the cycling equipment is only ±0.15 mV. The discharge curve will be very steep when $V_{bat}=2.8$ V at $t=t_{empty}$. For example, a slope of 0.04 mAh/mV can be inferred from the discharge curve at 381 mA in Figure 6.11. This means that the moment $t=t_{empty}$ is represented with high precision in the log file of the cycling equipment. However, the uncertainty in the time when the CAC register attains a certain value in the bq2050 log file will be 1 minute at maximum. This means that the maximum error ΔCAC in [mAh] that was made when E was derived from comparing the log files of the bq2050 and the cycling equipment was $(1/60) \cdot I$, in which current I is expressed in [mA].

The part of the error ΔQ_{out} that was made at the beginning of the determination of Q_{out} is equal to ΔCAC, because Q_{out} started on the basis of the content of the CAC register, i.e. CAC=0. The part of the error ΔQ_{out} that was made at $t=t_{empty}$ is negligible. This can be understood as follows. The discharge curve will be very steep at $t=t_{empty}$ and the voltage measurement accuracy is ±0.15 mV. This will lead to an error of 0.3 mV·0.04 mAh/mV=0.01 mAh in the case of a variation of 2·0.15 mV=0.3 mV in measured battery voltage. This is negligible compared to the part of the error ΔQ_{out} that was made at the start of Q_{out}. In summary, the error ΔQ_{out} will be assumed the same as ΔCAC in the tables below.

Results
The results of experiments 1 and 2 are summarized in Tables 6.1 and 6.2, respectively. The experiments started at the first row and ended at the last row in the tables. For example, a charge Q_{out} of 14 mAh was removed from the battery from the moment when the CAC register content had become zero until $t=t_{empty}$ at a discharge rate of 0.2 C at 25°C, see the first row, third column in Table 6.1. The LMD register had a value of 792 mAh at that moment. Hence, the error was negative then and its value was found from $E=(-14/792) \cdot 100\% = -1.8\%$. ΔE was calculated using (6.3), which yielded $\Delta E=1.8 \cdot ((2.5/14)+(0.5/792))=0.3\%$, in which $\Delta CAC=2.5$ mAh was found from $(1/60) \cdot 150=2.5$ mAh. This means that E=-1.8±0.3%. During the next discharge at a rate of 1 C, the content of the CAC register was 6 mAh at $t=t_{empty}$. The error was $(+6/792) \cdot 100\%= +0.8\%$ in this case, because the LMD register content remained unchanged. ΔE was found using (6.3): $\Delta E=1.6\%$, in which $\Delta CAC=(1/60) \cdot 700=11.7$ mAh was used. Therefore, E=0.8 ±1.6%.

Table 6.1: Errors (E) in estimation of empty battery by bq2050 in experiment 1

$I_{discharge}$ [A]	$I_{discharge}$ [C-rate]	$E \pm \Delta E$ at 25°C [%][1]	$E \pm \Delta E$ at -5°C [%][1]
0.15	0.2	-1.8±0.3	-6.7±0.4
0.7	1	+0.8±1.6	+2.3±1.7
2	2.7	+16.7±4.2	+57.8±4.8
0.7	1	0±1.5[2]	-86.2±1.7
0.15	0.2	-3.8±0.3	-15.6±0.4

Table 6.2: Errors (E) in estimation of empty battery by bq2050 in experiment 2

$I_{discharge}$ [A]	$I_{discharge}$ [C-rate]	$E \pm \Delta E$ at 25°C [%][1]
0.15	0.2	-574.0±7.7
2	2.7	+20.4±4.7
0.3	0.4	-2.6±0.7
0.3	0.4	-2.0±0.7
PD 2A/0.15A (1/8)[3]	0.5	-0.8±0.9
PD 2A/0.15A (1/8)[3]	0.5	-1.7±0.9

Note 1: $E = (Q_{out}$ after CAC=0)/LMD (-) or CAC$_{t=t_{empty}}$/LMD (+), see (6.3) for definition of ΔE

Note 2 (Table 6.1, fourth row): E=0 in this case, which leads to ΔE=0 according to (6.3). This is not realistic. Therefore, ΔE was defined as ΔCAC/LMD in this case, because the uncertainty in the real moment when the CAC register became zero is 1 minute.

Note 3: See Figure 6.10

Table 6.1 shows that the largest errors occur at large discharge currents and low temperatures. The estimation is indeed pessimistic for a discharge rate of 0.2 C, because of the negative error values, and below 5% at room temperature. However, the errors can become positive at discharge rates of 1 C and larger, especially at low temperatures. Positive errors should be avoided as much as possible, as stated above. The errors are also mostly negative in Table 6.2 for discharge rates smaller than 1 C with the exception of the first PD discharge, where the error can also be positive due to accuracy limitation of the derived E value. The error is positive for a discharge rate larger than 1 C. However, the error for the 150 mA current in the first row is a lot larger than in Table 6.1 for 150 mA at 25°C. This will be explained below.

Another interesting phenomenon can be observed in Table 6.2. The pulse-current yields reliable estimations, because the errors are very small. A low-pass filter with a bandwidth of only 0-8 Hz is used across the current sense resistor on the test PCB of the bq2050. Therefore, only the average current is measured. The result of experiment 2 shows that this is indeed allowed. This can also be understood from the battery models in chapter 4 and the simple battery model shown in Figure 6.3. All AC current will flow through the capacitors in Figure 6.3 when the frequency of the first harmonic of the current is high enough, and only the average current will flow through R_k and R_d. This is definitively the case for the current shown in Figure 6.10 for realistic time constants of 10 ms due to the charge-transfer processes and 1 minute or more due to the diffusion-controlled processes. This means that the charge-transfer and diffusion-controlled processes are determined by the average current. The current profile will only cause an additional voltage ripple around the discharge curve one obtains with the equivalent DC current of 381 mA. This voltage ripple equals $(2 A-150 mA)*R_{\Omega}$. This is illustrated in Figure 6.11.

The large error for the first 150 mA discharge current in Table 6.2 can be understood as follows. The battery was almost empty at the beginning of experiment 2. The first discharge is a 'valid' discharge and this led to an update of the LMD register to the capacity removed during the first discharge, which was only 70 mAh in experiment 2. The C-rate is expressed in terms of the capacity stored in the LMD register; see also the C-rate definition in chapter 3. This means that even 150 mA was larger than a 2 C-rate in experiment 2 according to the bq2050 definition. This means that the capacity was not updated further, because the EDV flags were not set for discharge rates higher than 2 C. This was solved by resetting the IC at the moment the battery was fully charged again after the discharging with 0.15 A.

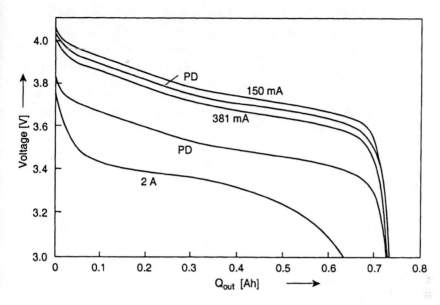

Figure 6.11: Measured discharge curves of the CGR17500 Li-ion battery. The top curve is for a DC discharge current of 150 mA. The third curve from the top is for a DC discharge current of 381 mA. The bottom curve is for a DC discharge curve of 2 A. The two remaining curves labelled *PD* form the discharge curve including voltage ripple for the current profile shown in Figure 6.10, which has an average value of 381 mA

Another discrepancy in the LMD update mechanism was discovered during the experiment at 25°C; see Table 6.1. The discharge current measured by the bq2050 during one of the discharge cycles in the initiation phase was somewhat higher than the real current. Therefore, the LMD register was updated to 792 mAh from the DCR register, which is higher than the actual battery capacity. As a result, the NAC register never reached the value of the LMD register again during the charging and none of the subsequent discharges were therefore 'valid'. Aging effects can no longer be taken into account in this case, because the LMD register will not be updated again.

6.2.4   Conclusions of the experiments

- At discharge rates lower than 0.5 C and at room temperature, the estimation error is smaller than 5%, which is acceptable in an application.
- The accuracy of the estimations is low during discharging at low temperature (-5°C) and at rates of 1 C and larger at all temperatures.
- Measuring pulse-shaped currents with a low bandwidth of 0-8 Hz yields reliable estimations of Cap_{rem}.
- The capacity updating mechanism does not work properly under certain conditions.

Hence, the main shortcomings of the system are estimations at large discharge currents and/or low temperatures and the capacity updating mechanism.

6.3   Direct measurements for Li-ion batteries: the EMF method

6.3.1   Introduction

The main problem in measuring a battery's voltage for SoC indication is the fact that the overpotentials depend on so many factors; see chapter 4 and section 6.1.2. An alternative would be to estimate the EMF only. The EMF is found to be an accurate indicator of the SoC for lead-acid batteries with little dependency on the battery's temperature and age [14]. This makes the EMF a very attractive candidate for SoC indication. This raises the question whether this is also the case for other types of batteries. EMF curves for Li-ion batteries will be described in this section [17]. Indications can be found in the literature that these curves' dependence on SoC is very reproducible [18]. This is confirmed by the measurements and the simulations discussed in this section. Various methods for assessing a battery's EMF will be discussed first in section 6.3.2. Note that the EMF is valid at equilibrium, which means that no external current flows. Therefore, SoC indeed relates to the charge inside the battery; see the definitions in section 6.1.1.

6.3.2   EMF measurement methods

Three methods are available for assessing the EMF:

• Voltage relaxation
• Linear interpolation
• Linear extrapolation

Voltage relaxation
Figure 6.3 shows that the battery voltage will eventually relax to the EMF value after current interruption. This may take a long time, especially when a battery is almost empty. Moreover, it is difficult to determine when the battery voltage has completely relaxed. One can try to estimate the EMF value on the basis of the voltage relaxation curve to speed up the estimation time without waiting until the battery voltage has fully relaxed [14]. Voltage relaxation measurements are performed under currentless conditions, as opposed to the two other methods described below.

As discussed in chapter 4, hysteresis is observable in the EMF curve. This means that the EMF value for a certain SoC depends on whether that SoC was reached by means of charging or by means of discharging. The effect of hysteresis is taken into account when the EMF curve is obtained through voltage relaxation after either charging or discharging steps.

Linear interpolation
The average battery voltage is determined from the battery voltages during charging and discharging with the same currents for linear interpolation. The basic concept of linear interpolation is illustrated in Figure 6.12, which shows that the battery voltage V_c is higher than the EMF during charging, whereas it is lower (V_d) during discharging. The average voltage obtained from voltages V_c and V_d is the EMF when $I_d=-I_c$ and the battery impedance is symmetrical. The battery temperature must moreover be the same when V_c and V_d are measured. Linear interpolation is applied in Figure 6.12 for two different SoC values, yielding two different EMF values. In order to determine the complete EMF-SoC curve, one can start with an empty

battery, charge the battery completely with current I_c and then discharge the battery completely again with current $I_d=-I_c$. The EMF curve is the average of the charging and discharging curves. The current values should not be taken to be too high, because at high I_c values the battery voltage will be charged in CV mode earlier, in which the battery voltage remains fixed and the current decreases, as described in chapter 5. Hence, the prerequisite for linear interpolation is not fulfilled. A different impedance value has been assumed in states 1 and 2 in Figure 6.12, as can be inferred from the different slopes of the interpolated lines. The battery impedance may be linearized and will then be symmetrical by default when sufficiently small values are used for I_c and I_d.

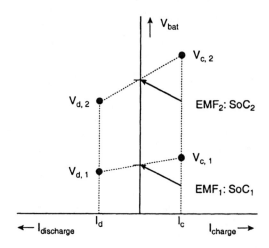

Figure 6.12: Basic concept of linear interpolation to assess the EMF voltage of a battery

It should be possible to apply a charge and a discharge current to access the EMF through linear interpolation. This becomes rather impractical when a battery is not attached to a charger, as will usually be the case in practical use. Moreover, hysteresis is ignored with linear interpolation, because a single EMF value is found for each SoC value. The occurrence of hysteresis will result in errors when the EMF curve, stored in the system's ROM, was determined by means of linear interpolation by the manufacturer and the EMF is determined using a different method during operation of the portable device, for example voltage relaxation. This should be taken into account in the system design.

Linear extrapolation
The battery voltages obtained with different currents with the same sign and at the same SoC are linearly extrapolated to a current of zero value in this case. The basic concept of linear extrapolation is illustrated in Figure 6.13. The linear extrapolation method is illustrated for discharge currents $I_{d,a}$ and $I_{d,b}$, where a larger discharge current leads to a lower voltage. Again, two different EMF values are determined at two different SoC values and the battery temperature must be the same during the determination of V_a and V_b. The use of more measurement data for extrapolation increases the accuracy. However, the larger the discharge currents, the larger the deviation from a straight line will be due to the non-linear overpotential-versus-

current characteristic. A polynomial fit through all the measured voltages should in such cases be used.

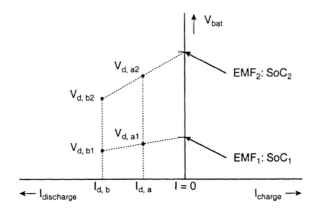

Figure 6.13: Basic concept of linear extrapolation to assess the EMF voltage of a battery

A manufacturer can use extrapolation to assess the complete EMF curve by successively charging and completely discharging a battery with various currents. The x-axes of the various discharge curves will be aligned when the battery is charged to the same SoC every time. This can be achieved by ending the CV mode with a minimum current, as discussed in chapter 5. Assessing a complete EMF curve through extrapolation takes a considerable amount of time, because the battery has to be fully charged and discharged several times.

    Linear extrapolation can also be used to assess the EMF in a portable product in practical use, as opposed to linear interpolation, because linear extrapolation can be realized simply by applying two discharge currents with different values and linearly extrapolating the accompanying battery voltages. However, it should be kept in mind that a steady-state situation has to occur after the application of each current. When this is not the case, this will influence the EMF value estimated in linear extrapolation. Hysteresis is taken into account with extrapolation. The EMF value valid for discharging will be found when discharge currents are used. This is another difference with respect to linear interpolation.

6.3.3   Measured and simulated EMF curves for the CGR17500 Li-ion battery

Measured EMF curves obtained with the aid of voltage relaxation and linear interpolation methods will be discussed first in this section [19]. The measurements were performed at three different temperatures, of 0°C, 25°C and 45°C, using three identical CGR17500 Li-ion batteries simultaneously, with the results obtained for the three batteries being averaged for each condition. The batteries were tested in over 600 charge/discharge cycles to investigate the influence of battery age on the EMF curves. Extrapolation was not considered in the measurements, because of the great amount of time involved in such cycle tests. All the batteries were activated to make sure that they were in the same condition before starting the experiments. The activation cycles included charging the batteries in CC/CV mode, at a 0.7 C charge rate in CC mode and a 4.1 V voltage in CV mode, with a minimum current of 35 mA. The batteries were discharged at a rate of 0.5 C.

The voltage relaxation measurements were performed by charging and discharging the batteries in small 15 mAh increments at a rate of 0.1 C. Each charge or discharge step was followed by a rest period of 30 minutes, after which the voltage was sampled. This voltage was assumed to be equal to the EMF. Experiments with rest periods of 720 minutes showed that the difference in voltage after 30 and 720 minutes of resting was only in the order of 2 mV, while the complete voltage relaxation was in the order of 25 mV. Therefore, the rest period of 30 minutes was assumed adequate to allow the battery voltage to relax almost completely. The batteries were charged and discharged in 48 steps. Charging continued at a rate of 0.1 C after the last charging step until the voltage of 4.1 V was reached. This resulted in a defined full state. Discharging continued at a rate of 0.1 C after the last discharging step until V_{EoD} (3 V) was reached, which led to a defined empty state.

The batteries were charged at a rate of 0.1 C until 4.1 V in the interpolation measurements. After a rest period of 30 minutes, the batteries were discharged at a rate of 0.1 C to 3 V. In order to average the results of the charging and discharging cycles, each nth discharge curve was averaged with each $(n+1)^{st}$ charge curve. A fully discharged battery was taken as a capacity reference to obtain an unambiguous capacity-axis definition. This means that the point at which the battery voltage dropped below 3 V was defined as 0 mAh. For example, the interpolated EMF value at 10 mAh was found by averaging the discharge voltage measured 10 mAh before the battery voltage dropped below 3 V with the charge voltage measured 10 mAh after the subsequent charging process started.

The maximum capacity was defined as the capacity obtained from the battery during the discharging to 3 V at a rate of 0.1 C in both the voltage relaxation and the interpolation measurements. The battery had to be charged at a rate of 0.1 C to 4.1 V before being discharged. This maximum capacity value was used to plot the EMF curves on a relative axis. A voltage measurement resolution of 1 mV has to be taken into account in the case of all voltage measurements obtained with the cycling equipment employed here. A comparison of the EMF curves obtained in voltage relaxation and interpolation is shown in Figure 6.14.

Figure 6.14 shows that the two curves obtained in voltage relaxation after charge and discharge steps are not identical. This is due to the hysteresis effect discussed above. The greatest hysteresis occurs at a capacity of roughly 30% and amounts to approximately 40 mV. The voltage difference is significantly larger than the measurement accuracy, which was the same as discussed in section 6.2.3, and cannot be explained by a too short relaxation time, as a relaxation time of 720 minutes only marginally moved the lines closer together. The interpolated EMF and the mathematical average of the voltage relaxation curves are practically identical. This inspires greater confidence in the consistency of the two measurement techniques. Only a small deviation occurs when the battery is almost empty, which is partly due to longer relaxation times during discharging. In this case 30 minutes rest time is not enough.

Figure 6.15 shows the EMF curves obtained in voltage relaxation after discharge steps at different temperatures. It illustrates that the temperature dependence of the EMF is zero at 50% SoC. At SoC values higher than 50%, the EMF decreases at lower temperatures. At values lower than 50%, the EMF increases at lower temperatures. The maximum error in the SoC in the temperature range from 0°C to 45°C occurs around 30% SoC and amounts to 8% when only the EMF curve at 25°C is taken into account in an SoC indication system. The dependence of the EMF on the temperature will not be taken into account in the remainder of this chapter.

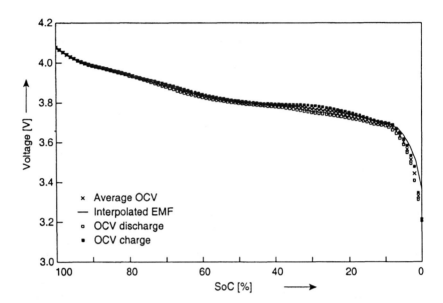

Figure 6.14: Comparison of the EMF curves obtained in voltage relaxation and interpolation measurements. The EMF values obtained in the voltage relaxation measurements are represented as squares, where OCV means Open-Circuit Voltage; measurement points were obtained after charging (+) and after discharging (□). The curve (x) represents the mathematical average of the (+) and (□) curves. The solid line represents the EMF curve obtained in interpolation. The x-axis shows SoC [%], normalized to maximum capacity (see text) and all measurements were performed at 25°C

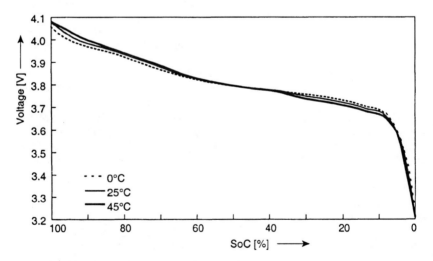

Figure 6.15: Measured EMF curves obtained in voltage relaxation after discharge steps at different temperatures: 0°C, 25°C and 45°C. The x-axis shows SoC [%], normalized to maximum capacity (see text)

Figure 6.16 shows the EMF curves obtained in interpolation at different temperatures. It is in agreement with the phenomena in Figure 6.15. This confirms the consistency between the two measurement methods. However, the SoC at which the temperature coefficient is zero differs. It occurs at 50% in Figure 6.15, but at 60% in Figure 6.16. It was found that it occurs at 70% in the EMF curve derived from voltage relaxation after charge steps; this has not been included in the figures. Hence, ignoring the hysteresis effect due to interpolation averages out the SoC value at which a zero temperature coefficient occurs. The EMF curves will differ at different temperatures when plotted on an absolute x-axis, because the absolute capacity obtained from a battery is lower at lower temperatures. On an absolute scale, a capacity of 718 mAh was found for 0°C, 739 mAh was found for 25°C and 755 mAh for 45°C.

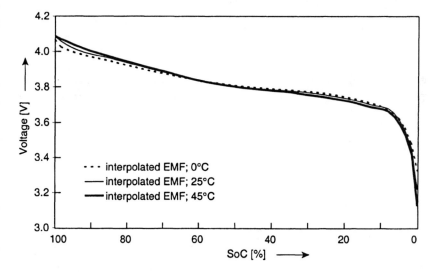

Figure 6.16: Measured EMF curves obtained in interpolation at different temperatures: 0°C, 25°C and 45°C. The x-axis shows SoC [%], normalized to maximum capacity (see text)

Figure 6.17 shows EMF curves interpolated for different charge/discharge cycles. The EMF curves are practically the same for all cycle numbers when plotted on the normalized capacity axis. The maximum difference is only 10 mV. The EMF curves plotted with an absolute capacity axis will differ, because the absolute battery capacity decreases when the battery becomes older. The absolute capacity values that were found for cycles 14, 289 and 615 were 733 mAh, 707 mAh and 660 mAh, respectively.

An EMF curve can also be simulated with the aid of the Li-ion battery model described in chapter 4. The EMF value can be simply obtained in simulations by adding the true equilibrium potentials of the two electrodes. The model is simply charged or discharged and the EMF is plotted versus SoC [%]. Figure 6.18 shows the simulation results for three different temperatures, of 0°C, 25°C and 45°C. The capacity obtained from the battery until the battery voltage dropped below 3 V was taken as the capacity reference when the model was discharged.

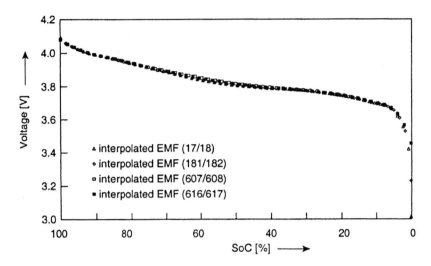

Figure 6.17: Measured interpolated EMF curves for different charge/discharge cycles at 25°C. Cycle 17/18 (Δ), cycle 181/182 (◇), cycle 607/608 (□), cycle 616/617 (■). The x-axis shows SoC [%], normalized to maximum capacity (see text)

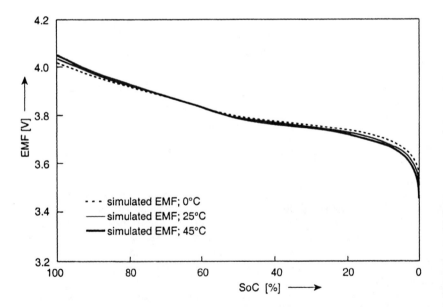

Figure 6.18: Simulated EMF curves for three different temperatures: 0°C, 25°C and 45°C

The simulated EMF curves shown in Figure 6.18 correspond very well with the measured interpolated curves in Figure 6.16. The general trend of a negative temperature coefficient at a low SoC and a positive temperature coefficient at a high SoC agrees with both Figures 6.15 and 6.16. The curves obtained in the interpolation of measurements and the simulated curves do not take into account the effect of hysteresis. Hysteresis was taken into account in the measured curves obtained with voltage relaxation in Figure 6.15. Therefore, the agreement between the results of the simulations and the interpolation measurements seems logical. The simulated EMF curve is inferred from $E^{eq}_{LiCoO_2}$- $E^{eq}_{LiC_6}$, see (Eq 4.78) and (4.87). The only temperature dependence that is taken into account in the present Li-ion battery model is in the terms $RT/n_{Li}F$. Apparently, this yields the correct temperature behaviour.

The influence of aging can be studied by simulating of the EMF curve with a 720 mAh and a 600 mAh maximum capacity. After normalization it was found that the two curves were exactly identical. (Eq 4.78) and (Eq 4.87) show that this is not surprising. Only the relative capacity in the form of x_{Li} occurs in both equations, and not the absolute battery capacity.

6.3.4   Conclusions

The measurements discussed in this section reveal that the EMF curve versus normalized capacity in [%] shows little dependence on temperature and number of cycles in the case of the investigated CGR17500 Li-ion battery. The results of the simulations with the Li-ion battery model developed in chapter 4 show the same behaviour. These results will be used in section 6.5 to define an alternative SoC indication system.

6.4   A simple mathematical model for overpotential description

In practice, current is drawn from a battery during discharging and overpotentials (η) occur; see Figure 6.3. A battery may appear empty to a user even if the estimated EMF value still estimates an SoC larger than 0%, because the battery voltage drops below V_{EoD}. This is illustrated in Figure 6.19, where SoC has been plotted on the horizontal axis to explain this effect, which can be translated to a time axis using the value of the discharge current. Discharging starts at A% SoC and the battery voltage drops with an overpotential η. As a result, the battery voltage drops below V_{EoD} at B% SoC. This means that the remaining capacity Cap_{rem} is zero at B% SoC; see the definitions in section 6.1.1. It should be possible to estimate this under the discharging conditions concerned by including a mathematical description of η in the system.

Simulations with the battery models presented in chapter 4 yield a first impression of the course of η under various conditions. This is a lot less time-consuming and a lot simpler than conducting measurements, because overpotentials at various discharge rates and various SoC values can be easily plotted after simulations that only take a few seconds to run. Figure 6.20 shows the overpotentials simulated with the Li-ion model using the parameter values given in chapter 4 for various discharge currents when the battery is discharged from a full state (100% SoC), half full state (50% SoC), 25% SoC and 10% SoC. For easy comparison of the overpotential curves at various discharge currents in one figure, Q_{out} has been plotted on the horizontal axis. Again, this axis can be translated to a time axis using the value of the discharge current. This has to be considered when relaxation processes are concerned. A Positive-Temperature Coefficient resistor (PTC) was placed in series with the battery, see chapter 4, but no safety IC resistance was considered for simplicity.

All the overpotential curves in Figure 6.20 represent similar behaviour. The overpotential quickly reaches a certain level in the initial stages of discharging, which level depends on the discharge current. This is attributable to the ohmic voltage drop and a relatively fast relaxation of the kinetic overpotentials. Then, a slow relaxation process occurs, after which the battery voltage again reaches a more or less stable level. Simulation results in Figure 4.53 illustrated that this relates to the build-up of a transport overpotential of Li^+ ions in the electrolyte. Finally, the overpotential increases again when the battery is almost empty. This is attributable to the increase in the kinetic and diffusion overpotentials due to diffusion limitation of Li^+ ions in the electrodes when the battery approaches the empty state. The increase in diffusion overpotential is dominant according to simulations; see Figure 4.53.

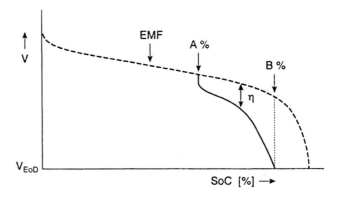

Figure 6.19: Schematic representation of EMF (dashed) and discharge (solid) curves leading to zero remaining capacity ($Cap_{rem}=0$) at B% SoC

For practical use in an SoC indication system, the behaviour shown in Figure 6.20 should be described in an equation that is as simple as possible and involves a minimum number of parameters. The parameter values should be easily obtainable in a limited number of measurements. In principle, such an equation can be inferred from the overpotential description used in the more complicated battery models described in chapter 4; see for example (4.92). In this chapter, a more or less empirical equation will be used for obtaining the first results of overpotential estimation in a proposed SoC indication set-up, which will be described in the next sections. This equation is given below [20]:

$$\eta(Q_{in},T,I,t) = \eta_{\Omega k}(T,I,t) + \eta_d(T,I,t) + \eta_q(Q_{in},T,I,t) =$$

$$I \cdot \left(R_{\Omega k}(T) + R_d(T) \cdot \left(1 - e^{\frac{-t}{\tau_d(T)}} \right) + \frac{U_q(I)E_q(T)}{Q_{in}(t)} \cdot \left(1 - e^{\frac{-t}{\tau_q(T)}} \right) \right) \qquad (6.4)$$

where η is the total overpotential (in [V]), which is composed of three contributions. Overpotential $\eta_{\Omega k}$ results from the combined effect of the ohmic and kinetic overpotentials, overpotential η_d is due to diffusion of Li^+ ions in the electrolyte and η_q describes the increase in overpotential when the battery becomes empty. I is the applied discharge current (in [A]), $R_{\Omega k}(T)$ is the temperature-dependent 'ohmic and kinetic' resistance (in [Ω]), $R_d(T)$ is the temperature-dependent 'diffusion' resistance (in [Ω]) and $\tau_d(T)$ is the temperature-dependent 'diffusion' time constant (in [s]). $R_{\Omega k}$ equals $R_\Omega + R_k$ in Figure 6.3, under the assumption that $R_k C_k$ is very small, and R_d and $\tau_d = R_d C_d$ can also be seen in Figure 6.3. The dimensionless function $U_q(I)$ is an inverse step function, with $U_q(I)=1$ for discharging ($I \leq 0$) and $U_q(I)=0$ for charging ($I >0$). Variable $E_q(T)$ is the temperature-dependent energy, normalized to the current I, which describes the increase in overpotential when the battery becomes empty (in [J/A]). Finally, $\tau_q(T)$ is a temperature-dependent time constant associated with the increase in overpotential in an almost empty battery. Parameter $E_q(T)$ is a measure of the energy that cannot be obtained from the battery when I increases. Hence, this parameter expresses the difference between Cap_{rem} and SoC. (6.4) must be expressed in differential form to describe the overpotential development under changing current conditions. The variable $Q_{in}(t)$ is the charge present inside the battery at time t, which means that Q_{in} expresses SoC in [C]. (6.4) illustrates that the overpotential will instantaneously attain a value $IR_{\Omega k}(T)$, after which the additional overpotential $IR_d(T)$ builds up with a time constant $\tau_d(T)$. The final term in (6.4) leads to an increase in overpotential due to a decrease in Q_{in} when the battery approaches the empty state. This occurs with time constant $\tau_q(T)$. The contribution of η_q to the total overpotential decreases drastically when an empty battery is charged, because the diffusion limitation of Li$^+$ ions in the electrodes vanishes. Therefore, $U_q(I)$ is made zero when a charge current is applied ($I >0$).

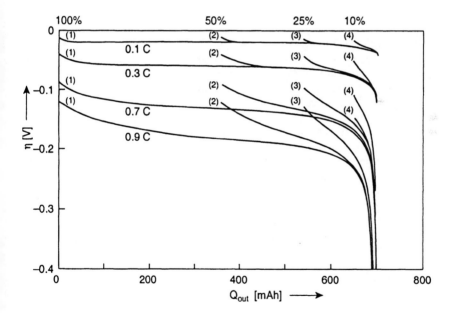

Figure 6.20: Simulated overpotentials η [V] versus charge obtained from the battery (Q_{out} [mAh]) for various discharge rates (0.1 C, 0.3 C, 0.7 C and 0.9 C) for a complete discharge (1), discharge from 50% SoC (2), discharge from 25% SoC (3) and discharge from 10% SoC (4) at 25°C. The Li-ion model described in chapter 4 was used in the simulations

Only five parameters occur in (6.4), some of which have an electrochemical meaning. All parameters can be found with the aid of curve fitting with battery discharge curves. However, the value of $R_{\Omega k}$ can also be inferred from an impedance diagram as the sum of R_{Ω} and R_{ct}; see Figure 6.4. The parameter R_d could in principle be inferred from an impedance diagram, but this is difficult in practice. The reason for this is that the shape of the impedance plot, of which the simplified version was shown in Figure 6.4, will depend on the nature of the diffusion process in practice. The measurements moreover have to be conducted at low frequencies leading to a change in the battery's condition, because the measurements take a long time. For the same reason, it is also difficult to infer parameter τ_d from an impedance plot. Parameter E_q can only be obtained by fitting with discharge curves obtained at different currents. Parameter τ_q can only be inferred from fitting (6.4) to experiments in which the current is interrupted when η_q has attained a significant value. This will lead to the occurrence of relaxation curves, which can be used to fit τ_q. All the parameters have to be derived at different temperatures.

All parameter values in (6.4) will be obtained by fitting measured complete battery discharge curves at various discharge currents in the remainder of this section. The use of impedance plots for obtaining some of the parameters will not be considered for simplicity. The measured overpotentials will be determined from the difference between discharge curves obtained at different discharge currents as an alternative to conducting tedious overpotential measurements by means of voltage relaxation. Figure 6.21 shows four experimental overpotential curves (solid). A CGR17500 Li-ion battery was discharged at various discharge rates including 0.1 C, 0.2 C, 0.4 C, 0.8 C and 1 C, at 25°C. Four differential curves were constructed, on the basis of 0.1 C as the smallest discharge rate, with the differential rates being 0.1 C (=0.2 C-0.1 C), 0.3 C, 0.7 C and 0.9 C.

The dashed curves in Figure 6.21 represent the calculated overpotentials that resulted from curve fitting of (6.4) to the measured differential overpotential curves. First, each of the four differential curves was fitted individually. The average value for the four fits was then calculated for each parameter. These average parameter values are revealed in the figure caption. The dashed curves shown in Figure 6.21 were calculated with these average parameter values. This illustrates that the fits are quite good, even with average parameter values. It shows that the overpotentials at various discharge currents can be fitted well in the case of a practical Li-ion battery with a single set of parameters. Parameter τ_q was chosen to be zero for simplicity, because its effect is negligible in the case of complete discharge curves. This can be understood from (6.4), where the exponential term in η_q has become zero already long before the term inversely proportional to Q_{in} starts to increase. This parameter becomes important when the overpotential behaviour of an almost empty battery has to be estimated and the current is often changed or interrupted in this region.

The measured differential curves shown in Figure 6.21 agree quite well with the simulated overpotential curves shown in Figure 6.20, which were obtained with the battery model described in chapter 4, in terms of order of magnitude and general shape. However, the relaxation process described by parameter τ_d in (6.4) was slower in the simulations than in the measurements. Care should be taken with the representation of the x-axis in [mAh], as explained above. The quantitative agreement between the results of the simulations and the measurements still needs improvement. It should be noted that the measurements shown in Figure 6.21 were not obtained with a high degree of accuracy. Therefore, new dedicated measurements were performed to derive the parameter values in (6.4). These measurements were performed at different temperatures, because the overpotentials depend strongly on temperature. A CGR17500 battery was discharged using two discharge currents at various temperatures. The discharge rates were 0.1 C and 0.5 C, which led to a single differential rate of 0.4 C. The resulting differential discharge curve was fitted at four temperatures of −10°C, 0°C, 25°C and 45°C. The result is shown in Figure 6.22. The parameter values obtained at different temperatures are listed in Table 6.3. Figure 6.22 shows very good fits of (6.4) with the measurements obtained at all the temperatures, especially at the moment when the overpotential begins to increase when the battery is empty. This is important, because (6.4) will be used to estimate the battery's remaining time of use.

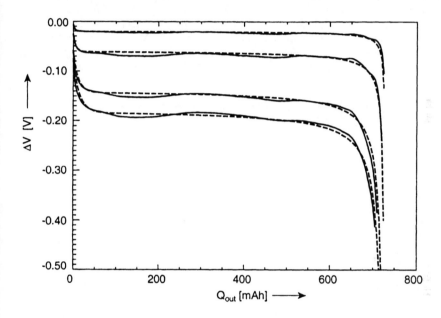

Figure 6.21: Experimental differential battery discharge curves in the case of a complete discharge obtained for differential rates 0.1 C, 0.3 C, 0.7 C and 0.9 C as a function of the charge obtained from the battery Q_{out} [mAh] at 25°C (solid curves). The dashed curves show the overpotentials calculated with the aid of (6.4) using parameter values valid at 25°C: $R_{\Omega k} = 0.11\ \Omega$, $R_d = 0.17\ \Omega$, $\tau_d = 74$ sec, $E_q = 28.8$ J/A and $\tau_q = 0$

Table 6.3 shows that $R_{\Omega k}$ decreases as the temperature increases. The kinetic part of $R_{\Omega k}$ will be larger at lower temperatures, because all electrochemical reactions proceed more slowly at lower temperatures. The ohmic part of $R_{\Omega k}$ decreases as the temperature decreases. Hence, the largest part of $R_{\Omega k}$ will be attributable to the kinetics of the reactions at −10°C. The reverse situation takes place at 45°C, where the ohmic part is dominant. The ohmic part of $R_{\Omega k}$ will be around 100 mΩ in practical batteries at 25°C, excluding the safety measures. The fits show that the increase in ohmic resistance as the temperature increases is not as high as the increase of the kinetic part as the temperature decreases, because $R_{\Omega k}$ is only 0.11 Ω at 45°C.

The diffusion coefficient increases with the temperature according to the Arrhenius relation; see (4.72). This leads to a lower diffusion overpotential, and hence to a lower value of R_d, at higher temperatures. Time constant τ_d also decreases with an increasing temperature, because it is derived from R_dC_d; see Figure 6.3. This can indeed found in Table 6.3. More energy can be obtained from the battery at higher temperatures because of the better kinetics and higher diffusion rates at higher temperatures. Therefore, E_q decreases as the temperature increases, as can indeed be seen in Table 6.3. Again, τ_q was chosen to be zero at all the temperatures. Table 6.3 shows that the parameters in Figures 6.21 and 6.22 are in the same order of magnitude at 25°C. The differences are partly due to the inaccuracy of the measurement results shown in Figure 6.21. Moreover, spread in battery behaviour has to be accounted for in parameter sets for different batteries of the same type. Many more measurements should be performed to obtain a better understanding of parameter spread between batteries. In addition, tests should be performed to investigate the influence of aging on the parameter values. This will not be considered here. The parameter values used in Figure 6.22 will be used in the remainder of this chapter.

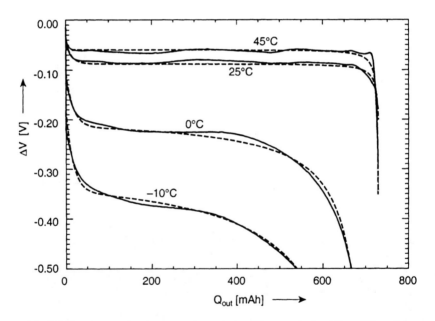

Figure 6.22: Experimental differential battery discharge curves obtained for a differential rate of 0.4 C as a function of the charge obtained from the battery Q_{out} [Ah] at $-10°C$, $0°C$, $25°C$ and $45°C$. The dashed curves show the overpotentials calculated with the aid of (6.4) using the parameter values listed in Table 6.3

Table 6.3: Values for the parameters in (6.4) for different temperatures; see Figure 6.22. The values obtained at 25°C in Figure 6.21 have been included for reference

Parameter:	–10°C	0°C	25°C	45°C	25°C (Figure 6.21)
$R_{\Omega k}$ [Ω]	0.45	0.29	0.15	0.11	0.11
R_d [Ω]	0.57	0.37	0.16	0.09	0.17
τ_d [s]	165	169	132	42	74
E_q [J/A]	468	216	7.2	7.2	28.8
τ_q [s]	0	0	0	0	0

6.5 Proposed set-up for State-of-Charge system

6.5.1 The algorithm

The results of sections 6.3 and 6.4 will be used in this section to derive a new set-up for an SoC indication system which is intended to overcome the shortcomings of the investigated bq2050 system. The proposed system is a combination of a direct-measurement system and a book-keeping system [20]. The EMF method described in section 6.3 is used for the direct measurement. The EMF curve, which is stored in the system, must be determined for the type of battery concerned using any of the methods described in section 6.3.2. The overpotential description of (6.4) is used for the book-keeping. The proposed algorithm operates in five different states:

- Initial state
- Equilibrium state
- Charge state
- Discharge state
- Transitional state

The system's estimations can be shown to the user in each state in the form of a value of SoC expressed in [%] and a remaining time of use t_{rem} available under the valid discharge conditions; see the definitions in section 6.1.1. The state diagram of the algorithm is shown in Figure 6.23.

The algorithm starts up in the *initial state*, when the battery is first connected to the SoC system. The algorithm then shifts to the appropriate state, depending on whether the battery is charged, discharged or in equilibrium. The *initial state* is re-entered only when the battery has been disconnected from and reconnected to the system. The SoC in the *initial state* is determined by measuring the battery voltage and translating this into an SoC value with the aid of the EMF curve. This is only an estimation, because there is no guarantee that the battery voltage is close to the EMF at the moment when the battery is connected to the system. However, at least an estimation of the initial SoC can be made, which is not possible with the Benchmarq system described in section 6.2.

A current smaller than or equal to I_{lim} is drawn from or flows into the battery and the battery voltage has relaxed from previous currents in the *equilibrium state*. The parameter I_{lim} is defined in the system. Its value should not be taken too large, because then the measured voltage will differ too much from the EMF. On the other hand, the *equilibrium state* will hardly be entered when the value of I_{lim} is taken too small. For example, the value of I_{lim} can be chosen a little higher than the current drawn by a cellular phone in the standby mode. The battery voltage will be close or equal to the EMF when the phone is in the standby mode and the battery voltage has relaxed from the current previously drawn in talk mode. For example, a current of 1 mA will only yield a deviation from the EMF of 300 μV at the battery terminals in the case of a realistic series resistance of 300 mΩ. The kinetic and diffusion overpotentials will be negligible under this condition. This means that in the *equilibrium state*, the SoC can be determined by means of the EMF method by simply measuring the battery voltage and translating it into an SoC value with the aid of the EMF curve. The validity of the '$|I| \leq I_{lim}$ and battery voltage stable' condition is checked before the *equilibrium state* can be entered. This is illustrated in Figure 6.23.

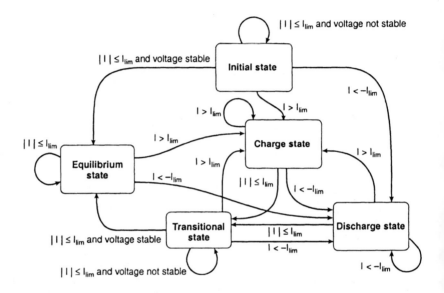

Figure 6.23: State diagram of the proposed SoC indication system

The algorithm resides in the *charge state* when the battery is charged with a positive current larger than I_{lim}. It is in the *discharge state* when the battery is discharged with a negative current larger than I_{lim} in absolute terms. The SoC is determined by coulomb counting in both states. In addition to the SoC, t_{rem} is also estimated in the *discharge state*. This will be described later in this section. A starting value of SoC for coulomb counting is always known when either the *charge* or *discharge state* is entered. This starting value is obtained with the aid of the EMF method when the state is entered from the *initial* or *equilibrium state*. Alternatively, the *charge state* can be entered from the *discharge state* and *vice versa*. In this case a starting value is also available.

The *transitional state* is entered during a change from either the *charge* or *discharge state* to the *equilibrium state*. The current is smaller than or equal to I_{lim} in this state. It is determined whether the *equilibrium state* may be entered or not, as in the *initial state*. This means that the voltage has to be stable or relaxed to allow a transition to the *equilibrium state*. The system shifts back into either the *charge* or the *discharge state* when this is not the case and an absolute current larger than I_{lim}

starts to flow again. Coulomb counting continues in the *transitional state* as long as the battery voltage has not fully relaxed.

Figure 6.24 shows the transitions between states in a practical situation. The figure was obtained in simulations using the battery model of Figure 6.3, with the complete overpotential being described by (6.4), and a measured EMF curve was entered in the model.

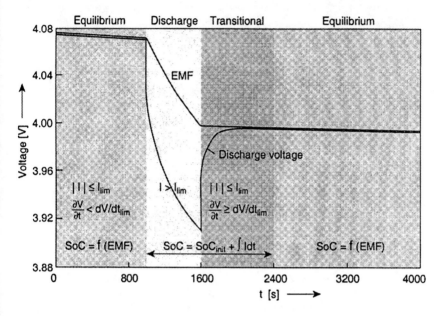

Figure 6.24: Simulated EMF (top curve [V]), discharge voltage (bottom curve [V]) and transitions between states as a function of time [s] in the proposed SoC indication system. The battery model shown in Figure 6.3 was used in the simulations, with the overpotential description according to (6.4) and an EMF curve derived from measurements

The algorithm is in the *equilibrium state* at $t=0$, because a current smaller than or equal to I_{lim} then flows out of the battery and the battery voltage is stable. The latter condition can be checked by calculating the derivative of the battery voltage over time and comparing it with the value of the parameter dV/dt_{lim}. The SoC is determined with the aid of the EMF method. A current larger than I_{lim} is drawn from the battery at $t=1000$ s and the battery voltage then drops. The algorithm shifts from the *equilibrium* to the *discharge state*. In addition to the SoC, the system also estimates t_{rem} on the basis of the SoC value in this state. The discharge current changes to a value smaller than or equal to I_{lim} at $t=1600$ s and the *transitional state* is entered. The SoC is determined by coulomb counting in both the *discharge state* and the *transitional state*. Coulomb counting continues in the *transitional state*, because a current smaller than or equal to I_{lim} still flows in this state. The battery voltage has relaxed by $t=2400$ s and the algorithm then changes to the *equilibrium state* again.

A first important aspect that must be borne in mind in the system is that the unit of SoC is [C] in coulomb counting, because the integral in time of the battery current is determined. The SoC is determined on a percentage scale in the EMF method; see section 6.3. Therefore, the content of the coulomb counter in the *charge* and *discharge state* has to be translated into a percentage scale, too. The maximum battery capacity Cap_{max} on an absolute scale is needed for this purpose. Some means of updating the value of Cap_{max} has to be taken into account to cope with capacity loss due to aging. A simple method for achieving this is illustrated in Figure 6.25.

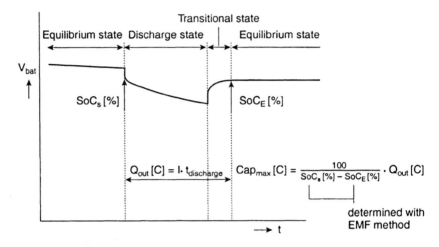

Figure 6.25: Simple method for updating Cap_{max} to take capacity loss into account

A necessary condition in Figure 6.25 is for the system to run through a sequence of states: *equilibrium state, discharge state, transitional state* and *equilibrium state*. The new value of Cap_{max} is simply calculated by relating the charge Q_{out} drawn from the battery mainly in the *discharge state* and a little in the *transitional state* to the difference in SoC (SoC_S-SoC_E) before and after discharging as derived from the EMF method. The method can be made more complex by enforcing a minimum value of Q_{out} for the update to be valid. Moreover, only small changes in Cap_{max} should be allowed. Finally, the update mechanism should be allowed to occur only under 'standard' conditions, for example a discharge rate that is not higher than 1 C and a temperature that is within the range from 10°C to 40°C. A similar updating mechanism can be implemented during charging.

A second important aspect to be borne in mind in the *discharge state* was illustrated in Figure 6.19: Cap_{rem} may differ from the SoC during discharging. Both Cap_{rem} and t_{rem} can be estimated under the prevailing discharge condition and from the present SoC value. Note that the SoC value itself will be updated continuously with the aid of coulomb counting in the *discharge state* and will remain available at all times. In the present algorithm, the time at which the voltage will drop below V_{EoD}, or EMF-η=V_{EoD} is estimated. The difference between that time and the present time yields the remaining time of use t_{rem}. To estimate the time at which EMF-η=V_{EoD}, the future overpotential development must be estimated under the prevailing discharge conditions by increasing the time and decreasing the value of Q_{in} in (6.4). This is achieved by expressing (6.4) in a differential form and by calculating the anticipated changes $d\eta$ in time dt. At the same time, the future EMF development must be estimated from Q_{in}/Cap_{max}. The estimation of future values of η and EMF can be seen as a 'fast-forward' anticipation of battery behaviour under the prevailing discharge condition. The time at which the battery voltage will drop below V_{EoD} has to be re-estimated when the discharging condition changes. This leads to an updated value of t_{rem}. The future overpotential may also be estimated for a condition which does not currently apply. It may be a nice feature to show t_{rem} under different conditions. For example, a cellular phone may display the remaining standby time, the remaining talk time under the current conditions and the remaining talk time at low temperatures [20].

6.5.2   Comparison with the bq2050 system

The proposed system can be compared with the bq2050 system; see section 6.2.

- The SoC system proposed in this section can be regarded as a book-keeping system, which is calibrated each time the battery relaxes to equilibrium. This is achieved by resetting the SoC value calculated with the aid of the book-keeping system to the value obtained with the EMF method. More calibration points will be encountered during use than is the case with the bq2050 system, which is only calibrated when the battery voltage drops below the EDV threshold during a 'valid' discharge. The periods for which coulomb counting is applied will be shorter in the case of the proposed system and hence the errors will accumulate over shorter time periods.
- The two systems' methods for expressing the difference between the SoC and Cap_{rem} during discharging are comparable. A correction factor for the coulomb counter is used in the bq2050 during discharging to calculate the CAC from the NAC register depending on the discharging conditions. Both NAC and CAC remain available to the user. The coulomb counter (NAC) reflects the SoC, while a corrected counter (CAC) shows Cap_{rem} for the current discharging conditions. In the proposed system, t_{rem} is estimated from the SoC, with both variables remaining available to the user. The way in which Cap_{rem} or t_{rem} is derived from the SoC value changes with the discharging conditions in the case of the two systems. However, a difference between the two systems concerns the method itself that is used to estimate Cap_{rem} from the SoC. Only a limited number of correction factors is used in the bq2050 system, which depend on the temperature and current. A fast-forward estimation of battery behaviour is performed for each condition in the proposed system. This yields different behaviour for a whole range of currents and temperatures.
- The updating mechanism of Cap_{max} presented in this section differs from the updating mechanism used in the bq2050. A 'valid' discharge should take place with the bq2050 system before the value of Cap_{max} (called LMD in the bq2050) can be updated. Cap_{max} can be updated more frequently in the proposed mechanism, for example every time the condition illustrated in Figure 6.25 is satisfied.
- The system proposed in this section estimates the initial SoC by itself in the initial state. The bq2050 is a pure book-keeping system, so the initial SoC has to be programmed by the user. Problems may occur, depending on the initial SoC and also on the subsequent discharging and charging conditions. This was shown in section 6.2, in which a wrong capacity update occurred on the basis of the initial condition and the subsequent discharging.
- Self-discharge is taken into account in different ways in the two systems. The coulomb counter counts down at a speed proportional to its contents and temperature in the bq2050 system. This is not necessary in the system proposed

here. The self-discharge will eventually lead to a decrease in EMF in equilibrium. As a result, the SoC will be automatically updated.

6.5.3   Comparison with systems found in the literature

System 1
The concept of using book-keeping when current flows and the EMF method when the battery voltage has stabilized has also been described for lead-acid batteries [14]. The system was applied in field tests using electric wheelchairs and yielded satisfying results. A pronounced relationship between the EMF and the SoC in [%] exists for lead-acid batteries, which is even linear. Two interesting differences with respect to the system proposed in this section can be found in the system proposed in [14].

First of all, a different correction is applied to the coulomb counter during discharging in the wheelchair system. A linear relation is assumed between the discharge current I and the battery capacity removed per sampling period $I \cdot t \cdot C$, in which C is a correction factor. The difference between Cap_{rem} estimated with the aid of book-keeping and the SoC estimated using the EMF method is determined when the battery is in equilibrium. C is modified on the basis of this difference. This means that the effects of discharging efficiency and capacity loss are combined. These two effects are separated in the system proposed here. The 'correction factor' in the form of an estimated future overpotential behaviour is updated every time the condition changes, even when the battery does not return to equilibrium. A separate updating mechanism of Cap_{max} leads to a better insight into the actual value of Cap_{max}, which can serve as battery health monitor.

A second difference concerns the transition to *equilibrium state*. The algorithm in the system proposed here simply waits until the battery voltage has stabilized. This can take a considerable time, depending on the battery's condition. An elegant approach was adopted in the wheelchair system. The EMF value is estimated from an assumed relaxation profile of the battery voltage when the discharge current is interrupted. This allows calibration without the necessity to wait until the battery voltage has stabilized. It would be interesting to further investigate the possibilities of including this in the proposed system, because an overpotential description is available. This approach was however not adopted in the experiments described in the next section.

System 2
An empirical function describing the entire battery discharge voltage is used for lead-acid batteries in [15]. The remaining time of use until V_{EoD} is reached is calculated under the current discharge conditions on the basis of this empirical function. No distinction is made between the EMF and the overpotential. It is unclear how the system behaves when no or little current flows, because the empirical function is obtained by fitting discharge curves for significant currents. The concept of using a description of battery voltage during current flow to estimate t_{rem} is similar to the system described in this section. The parameters in the empirical function in [15] are updated on the basis of a comparison of the battery voltage estimated by the empirical function and the measured battery voltage. This adaptivity is an interesting subject for further research for the system described here; the parameters in (6.4) could then be updated on the basis of measurements. This was not taken into account in the experiments described in the next section.

6.6   Experimental tests with the system proposed in section 6.5

6.6.1   Introduction

Some preliminary test results obtained with the system proposed in the previous section will be described in this section. Again, a CGR17500 battery was used. The main aim of this section is:

- to investigate the accuracy of the estimation of t_{rem} under different load conditions and at different temperatures. Moreover, the behaviour of the system when load changes occur during discharge will be investigated;
- to investigate the calibration of the book-keeping system in *equilibrium state*;
- to investigate the capacity updating mechanism proposed in Figure 6.25.

6.6.2   Set-up of the experiments

Two experiments have been performed. The battery was initialized with a standard regime before each experiment and at each temperature. This regime comprised three charge-discharge cycles, using a 500 mA current in CC mode and a 4.1 V voltage in CV mode, with a minimum current of 35 mA, in the charge regime. The battery was discharged at a 0.5 C-rate to a voltage of 2.8 V. A 30-minute rest period was allowed between the charging and discharging. Both experiments were performed at 25°C and at 0°C. This means that the initiation was performed four times.

Experiment 1
Experiments similar to those carried out with the bq2050 were carried out in the first experiment to facilitate comparison with the results presented in section 6.2. The charge regime was the same as during initiation. A rest period of 30 minutes was allowed between the charging and discharging. A rest period of 30 minutes after discharges to 2.8 V at rates higher than 0.2 C was followed by another discharge at a 0.2 C-rate to 2.8 V and another 30-minute rest period. This was done to ensure an empty battery. The battery was successively discharged at rates of 0.2 C, 1 C, 2.7 C, and with the pulse current defined in Figure 6.10. Besides for comparison with the results presented in section 6.2, the results of the first experiment also serve the first main aim mentioned above.

Experiment 2
Successive charge and discharge cycles, separated by rest periods of 30 minutes, were applied again in the second experiment. The charge cycles were the same as in the first experiment. The same remark as above holds for discharging to 2.8 V at rates larger than 0.2 C. Three different discharge regimes were successively applied:

1.   Discharge with I_1=0.2 C-rate for 60 minutes, followed by a discharge with I_2=2.7 C-rate, I_2>I_1, to 2.8 V.
2.   Discharge with I_2 for 10 minutes, followed by a discharge with I_1 to 2.8 V.
3.   Discharge at a 0.013 C-rate (10 mA) for 15 minutes, followed by a discharge at a 0.5 C-rate for 30 minutes, another discharge at a 0.013 C-rate for 30 minutes, and finally a 0.5 C-rate discharge to 2.8 V.

Discharge regimes 1 and 2 yield information on the first aim with respect to load changes. A discharge with a small current is followed by a discharge with a high current in discharge regime 1. The reverse occurs in discharge regime 2. Discharge regime 3 serves the second and third aims in this section. It leads to the situation illustrated in Figure 6.25. Therefore, Cap_{max} can be derived as discussed above. Moreover, the book-keeping part is calibrated at SoC_E. Regime 3 was repeated twice, with an intermittent full discharge at a rate of 0.1 C. The latter discharge was incorporated between the two 3 regimes to derive the real value of Cap_{max} at that moment.

The experiments were organized in time as follows. First, the order *standard regime, experiment 1, standard regime, experiment 2*, was performed at 25°C. The algorithm continuously monitored the SoC throughout this period. Then the same order was repeated at 0°C. Again, the algorithm continuously monitored the battery.

6.6.3   Experimental results

The parameters obtained in Figure 6.22 were used in (6.4); see Table 6.3. A value of 720 mAh was used for Cap_{max}, which is the nominal battery capacity. To allow fair comparison with the results obtained with the bq2050, the measured value of Cap_{max} was not used. The values of I_{lim} and dV/dt_{lim} in the algorithm were chosen to be 20 mA and 15 µV/s, respectively.

Results of experiment 1
The results of experiment 1 are summarized in Table 6.4. The remaining time of use under the prevailing discharge conditions was estimated by the algorithm at three moments during each discharge. This estimated time was compared with the actual time from that moment until the battery voltage dropped below 2.8 V for the first time. The first moment was chosen to be the moment when the discharge current was applied. The second moment was chosen somewhere halfway through the discharging, and the last moment was chosen at which either the estimated or the real remaining time of use became zero. Estimated times of less than 10 seconds were regarded as zero.

In order to allow fair comparison with the errors calculated in the case of the bq2050 in section 6.2, the errors E were calculated as follows:

$$E = \left(\frac{t_{i,estimated} - t_{i,actual}}{t_{1,actual}} \right) \cdot 100\% \tag{6.5}$$

where $i=1,2$ or 3. This means that the difference between estimation and reality is expressed relative to the full discharge time $t_{1,actual}$ with the current for which the remaining time of use is estimated. This is indeed a fair comparison with the results presented in section 6.2, because the difference between the CAC register (estimation) and the real remaining capacity was expressed relative to the LMD register in that section, with the LMD being a measure of the full capacity. The accuracy ΔE with which E can be determined will be discussed further below.

Table 6.4: Results of experiment 1: Estimated and actual remaining times of use at three moments during discharging at various discharge rates

(a) 25°C

$I_{discharge}$ [C-rate]	$t_{1,estimated}$ [s]	$t_{1,actual}$ [s]	Δt [s]	$E \pm \Delta E$ [%]	$t_{2,estimated}$ [s]	$t_{2,actual}$ [s]	Δt [s]	$E \pm \Delta E$ [%]
0.2	17694.1	18141.3	-447.2	-2.47±0.01	8160.5	8612.1	-451.6	-2.49±0.01
1	3564	3643.2	-79.3	-2.18±0.01	1702.7	1772.3	-69.6	-1.91±0.01
2.7	1244.2	1030.4	213.8	20.75±0.03	728.9	513.8	215.1	20.88±0.03
PD[1]	14805	15338.2	-533.2	-3.48±0.01	7028.6	7426.3	-397.7	-2.59±0.01

$I_{discharge}$ [C-rate]	$t_{3,estimated}$ [s]	$t_{3,actual}$ [s]	Δt [s]	$E \pm \Delta E$ [%]
0.2	0	454.7	-454.7	-2.51±0.01
1	0	80.5	-80.5	-2.21±0.01
2.7	207.9	0	207.9	20.18±0.03
PD[1]	0	526.3	-526.3	-3.43±0.01

(b) 0°C

$I_{discharge}$ [C-rate]	$t_{1,estimated}$ [s]	$t_{1,actual}$ [s]	Δt [s]	$E \pm \Delta E$ [%]	$t_{2,estimated}$ [s]	$t_{2,actual}$ [s]	Δt [s]	$E \pm \Delta E$ [%]
0.1[2]	34092.1	33539	553.1	1.65±0.01	17106.4	16568	538.4	1.61±0.01
1	3060	3194.2	-134.2	-4.20±0.02	1439	1555.9	-116.9	-3.66±0.02
2.7	115.4	88.9	26.5	29.81±0.37	69.3	42.7	26.6	29.92±0.37
PD[1]	13952.6	13877.2	75.4	0.54±0.01	6535.2	6471.3	63.9	0.46±0.01

$I_{discharge}$ [C-rate]	$t_{3,estimated}$ [s]	$t_{3,actual}$ [s]	Δt [s]	$E \pm \Delta E$ [%]
0.1	542.1	0	542.1	1.62±0.01
1	0	150.2	-150.2	-4.70±0.02
2.7	26.7	0	26.7	30.03±0.37
PD[1]	0	169.3	-169.3	-1.22±0.01

Note 1: See Figure 6.10. However, due to an error in the programming of the cycling equipment, the duty cycle of the 2A pulses was 1/80 instead of 1/8. This led to an average current of 173 mA
Note 2: The result at a 0.1 C-rate has been included in this table because of measurement problems encountered in the cycle at a rate of 0.2 C at 0°C

Accuracy considerations

On the basis of (6.5), the following expression can be derived for the accuracy ΔE with which E can be determined. The error can again be expressed by $E \pm \Delta E$:

$$\Delta E = |E| \cdot \left(\frac{\Delta t_{i,estimated}}{|t_{i,estimated} - t_{i,actual}|} + \frac{\Delta t_{i,actual}}{|t_{i,estimated} - t_{i,actual}|} + \frac{\Delta t_{1,actual}}{t_{1,actual}} \right) \qquad (6.6)$$

where $\Delta t_{i,estimated}$ and $\Delta t_{i,actual}$ are the accuracies with which the estimated and actual remaining times of use can be determined, respectively. We now have to estimate these accuracies, where $\Delta t_{i,actual} = \Delta t_{1,actual} = \Delta t_{actual}$ can be assumed. Let us first consider $\Delta t_{i,estimated}$. This time is inferred from $t_{empty} - t_p$, with the battery being empty at $t = t_{empty}$ ($V_{bat} = V_{EoD}$) and t_p being the time at which the estimation is made. The time t_{empty} was determined internally with high precision from the stored EMF curve and the overpotential function of (6.4). It has therefore been assumed that there is no measurement error in the determination of this time. The time t_p is inferred from a time point in the log file of the cycling equipment. These times have an accuracy of

±0.06 s. This means that $\Delta t_{i,estimated}$ has been assumed 0.06 s in the tables in this section.

The actual remaining time of use is also found from t_{empty}-t_p. The same remarks with respect to accuracy as discussed above hold for t_p. However, now the time t_{empty} was inferred from measurements instead of internal calculations. As discussed above, t_{empty} is the moment when the battery voltage has dropped below 2.8 V. The cycling equipment stored measured voltage values when they had changed by 2 mV. This means that in worst case, t_{empty} was determined when V_{bat}=2.8 V-2 mV. Assuming a slope of the discharging curve of 0.04 mAh/mV as before, the error expressed in [hr] can now be found from ((0.04 mAh/mV)·2mV)/I, in which I is expressed in [mA]. The total error in the actual remaining time of use Δt_{actual} expressed in [s] can now be found from (0.08 mAh/I)·3600+0.06.

Results of experiment 2
The results of experiment 2 for discharge regimes 1 and 2 are summarized in Table 6.5. Again, the accuracy of the t_{rem} estimations was determined at three moments. This time, the first moment was chosen to be the moment at which the discharge current changed, which was from I_1 to I_2 in discharge regime 1 and from I_2 to I_1 in discharge regime 2. The second moment was chosen halfway through the discharging with the changed current (I_2 in regime 1 and I_1 in regime 2). The third moment was chosen to be the moment at which either the estimated or actual time became zero, with estimated times smaller than 10 seconds again being regarded as zero.

Table 6.5: Results of experiment 2: Estimated and actual remaining times of use at three moments during discharging with discharge regime 1 and discharge regime 2

(a) 25°C

$I_{discharge}$ [C-rate]	$t_{1.estimated}$ [s]	$t_{1.actual}$ [s]	Δt [s]	E±ΔE [%]	$t_{2.estimated}$ [s]	$t_{2.actual}$ [s]	Δt [s]	E±ΔE [%]
Reg. 1: 0.2→2.7	985.1	789.5	195.6	24.78±0.04	686.1	494.3	191.8	24.29±0.04
Reg. 2: 2.7→0.2	9643.6	10106.5	-462.9	-4.58±0.02	4933.8	5425.9	-492.1	-4.87±0.02

$I_{discharge}$ [C-rate]	$t_{3.estimated}$ [s]	$t_{3.actual}$ [s]	Δt [s]	E±ΔE [%]
Reg. 1: 0.2→2.7	195.2	0	195.2	24.72±0.04
Reg. 2: 2.7→0.2	0	449.6	-449.6	-4.45±0.02

Table 6.5 (continued): Results of experiment 2: Estimated and actual remaining times of use at three moments during discharging with discharge regime 1 and discharge regime 2

(b) 0°C

$I_{discharge}$ [C-rate]	$t_{1,estimated}$ [s]	$t_{1,actual}$ [s]	Δt [s]	$E \pm \Delta E$ [%]	$t_{2,estimated}$ [s]	$t_{2,actual}$ [s]	Δt [s]	$E \pm \Delta E$ [%]
Reg. 1: $0.2 \rightarrow 2.7$	57.1	55.8	1.3	2.33±0.48	29.2	27.6	1.6	2.87±0.48
Reg. 2: $2.0^1 \rightarrow 0.2$	10835.7	10551.1	284.6	2.70±0.02	5809.6	5536.1	273.5	2.59±0.02

$I_{discharge}$ [C-rate]	$t_{3,estimated}$ [s]	$t_{3,actual}$ [s]	Δt [s]	$E \pm \Delta E$ [%]
Reg. 1: $0.2 \rightarrow 2.7$	1.8	0	1.8	3.23±0.48
Reg. 2: $2.0^1 \rightarrow 0.2$	289	0	289	2.74±0.02

Note 1: Due to measurement problems encountered in regime 2 at 0°C, an alternative cycle with a change from a 2 C-rate instead of a 2.7 C-rate was used in this table

The results of discharge regime 3 of experiment 2 are summarized in Table 6.6. The variables necessary to calculate Cap_{max} are listed for both applications of the regime. The value of Cap_{max} derived from the intermediate discharging at a rate of 0.1 C has been included for comparison. The value of SoC in *transitional state*, just before the battery entered the *equilibrium state* (SoC_t), and the first value in the equilibrium state (SoC_E) are listed for the sake of the second aim in this section. The difference between SoC_t and SoC_E illustrates the effect of calibration.

Table 6.6: Results of discharge regime 3 of part 2: SoC_S, SoC_E, Q_{out} and calculated Cap_{max} (see also Figure 6.25), Cap_{max} derived from full discharge at a rate of 0.1 C, and SoC in the transitional state just before the battery entered the equilibrium state (SoC_t)

(a) 25°C

Application of reg.3	SoC_S [%]	SoC_E [%]	Q_{out} [C]	$Cap_{max. est.}$ [mAh][1]	$Cap_{max. meas.}$ [mAh][2]	SoC_t [%]
First	100	74.3	651.2	702.9	735.1	74.9
Second	100	74.2	651.1	701.0	735.1	74.9

(b) 0°C

Application of reg.3	SoC_S [%]	SoC_E [%]	Q_{out} [C]	$Cap_{max. est.}$ [mAh][1]	$Cap_{max. meas.}$ [mAh][2]	SoC_t [%]
First	97.1	68.0	654.1	625.7	670.9	71.9
Second	93.0	65.0	655.3	650.1	670.9	67.8

Note 1: Estimated from SoC_S, SoC_E and Q_{out}; see Figure 6.25
Note 2: Measured from full discharge at a rate of 0.1 C

6.6.4 Discussion of the results

Experiment 1

Table 6.4 shows that the absolute error Δt remains constant throughout the complete discharging time. This was to be expected from the applied algorithm, because the conditions do not change from t_1, through t_2 until t_3. As a result, the time when the battery is estimated to be empty does not change, and hence the estimated times $t_{i,estimated}$ decrease with the actual remaining times of use $t_{i,actual}$. The results shown in Table 6.4 can be compared with the results obtained with the bq2050 presented in Table 6.1. This has been done in Table 6.7.

Table 6.7: Comparison of errors in empty estimation with the bq2050 and the SoC algorithm proposed in section 6.5 under identical circumstances at 25°C and similar circumstances at low temperatures (-5°C in the case of bq2050 and 0°C in the case of the algorithm presented in section 6.5)

I [C-rate]	E±ΔE of bq2050 at 25°C [%]	E±ΔE of new algorithm at t_3 at 25°C [%]	E±ΔE of bq2050 at -5°C [%]	E±ΔE of new algorithm at t_3 at 0°C [%]
0.2	-1.8±0.3	-2.51±0.01	-6.7±0.4	1.62±0.01[1]
1	0.8±1.6	-2.21±0.01	2.3±1.7	-4.70±0.02
2.7	16.7±4.2	20.18±0.03	57.8±4.8	30.03±0.37
PD	-1.3±0.9[2]	-3.43±0.01[3]	Not available	-1.22±0.01[3]

Note 1: This measurement was performed at a rate of 0.1 C instead of a rate of 0.2 C
Note 2: See Figure 6.10; this error is the average of the errors in both pulse discharges in Table 6.2
Note 3: See Figure 6.10; a duty cycle of 1/80 instead of 1/8 was used

Inspection of Table 6.7 shows that the estimations obtained with the bq2050 yield smaller errors than those obtained with the new algorithm at 25°C at rates of 1 C and smaller. Moreover, the estimation of t_{rem} obtained with the new system has not improved at the large discharge rate of 2.7 C compared to the Cap_{rem} estimations obtained with the bq2050. However, the error in the estimation obtained with the new algorithm is still lower than 5% and also negative, i.e. pessimistic, at rates of 1 C and lower encountered in practical portable devices. This will usually be acceptable. It should also be noted that better results will be obtained with the proposed algorithm once the real value of Cap_{max} of 735 mAh (see Table 6.6) has been introduced instead of the nominal capacity of 720 mAh. The estimations at low temperatures and large discharge currents obtained with the new algorithm are considerably better, although the estimation error is still positive and large, hence annoying for a user in practice. The error in the estimation is again smaller than 5% at discharge rates of 1 C and lower.

Experiment 2
The results presented in Table 6.5 illustrate the effect of previous discharge currents on the estimations when the current is changed during discharging. Some of the results given in Tables 6.5 and 6.6 have been included in Table 6.8 to further clarify this effect.

Table 6.8: Errors in estimated remaining time of use for full (Table 6.4) and partial (Table 6.5) discharges at rates of 0.2 C and 2.7 C at 25°C and 0°C

Temperature [°C]	E±ΔE of partial discharge at 0.2 C-rate [%][1]	E±ΔE of full discharge at 0.2 C-rate [%][2]	E±ΔE of partial discharge at 2.7 C-rate [%][1]	E±ΔE of full discharge at 2.7 C-rate [%][2]
25	-4.45±0.02	-2.51±0.01	24.72±0.04	20.18±0.03
0	2.74±0.02	1.62±0.01[3]	3.23±0.48	30.03±0.37

Note 1: See Table 6.5
Note 2: See Table 6.4
Note 3: Discharge rate of 0.1 C instead of 0.2 C

Table 6.8 shows that the error in the estimation increases somewhat, but not dramatically, for partial discharges using a certain current. For example, the error is -2.51% when the battery is completely discharged at a rate of 0.2 C at 25°C. This error increases to -4.45% for a partial discharge at a rate of 0.2 C when the battery has first been discharged at a large rate of 2.7 C for 10 minutes. This error is hence still pessimistic and smaller than 5%. This inspires greater confidence in the applied approach of overpotential estimation, because apparently the overpotential relaxation resulting from the application of the 2.7 C discharge rate is taken into account correctly. The estimation of t_{rem} for a partial discharge at a rate of 2.7 C at 0°C is much better than that at full discharge. More experiments will have to be carried out before a definite conclusion can be drawn concerning this improved accuracy for partial discharges with large currents at low temperatures.

The results shown in Table 6.6 show that the capacity updating mechanism proposed in Figure 6.25 yields estimations of Cap_{max} that are lower than the actual value. This can be understood from the fact that SoC_E is lower than SoC_t in all cases. The battery was in a state of equilibrium when the first discharge at a rate of 0.5 C was started. This means that SoC_S was based on the EMF method. The charge Q_{out} represents the charge that was drawn from the battery when the battery was in the *discharge state* and subsequently in the *transitional state*. The latter state was entered when the discharge current at a 0.5 C-rate was interrupted after 30 minutes. The SoC was updated by means of coulomb counting in these two states. A value of 720 mAh was used for Cap_{max} to translate the absolute values of SoC, calculated by coulomb counting, into a relative SoC in [%]. This value can be recalculated from Table 6.6 by relating the difference between SoC_S and SoC_t to Q_{out}, which means that SoC_t replaces SoC_E in the capacity updating equation given in Figure 6.25. The battery voltage relaxed and the current of 10 mA was lower than I_{lim} (20 mA) in the period of 30 minutes after the first application of the 0.5 C discharge rate. The *equilibrium state* was re-entered during this period after the change in battery voltage in time dropped below 15 µV/s. The value of SoC_E was the first SoC value that was again based on the EMF method. This value was somewhat lower than SoC_t in all cases, because the battery voltage had not fully relaxed. This effect was more pronounced at low temperatures, because the overpotential that had to relax was larger ($R_{\Omega k}$ and R_d were larger) and therefore also the associated time constants were larger. Therefore, the difference between SoC_t and SoC_E is larger at 0°C in Table 6.6. As a result, the difference between the estimated and measured Cap_{max} value was larger at 0°C; see the capacity updating equation given in Figure 6.25. However, the fact that Cap_{max} is lower at lower temperatures is indeed revealed by the algorithm. An improvement in the conditions under which the *equilibrium state* may be re-entered should lead to more accurate estimations of Cap_{max}.

The effect of calibration is revealed by the difference between SoC_t and SoC_E in Table 6.6. The calibration is not fully accurate, because the battery voltage has not fully relaxed. The value of SoC increased a little further once the *equilibrium state* was re-entered in the experiments, because the battery voltage relaxed further. Due to the relatively long rest periods of 30 minutes applied in the experiments, the effect of this 'premature' calibration on the accuracy of the estimations was limited. The effect cannot be seen at all throughout the *equilibrium state*, when the battery voltage has fully relaxed until the next *charge* or *discharge state* is entered.

6.6.5 Conclusions of the experiments

The main aim of this section was to investigate the accuracy of the estimations of remaining time of use under various conditions, including changing load conditions. In addition, the proposed capacity updating mechanism was investigated. No definite conclusions can be drawn, because only few experiments have been carried out.

The estimation errors obtained with the new algorithm are only slightly larger than those obtained with the existing bq2050 system. The estimation for discharges with large currents at low temperatures is better with the new system than with the bq2050. Especially the estimation for a partial discharge at a 2.7 C-rate at 0°C has an acceptable accuracy. The accuracy of estimations at large currents at room temperatures is not better than that obtained with the bq2050 system and is still too low. This can be improved by investigating the fit of (6.4) with measurements with large discharge currents. The parameters for (6.4) that were used in the experiments were derived only at discharge rates smaller than 1 C. The new system works well when load changes occur, because the differences in the accuracy of estimations for full and partial discharges are limited. The capacity updating mechanism needs improvement, which should aim mainly at improving the conditions under which the *equilibrium state* may be re-entered. This will also have a beneficial effect on the system's calibration. However, problems with the capacity updating mechanism of the kind described for the bq2050 system in section 6.2 cannot occur with the new system.

The expected advantages of the new system over the bq2050 were listed in section 6.5.2 and included (i) improved accuracy due to more calibration points, (ii) an improved capacity updating mechanism and (iii) a realistic estimation of the initial SoC. Unfortunately, it is not possible to draw solid conclusions with respect to these possible advantages on the basis of the few experiments. Experiments should be performed with both systems under many more conditions than presented in this book to quantify these advantages in practice. Both systems should eventually be tested in field tests.

6.7   Conclusions

The subject of SoC indication has been discussed in this chapter. Three basic methods for SoC indication were identified in sections 6.1 and 6.2 and their main characteristics and problems have been discussed. The main problem involved in direct-measurement systems is the inclusion of all possible battery and usage conditions in the function f_i^d, which links the measured battery variable to the SoC. The main problem in book-keeping systems is to define reliable calibration opportunities that occur often enough during the battery's use.

The focus in sections 6.3 and 6.4 was on Li-ion batteries. Two usable phenomena for SoC indication were identified. First of all, the EMF method was discussed as a possible candidate for direct measurement. Measurements and simulations show that the EMF curve does not depend on many parameters. It is not dependent on aging of the battery and its dependence on temperature is limited. Secondly, a simple way of describing overpotentials has been derived from simulations, which can be used to determine the remaining capacity under various discharge conditions. The battery model described in chapter 4 greatly enhances understanding of both the EMF method and the overpotential description. The simulations show good agreement with practical conditions. Hence, the use of the battery models described in chapter 4 in designing Battery Management Systems has again been proven useful.

A new SoC indication system was proposed and discussed in sections 6.5 and 6.6, in which information presented in the previous sections was applied. The system is a combination of direct measurement by means of the EMF method and book-keeping using a simple description of overpotentials. The book-keeping part is calibrated through direct measurement in the form of the EMF method. Hence, the benefits of both methods are combined. The results of some preliminary tests performed with the new system have been discussed. The accuracy of estimations under various conditions, including load changes, is satisfactory. In comparison with the results obtained with the existing bq2050 book-keeping system, mainly the estimations at large discharge currents and low temperatures are better with the new system. However, the estimations obtained with the bq2050 system at moderate currents and at room temperature are better than those obtained with the proposed system.

Many more tests will have to be carried out in the future to investigate the new system in greater detail. For example, tests with partial charges and discharges with an almost empty battery should be performed. The parameter τ_q in (6.4) should be given a value for this purpose. The results of the capacity updating mechanism will improve when a more accurate EMF estimation can be made on re-entry of the *equilibrium state*. Apart from experimenting with different values of dV/dt_{lim}, one should also consider the approach adopted in [14], where the EMF value is estimated from the voltage relaxation curve. An improvement in this area will have a positive effect not only on the capacity updating mechanism, but also on the accuracy of the system's calibrations. Finally, adding adaptivity to the system to update the parameters in (6.4) should be considered. This adaptivity should improve the system's capability of coping with aging effects and parameter spread for different batteries of the same type.

6.8   References

[1]   D. Linden, *Handbook of batteries*, Second edition, McGraw-Hill, New York, 1995

[2]   G.M. Barrow, *Physical Chemistry*, Fourth Edition, McGraw-Hill, Tokyo, Japan, 1979

[3]   Sony Infolithium webpage: http://www.sel.sony.com/SEL/consumer/ handycam/ worry-free/infolithium.html

[4]   F. Huet, "A Review of Impedance Measurements for Determination of the State-of-Charge or State-of-Health of Secondary Batteries", *J. Power Sources*, vol. 70, no. 1, pp. 59-69, January 1998

[5]   A.J. Bard, L.R. Faulkner, *Electrochemical Methods: Fundamentals and Applications*, John Wiley&Sons, New York, 1980

[6]   W.S. Kruijt, H.J. Bergveld, P.H.L. Notten, *Electronic-Network Model of a Rechargeable Li-ion Battery*, Nat.Lab. Technical Note 265/98, Philips Internal Report, October 1998

[7]   M. Kim, E. Hwang, "Monitoring the Battery Status for Photovoltaic Systems", *J. Power Sources*, vol. 64, no. 1, pp. 193-196, January 1997

[8]   W. Cantor, "Monitoring Lead-Acid Batteries in UPS Systems", *Electronic Design*, pp. 123-128, January 12, 1998

[9]   J. Alzieu, H. Smimite, C. Glaize, "Improvement of Intelligent Battery Controller: State-of-Charge Indicator and Associated Functions", *J. Power Sources*, vol. 67, no. 1&2, pp. 157-161, July/August 1997

[10] Benchmarq Microelectronics, bq2050 Lithium-Ion Power Gauge IC, *Datasheet*, June 1995

[11] P.H.L. Notten, H.J. Bergveld, W.S. Kruijt, "Battery Management System and Battery Simulator", *Patent US6016047, WO9822830, EP0880710*, filed 11 October 1997

[12] C.E. Barbier, H.L. Meyer, B. Nogarede, S. Bensaoud, "A Battery State-of-Charge Indicator for Electric Vehicle", *Proc. of the Institution of Mechanical Engineers*, Automotive Electronics, ImechE 142, pp. 29-34, May 1994

[13] P. Lürkens, W. Steffens, "Ladezustandsschätzung von Bleibatterien mit Hilfe des Kalman-Filters", *etzArchiv*, Bd. 8, H. 7, pp. 231-235, 1986

[14] J.H. Aylor, A. Thieme, B.W. Johnson, "A Battery State-of-Charge Indicator for Electric Wheelchairs", *IEEE Trans. on industrial electronics*, vol. 39, no.5, pp. 398-409, October 1992

[15] A.M. Pesco, R.V. Biagetti, R.S. Chidamber, C.R. Venkatram, "An Adaptive Battery Reserve Time Prediction Algorithm", *Proc. of the 11th international telecommunications energy conference*, Session 6.1, pp. 1-7, Firenze, Italy, October 15-18, 1989

[16] O. Gerard, J.N. Patillon, F. d'Alche-Buc, "Neural Network Adaptive Modelling of Battery Discharge', *Lecture Notes in Computer Science*, vol. 1327, pp. 1095-1100, 1997

[17] F.A.C.M. Schoofs, W.S. Kruijt, R.E.F. Einerhand, S.A.C. Hanneman, H.J. Bergveld, "Method of and Device for Determining the Charge Condition of a Battery", *Patent W0062086, EP0002787*, filed 29 March 2000

[18] J. Molenda, A. Stoklosa, T. Bak, "Modification in the Electronic Structure of Cobalt Bronze Li_xCoO_2 and the Resulting Electrochemical Properties", *Solid State Ionics*, vol. 36, no. 1&2, pp. 53-58, October 1989

[19] R.E.F. Einerhand, S.A.C. Hanneman, *State-of-Charge Indication Method: Li-ion Batteries*, Nat.Lab. Technical Note 190/98, Philips Internal Report, September 1998

[20] H.J. Bergveld, H. Feil, J.R.G. van Beek, "A Method of Predicting the State-of-Charge as well as the Use Time Left of a Rechargeable Battery", *Patent application PHNL000675EPP*, filed 30 November 2000

Chapter 7
Optimum supply strategies for Power Amplifiers in cellular phones

The market for cellular phones is growing rapidly and cellular phones rank among the most prominent portable devices. All these phones are battery-powered and long use times on a single battery charge are required. The Power Amplifier (PA) draws most of the current from the battery in talk mode. Therefore, a reduction in this current will lead to a significant increase in talk time. This chapter covers powering the PA with the optimum supply voltage for each output power, which leads to a decrease in supply current. As described in chapter 1, this is one of the basic tasks of a Battery Management System.

Section 7.1 gives some basic information on cellular systems and describes the trends in these systems. These trends lead to an increased desirability of reductions in the PA supply current. The concept of efficiency control enabling such reductions in supply current will be introduced in section 7.2. A DC/DC converter is needed for efficiency control. Different ways of translating one DC voltage into another DC voltage will be discussed in section 7.3. Simulation models will be developed for a DC/DC converter and a PA in section 7.4. Simulations can be performed with these models and the Li-ion battery model developed in chapter 4 to define the theoretical benefits of efficiency control. These simulations will be described in section 7.5. Section 7.6 discusses PA measurements to verify the theoretical benefits derived in section 7.5. The battery model described in chapter 4 will be used to estimate talk time improvements in a complete cellular phone. Results of measurements of applying efficiency control in a complete cellular phone will be discussed in section 7.7. Conclusions will be drawn in section 7.8.

7.1 Trends in cellular systems

Cellular phones are used in cellular systems [1],[2], which form part of the large group of wireless systems. A cellular system is organized in the form of cells, which complement each other and each cover a certain geographical area. One base station is present in each cell, which can accommodate several handsets. Information to and from these handsets is transmitted in the form of modulated signals. The base station controls the links with the various handsets. Communication from handset to base station is referred to as the up-link, while communication in the other direction is referred to as the down-link.

Many different types of cellular systems exist nowadays. Each cellular system is characterized by the employed frequency band, the applied modulation technique and by the way in which the available channels are distributed over the handsets. The frequency band is divided into several channels, whose number and width depend on the cellular system. Each handset is allowed to use a certain channel. Data is transmitted by means of signals, which are modulated around the centre channel frequency. Most of the current cellular systems are based on digital modulation techniques, as opposed to conventional analogue modulation techniques, such as AM (Amplitude Modulation) and FM (Frequency Modulation). The main reason for using digital modulation techniques is the greater channel capacity and signal integrity in comparison with that available with analogue techniques. This increases the quality of the established links between the handsets and the base station. Moreover, digital modulation provides compatibility with digital data services and higher data security [1]-[3].

Data in digital modulation techniques is represented by one- or multiple-bit symbols. These symbols are transmitted by modulating a carrier, with, for example, the amplitude and phase of the transmitted signal reflecting the symbol's value. Examples of digital modulation techniques include QPSK (Quadrature Phase Shift Keying), FSK (Frequency Shift Keying), MSK (Minimum Shift Keying) and QAM (Quadrature Amplitude Modulation). A detailed description of digital modulation methods is beyond the scope of this book. A comprehensive introduction to digital modulation can be found in [3].

Channel allocation to separate the different users of the available spectrum is organized by means of multiplexing methods in all communication systems. In cellular systems, multiplexing is already achieved by geographical separation. Two handsets are allowed to transmit in the same channel when they are present in different cells, because of the inherent geographical distance between them. Apart from geographical multiplexing as applied in cellular systems multiplexing methods based on frequency, time and code are also applied in communication systems. These multiplexing methods are denoted as access methods. The three main access methods are FDMA (Frequency-Division Multiple Access), TDMA (Time-Division Multiple Access) and CDMA (Code-Division Multiple Access) [1]-[3].

Each transmitter and each receiver uses a different frequency in FDMA systems. Therefore, each transmitter or receiver is identifiable by the frequency it uses. In TDMA systems, data is communicated in bursts, which are separated in time. The minimum time unit is a time slot. When handset A transmits data in a certain channel in time slot N, handset B can transmit in the same channel in time slot $N+1$. As a result, each channel is shared by several handsets and each handset is identifiable by the time at which it sends or receives data. Sharing of the same channel by various handsets is realized in a different way in CDMA systems. Here, several handsets transmit their data in the same channel simultaneously and a unique code is added to each signal. The receiving party correlates the received signal with the unique code of the other party, with whom communication is in progress. As a result, only the signal of interest is derived from the received signal. In this case, the added code enables identification of transmitters and receivers.

In addition to the access methods described above, two different duplex methods are available. The transmitter and the receiver are multiplexed on the same frequency in TDD (Time Division Duplex) systems, as opposed to TDMA systems, in which several transmitters or receivers are multiplexed. A simple example is a two-way radio, with which one has to press a button to talk and release it to listen. The transmitter and receiver use separate frequencies or frequency bands in FDD (Frequency Division Duplex) systems. Some examples of cellular systems available on today's market are summarized in Table 7.1.

A simplified transmit path from base-band to antenna in a cellular phone is shown in Figure 7.1. The information that has to be transmitted originates in the base-band part in the form of bits. These bits are translated into a modulated data stream, which is up-converted to the channel frequency f_c by a Radio-Frequency (RF) mixer. The PA amplifies this signal to the antenna, where some means of filtering and impedance transformation is present between the output of the PA and the antenna. This leads to a loss of power between the PA and the antenna, which is usually referred to as insertion loss. Important issues in Figure 7.1 within the scope of this book are power control and the demands on signal quality at the antenna.

Table 7.1: Overview of cellular systems on today's cellular telephony market [3]

	GSM900	NADC	CDMA	DCS1800
Geography	Europe	North America	North America, Korea, Japan	Europe
Introduction	1992	1992	1995-1997	1993
Frequency Range	935-960 MHz down 890-915 MHz up	869-894 MHz down 824-849 MHz up	824-849 MHz (down, US) 869-894 MHz (up, US) 832-834, 843-846, 860-870 MHz (down, Japan) 887-889, 898-901, 915-925 MHz (up, Japan)	1710-1785 MHz down 1805-1880 MHz up
Access Method	TDMA/ FDMA	TDMA/ FDMA	CDMA	TDMA/ FDMA
Duplex Method	FDD	FDD	FDD	FDD
Modulation	0.3 GMSK (1 bit/symbol)	π/4 DQPSK (2 bits/symbol)	Handset: QPSK Base station: OQPSK (1 bit/symbol)	0.3 GMSK (1 bit/symbol)
Mobile output power	3.7 mW-2 W	2.2 mW-6 W	10 nW-1 W	250 mW-2 W
Channel spacing	200 kHz	30 kHz	1.23 MHz	200 kHz
Number of channels	124 freq. ch., 8 timeslots per ch. (992)	832 freq. ch., 3 users per ch. (2496)	20 (US), 12 (Japan)	3000-6000

The base station orders the handset to transmit at a certain output power level for power control. The control range and the step size depend on the cellular system. Table 7.1 shows different output power ranges for different cellular systems. Moreover, the power control is organized in the form of a protocol, which also depends on the cellular system. This protocol specifies the organization in time of the output power changes. For example, it specifies how often output power changes may occur, and how long a transition from one output power to another takes. An example for GSM (Global System for Mobile communication) will be given in section 7.7. The output power of the PA has to be ramped up and down during a transmit burst in TDMA systems, such as GSM, according to a specified power-time template. This is referred to as the ramping specification, which prevents unwanted signals in other parts of the frequency spectrum.

Power control is realized in a closed loop in most cellular phones, because of the stringent demands made on it. This is shown in Figure 7.1, where the measured output power P_{meas} is forced to be equal to the reference output power P_{ref}, dictated by the base-band part, in a closed control loop. The base-band part obtains information on the required value of P_{ref} from the base station. The bandwidths of these loops can be up to 1 MHz, because the ramping times in current cellular systems are in the range of tens of microseconds. The bandwidth and temperature stability of this loop have to be high enough to be able to control the output power to the right level, irrespective of PA tolerances and temperature variations.

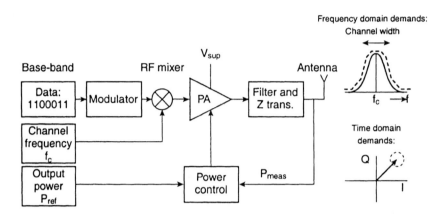

Figure 7.1: Simplified transmit path in a cellular phone from base-band to antenna

Stringent demands apply to the quality of the signal transmitted at the antenna. The demands in the frequency domain limit the power transmitted in neighbouring channels and further away from the carrier frequency f_c. This prevents violation of communication outside the own frequency channel. It is implemented by a spectral mask for the neighbouring or adjacent channels, which defines the allowed Adjacent Channel Power Ratio (ACPR). An example is shown in the top right part of Figure 7.1. A similar spectral mask exists over a much wider frequency range, which defines how much power may be transmitted at distances further away from f_c, for example in frequency bands of other cellular systems.

Similar quality demands apply in the time domain on the basis of a maximum allowed Bit Error Rate (BER) for each cellular system. The receiver on the other side may not be able to distil the correct symbols when a transmitter violates the demands on the signal quality in the time domain, and the BER may be too high. An example of a demand specified in the time domain is the vector error, which is the allowed deviation in amplitude and phase of a vector that represents a certain symbol in the I/Q plane [3]. This is shown in the bottom right part of Figure 7.1. Several trends can be distinguished in cellular systems, which are summarized below:

Trend 1: Higher ON/OFF ratio of the PA

The first trend is towards a higher ON/OFF ratio of the PA. In the European GSM system, the ON/OFF ratio increases from 1/8 in the case of conventional GSM system to 3/8 in the case of newer generations of the system, such as the EDGE (Enhanced Data rates for GSM Evolution) system. This means that three out of eight instead of one out of eight time slots are used for data transmission. The third generation (3G) UMTS (Universal Mobile Telecommunication System) will be based on the CDMA principle, with the PA always being ON. The PA current consumption will have a greater impact on the talk time when it is ON more often.

Trend 2: Non-constant envelope modulation techniques

The second trend is towards non-constant envelope digital modulation techniques. The reason for this is the more efficient use of the available spectrum for these techniques, which is needed to accommodate the increasing number of phones and their increasing functionality. For example, the conventional GSM system uses GMSK (Gaussian Minimum Shift Keying); see Table 7.1. This is a constant-envelope modulation method with which the amplitude of the modulated signal remains constant. 8PSK modulation will be applied for the EDGE system, leading to a non-constant envelope signal. The 3G UMTS system will use QPSK modulation, which is also a non-constant envelope modulation method due to the raised cosine filtering for band compression.

The transmit path has to be linear with non-constant envelope modulation, because information is present in the amplitude of the transmitted signal. A simple way of increasing the linearity of a PA is back off its output power or to drive the PA in a less efficient but more linear class, such as class A. This leads to an increase in the PA's current consumption, and hence to a decrease in talk time, as will be shown in later sections.

Trend 3: Lower average output power of handsets

A third trend is towards a lower average output power, because of the increasing number of cells and the associated decrease in cell size. Each cellular system is characterized by a certain maximum output power, as was shown in Table 7.1. The PAs are designed to be able to transmit at this maximum output power. The PA is only used optimally at this maximum output power, as will be shown in the next section. Assuming a fixed supply voltage, the PA will draw significantly more current than is necessary at lower output powers. Hence, although a decrease in the average output power transmitted by the handsets leads to less overall power consumption, the PA is no longer used optimally and draws more current than is necessary.

As will be clear from the three trends described above, a reduction of a PA's supply current is becoming increasingly important. One possible way of achieving this will be described in the next section.

7.2   The efficiency control concept

The efficiency of a PA (η_{PA}) is in this book defined as:

$$\eta_{PA} = \frac{P_{out}}{P_{sup}} = \frac{P_{out}}{V_{sup}I_{sup}} \qquad (7.1)$$

where P_{out} is the radiated output power and P_{sup} is the power drawn at the supply pin to enable the amplification. An increase in the PA efficiency for a certain output power P_{out} leads to the desired reduction in the PA supply current I_{sup} drawn from the supply voltage V_{sup} as illustrated by (7.1).

Voltage V_{sup} is either the battery voltage or the output voltage of a DC/DC converter in present-day cellular phones. In the latter case, the supply voltage of the

PA remains constant during the discharging of the battery. In the former case, the supply voltage of the PA varies during the discharging, according to the battery discharge curve. An optimum supply voltage leading to an optimum efficiency exists for each output power. This will be discussed later in this section. The efficiency control concept [4] involves the use of this optimum PA supply voltage for each output power. This ensures that the PA is always used at the maximum possible efficiency. For a better understanding, the next section gives some more detailed information on PAs and their efficiency.

7.2.1   Basic information on Power Amplifiers

A simple representation of a PA is shown in Figure 7.2 [5],[6]. This simple representation holds for the PA classes (A, AB, B and C) considered in this book. The PA is represented by its final stage only. Several stages will be used in series in a real PA to ensure the proper power amplification. The transistor shown is bipolar, but FET devices are also used in practice. When the collector voltage of the transistor approaches 0 V saturation will introduce non-linear effects. The DC bias network at the base of the illustrated transistor has been omitted for simplicity. The RF signals are represented by simple non-modulated sine waves with frequency f_{RF}. The LC output filter formed by C_{block}, C_{filter} and L_{filter} resonates at f_{RF} and is assumed to be ideal, which means that its quality factor Q is so high that the voltage across C_{filter} and L_{filter} is always sinusoidal with frequency f_{RF}, even if the current through it is non-sinusoidal. Non-sinusoidal currents will arise from distortion due to the large-signal behaviour of the transistor. The assumption of a high Q implies narrow-band operation [5]. Finally, the resistor $R_{antenna}$ represents the antenna load.

Coil L_{bias}, often referred to as 'Big Fat Inductor', is a short-circuit in the case of DC current and hence the DC voltage at the collector of the transistor equals V_{sup}. L_{bias} is an open-circuit in the case of AC (RF) signals, which means that the actual voltage at the collector can be higher than V_{sup}. The maximum amplitude of the AC signal at the collector is V_{sup}, because the minimum voltage at the collector is 0 V and the DC voltage is V_{sup}. Therefore, the maximum voltage at the collector is $2V_{sup}$. The blocking capacitor C_{block} prevents DC power consumption in $R_{antenna}$ and is a short-circuit for RF signals. The ideal LC filter suppresses higher harmonics of the input signal, which are caused by distortion as discussed earlier. As a result, only the signal with the fundamental RF frequency f_{RF} is present across $R_{antenna}$.

A well-known theorem states that maximum power transfer occurs when the load impedance matches the source impedance. The efficiency will be only 50% in this case. Such an efficiency is unacceptable for the output power levels encountered in cellular phones. Therefore, PAs are generally designed to supply a specified output power to a load with the highest efficiency possible [5]. Hence, the objective is to obtain maximum output power, not maximum power gain. Assume that the load resistance observed at the collector in Figure 7.2 is R_{load}. Maximum output power occurs when a full signal swing is present across R_{load}, which is a sinusoidal voltage with an amplitude V_{sup}. Therefore, maximum power is found from:

$$P_{out,max} = \frac{V_{sup}^2}{2R_{load}}$$
(7.2)

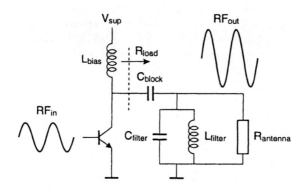

Figure 7.2: Simple representation of a PA [5],[6]

The antenna impedance $R_{antenna}$ is fixed at 50 Ω in most cellular phones. The maximum output power at the antenna in, for example, a GSM phone is 2 W; see Table 7.1. For a typical supply voltage of 3.6 V, which results from a direct connection of the PA supply pin to a Li-ion battery, the load at the collector R_{load} has to be 3.24 Ω to achieve the maximum required output power of 2 W; see (7.2). Therefore, an impedance transformation from 3.24 Ω to 50 Ω is needed. This is realized by the LC-filter in Figure 7.2. C_{block} resonates with L_{filter} at the signal frequency f_{RF}, while C_{filter} and L_{filter} resonate at $3f_{RF}$, i.e. the third harmonic of the input frequency. Series resonance of C_{block} and L_{filter} will lead to the necessarry low impedance R_{load} of 3.24 Ω at f_{RF}. The parallel resonance of C_{filter} and L_{filter} leads to the high impedance of 50 Ω seen at R_{load}. This leads to a suppression of the third harmonic in the output signal. An impedance transformation network was also shown in Figure 7.1. The impedance transformation network in Figure 7.2 has been assumed lossless. However, due to series resistances of the inductors and capacitors, a power loss of at least 1 dB will occur in practice. The PA has to transmit at least 1 dB more than the desired output power at the antenna, because the output power at the antenna is specified in the system.

Four classic PA classes are encountered in the literature [5]-[7], notably class A, class AB, class B and class C. The difference between these classes arises from the bias conditions. The transistor conducts 100% of the time in class A operation, it only conducts 50% of the time in class B operation, between 50% and 100% of the time in class AB operation and less than 50% of the time in class C operation. As a result, the class C PA is the most efficient, followed by the class B and the class AB PAs. The class A PA is the least efficient. This can be understood by considering that power is dissipated in the transistor when current flows while the collector voltage is larger than zero. The basic equation for the collector current for all classes is given by:

$$I_C(t) = \max\left(I_{DC} + i_{RF}\sin(\omega_{RF}t),0\right) \tag{7.3}$$

where I_{DC} is the current through L_{bias} and $i_{RF}sin(\omega_{RF}t)$ is the current through C_{block}. The different classes depend on the value of I_{DC}, while I_C remains greater than or equal to zero. The collector currents of the different PA classes are depicted in Figure 7.3.

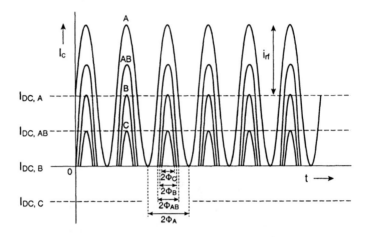

Figure 7.3: Collector currents of different PA classes

The conduction angle 2Φ is shown in Figure 7.3 for all classes and for an arbitrary output power. The maximum theoretical efficiency $\eta_{max,theory}$ occurs at maximum output power $P_{out,max}$ for all classes and can be calculated using (7.1) [5], where I_{sup} is the DC content of the collector current I_C. The output power follows from the fundamental component of the current i_c, because this fundamental component is the only current allowed to flow in $R_{antenna}$. The maximum theoretical efficiency $\eta_{max,theory}$ is then found to be:

$$\eta_{max,theory} = \frac{2\Phi - \sin(2\Phi)}{4(\sin(\Phi) - \Phi\cos(\Phi))} \tag{7.4}$$

In practice, the maximum attainable efficiency η_{max} at maximum output power $P_{out,max}$ will be somewhat lower than the theoretical maximum efficiency $\eta_{max,theory}$. This is amongst others caused by non-zero minimum collector voltage and inevitable losses in interconnect and filter components. A summary of the PA classes A, AB, B and C is given in Table 7.2.

Table 7.2: Summary of PA classes A-C: Conduction angle 2Φ [rad], maximum theoretical efficiency $\eta_{max,theory}$ [%] and linearity class

Class:	2Φ [rad]:	$\eta_{max,theory}$ [%]:	Linearity class
A	2π	50	++
AB	$\pi < 2\Phi < 2\pi$	$50 < \eta_{max} < 78.5$	+
B	π	78.5	-
C	$2\Phi < \pi$	$78.5 < \eta_{max} < 100^1$	$-^2$

Note 1: The maximum efficiency of 100% for class C occurs at zero output power
Note 2: Class C PAs are not often encountered in practice because they demand substantial filtering due to their non-linearity. This leads to an expensive system solution

The efficiency will be lower than η_{max} at output powers lower than $P_{out,max}$. For each of the classes in Table 7.2, the efficiency expressed in P_{out}, V_{sup} and η_{max} is given by:

$$\eta_{PA} = \frac{\sqrt{2P_{out}R_{load}}}{V_{sup}} \times \eta_{max} \qquad (7.5)$$

Substitution of (7.2) into (7.5) shows that the efficiency is indeed η_{max} at $P_{out}=P_{out,max}$. The value of η_{max} can be found from PA data sheets and has an upper limit of $\eta_{max,theory}$ for each PA class. Moreover, the efficiency is proportional to the root of the output power P_{out}. In practice, the value of V_{sup} will be chosen such that, given the value of the transformed antenna load impedance observed at the collector (R_{load}), maximum output power can be transmitted; see (7.2). The efficiency will be lower than η_{max} when V_{sup} has the same value at output powers lower than $P_{out,max}$. This can be understood from the way in which power is controlled in practice. Either the bias of the PA is changed or a linear regulator in the supply line ensures a lower voltage at the collector/drain of the output transistor. The difference between these two power control methods is illustrated in Figure 7.4. The voltage V_{con} denotes the control voltage needed to establish a variation in output power. Only the final stages of the PAs have been depicted for simplicity.

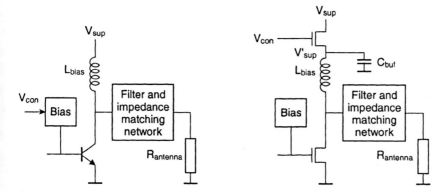

Figure 7.4: Two different approaches to power control in PAs: (a) bias control for bipolar PAs, (b) drain control by means of a linear regulator for FET PAs

The bias control method shown in Figure 7.4a is used for bipolar PAs. The result is an output voltage with less than full swing for output powers lower than $P_{out,max}$. This leads to an efficiency lower than η_{max}, because η_{max} is only attained at full swing. The power control method with the linear regulator is used for FET PAs. In that case, a FET switch has to be present in the supply line, as indicated in Figure 7.4b, to switch the device off completely, because most FET devices are 'normally-on'. Therefore, power control is realized on the drain-side of the FET by using the switch as a linear regulator. This results in a voltage V'_{sup} lower than V_{sup}, as indicated in the figure, and hence a lower output power. Although the voltage across R_{load} observed at the input of the impedance matching network will in this case have an amplitude of V'_{sup}, which is full swing, the efficiency is still lower than maximum. The reason for this is that the efficiency of the linear regulator decreases at lower values of V'_{sup} and a fixed value of V_{sup}. More information on linear regulators will be given in section 7.3.

Other PA classes in which the output transistor is used as a switch are also encountered in the literature [5]-[7]. These PA classes are D, E and F. The theoretical efficiency of the PA is 100% in these cases, because an ideal switch has zero voltage across it when current flows, whereas zero current flows when the switch does not conduct. The efficiency is lower than 100% in practice. These PAs are highly non-linear, resulting in more filtering at the output. This leads to a more expensive system. Therefore, PAs of classes D, E and F have not yet been used in cellular phones. The remainder of this chapter will concentrate on class A, AB and B type PAs.

7.2.2   Optimum supply voltage for optimum efficiency

In order to have an efficiency of η_{max} at all output powers, V_{sup} should be made dependent on P_{out}. This is the essence of efficiency control and applies to both bipolar and FET PAs. The optimum value $V_{sup,opt}$ as a function of P_{out} can be obtained from the general equation (7.5):

$$V_{sup,opt} = \sqrt{2 P_{out} R_{load}} \qquad (7.6)$$

where R_{load} reflects the transformed antenna impedance again. When the supply voltage for maximum output power $P_{out,max}$ is named V_{nom}, which is the nominal supply voltage described in the data sheet of the PA, (7.6) can be rewritten as:

$$V_{sup,opt} = \sqrt{\frac{P_{out}}{P_{out,max}}} \cdot V_{nom} \qquad (7.7)$$

using (7.2). Some means of DC/DC conversion is needed for efficiency control to translate the battery voltage into the desired value of $V_{sup,opt}$. The advantages of efficiency control are listed below:

- The maximum efficiency of the PA, as specified in the data sheet, is maintained over a wider range of output powers, as opposed to only at maximum output power.
- The application of a well-defined supply voltage for each output power may make it easier to design the PA. The reason for this is that the PA designer now

only has to take a limited range of supply voltages into account during the design, as opposed to a whole range of supply voltages when the PA is connected directly to the battery. This advantage already holds when a DC/DC converter with a fixed output voltage is used.

- The PA designer can choose the optimum nominal supply voltage when a DC/DC converter is present in the system anyway. Impedance transformation is needed to be able to transmit at the desired output power, as was explained in section 7.2.1. (7.2) shows that the higher the nominal supply voltage for maximum output power is chosen to be, the higher the load observed by the PA will be allowed to be and the lower the ratio of the 50 Ω antenna impedance and the impedance R_{load} observed at the PA side will become. This lowers the losses in the impedance transformation network, because the parasitic resistances in this network are now smaller relative to R_{load}.

The implementation of efficiency control in a cellular phone does not imply drastic changes to the transmit architecture. An example for a GSM phone will be given in section 7.7.

The transmit path needs to be sufficiently linear when the cellular system employs non-constant envelope modulation; see section 7.1. Full swing RF signals at the output of the PA are then not allowed, because saturation of the output transistor will lead to an unacceptable level of non-linearity. In that case, a somewhat higher value than that found from (7.6) and (7.7) has to be chosen for $V_{sup,opt}$. This will lead to a smaller improvement in efficiency. A quantitative example will be given in section 7.6, which shows that a considerable improvement can still be realized in practice.

7.3   DC/DC conversion principles

In general, two approaches exist for converting one voltage into another. The first approach involves a time-continuous circuit with a dissipative element, whereas the second approach involves a time-discrete circuit with an energy-storage element. The first approach is only suitable for converting a higher voltage into a lower voltage, which is down-conversion, whereas the second approach enables up- and down-conversion. The two approaches are illustrated in Figure 7.5, in which the battery voltage V_{in} is converted into V_{out}.

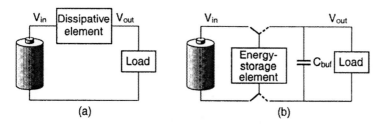

Figure 7.5: Two approaches to convert a battery voltage into another voltage. (a): Time-continuous with dissipative element ($V_{in} > V_{out}$ only) (b) Time-discrete with energy-storage element

The dissipative element in Figure 7.5a remains connected between the battery and the load. The efficiency of the voltage conversion will always be lower than 100%, because of its dissipative nature. The energy-storage element in Figure 7.5b is first connected to the battery to store energy, after which it is connected to the load to supply this energy. The efficiency of the voltage conversion process is 100% in the theoretical case in which no energy is lost in the energy-storage element and switches. The type of time-discrete voltage converter employed and the characteristics of the employed components will determine the efficiency in practice. An energy buffer C_{buf} is necessary, because of the time-discrete nature of converters of this type.

7.3.1   Linear voltage regulators

The configuration of Figure 7.5a is commonly known as a linear voltage regulator [8]. A more detailed schematic representation of a linear voltage regulator is shown in Figure 7.6. It consists of a transistor, which may be of any type, controlled by a regulator, which compares a fraction of V_{out} with a reference voltage V_{ref}. Linear regulators vary in the used type of transistor and its drive circuit. The transistor is operated in the saturation region in the case of FETs and the linear region in the case of bipolar transistors. This means that the output current I_{out}, and hence V_{out}, will hardly change when V_{in} changes. However, a certain minimum voltage difference $V_{in}-V_{out}$, which is the dropout voltage, has to be present for proper operation. A simple resistor represents the load in Figure 7.6.

Figure 7.6 illustrates that the maximum efficiency η is the ratio of V_{out} and V_{in}. A decrease from this maximum value is caused by the current consumption of the linear voltage regulator itself, including the opamp, voltage reference and current through R_1 and R_2. Hence, I_{out} is smaller than I_{in}. The efficiency of the linear voltage regulator will be higher in the case of smaller differences between V_{in} and V_{out}, as can be understood from the ratio. Therefore, the term Low DropOut regulator (LDO) is often encountered in practice when a low voltage drop can be accommodated, the term 'low' being a relative term.

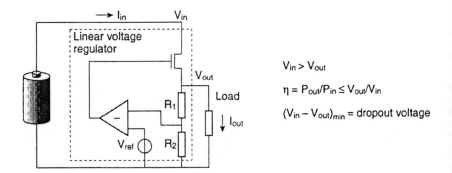

Figure 7.6: Basic schematic representation of a linear voltage regulator

Linear voltage regulators are encountered in portable devices when the difference between the battery and load supply voltages is not too high, because the efficiency will then still be acceptable. An advantage of linear regulators is that they are cheap and small because they do not need an energy-storage element, which usually takes up quite some volume. Linear regulators are often used as filters, because variations in V_{in} are suppressed in V_{out}. Although not included in Figure 7.6, small capacitors with values specified in the data sheet are added to the input and output of a linear voltage regulator in practice. The output capacitor improves the response to load changes and usually implements regulator loop frequency compensation.

7.3.2   Capacitive voltage converters

A converter as shown in Figure 7.5b has to be used when there is a large difference between the battery voltage and the voltage needed by the load, or when the load voltage should be higher than the battery voltage. The energy-storage elements can either be capacitive or inductive. The basic principle of a capacitive voltage converter is shown in Figure 7.7. Such a configuration is commonly referred to as a charge pump. The switches are operated from a non-overlapping clock signal with periods Φ_1 and Φ_2. The capacitors are connected in parallel to the battery for the up-converter for period Φ_1 and each capacitor is charged to the battery voltage. The capacitors are connected in series for period Φ_2, added to the battery voltage, and connected to the output buffer capacitor C_{out}. At no load, this leads to an output voltage $V_{out}=(n+1)\cdot V_{in}$, where n is the number of capacitors.

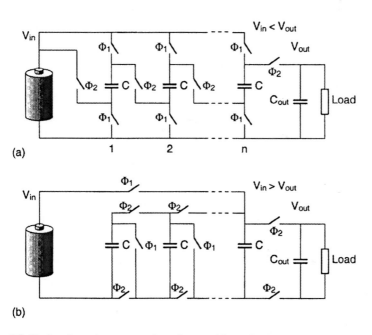

Figure 7.7: Basic schematic representation of a capacitive voltage converter: (a) Voltage up-converter (b) Voltage down-converter

The capacitors are connected in series for the down-converter for period Φ_1 and the total series connection is charged to the battery voltage. The capacitors are connected in parallel for period Φ_2 and connected to the output buffer capacitor C_{out}. This leads to an output voltage $V_{out}=V_{in}/n$ at zero load, with n being the number of capacitors. By combining parallel and series connections for periods Φ_1 and Φ_2, non-integer conversion factors can be realized.

Efficiency considerations

Consider the charging of an ideal capacitor by a voltage source V through a switch S with an on-resistance R_S. This is illustrated in Figure 7.8, from which the following equation can be derived for the energy $E_C(t)$ stored in C as a function of time t:

$$E_C(t) = \int_0^t V_C(t)I(t)dt = \frac{1}{2}CV^2 \cdot \left(e^{\frac{-2t}{\tau}} - 2e^{\frac{-t}{\tau}} + 1 \right) \qquad (7.8)$$

And for the energy $E_R(t)$ dissipated in R_S as a function of time t:

$$E_R(t) = \int_0^t V_R(t)I(t)dt = \frac{1}{2}CV^2 \cdot \left(1 - e^{\frac{-2t}{\tau}} \right) \qquad (7.9)$$

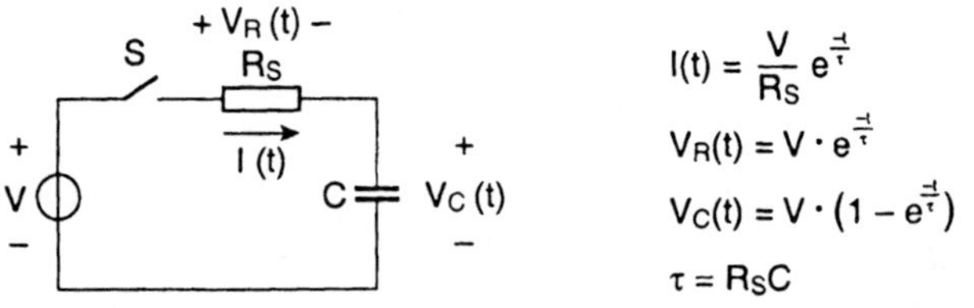

$$I(t) = \frac{V}{R_S} e^{\frac{-t}{\tau}}$$

$$V_R(t) = V \cdot e^{\frac{-t}{\tau}}$$

$$V_C(t) = V \cdot \left(1 - e^{\frac{-t}{\tau}} \right)$$

$$\tau = R_S C$$

Figure 7.8: Charging a capacitor C by a voltage source V through a switch S with an on-resistance R_S

$E_C(t)$ and $E_R(t)$ both become $E_{max} = \frac{1}{2}CV^2$ when t approaches infinity. This means that, irrespective of the value of R_S, equal amounts of energy are stored and dissipated. This also holds when R_S is zero or is not constant. The latter case occurs, when the capacitor C is charged through a transistor. Table 7.3 shows the course in time of $E_C(t)$ and $E_R(t)$. For clarity, these energies have been normalized to the maximum energy E_{max}.

Table 7.3: Normalized energies $E_C(t)$ and $E_R(t)$ and normalized voltage $V_C(t)/V$ at normalized times t/τ

t/τ	V_C(t)/V [%]	E_C(t)/E_max [%]	E_R(t)/E_max [%]
0.25	22.4	5	39.7
0.38	31.6	10	53.3
0.69	50	25	75
1	63.2	40	86.5
1.23	70.7	50	91.4
2.97	94.9	90	99.7
3.68	97.5	95	99.9

Table 7.3 shows that storing the first 5% of E_{max} causes an energy loss of 39.7% of E_{max} in R_S. However, when the same amount of energy is stored starting at 90% (90% → 95% of E_{max}), the energy loss is only 0.2% of E_{max}. Hence, although the

energy storage occurs fast at the beginning of charging, because it only takes 0.25τ to store the first 5% of E_{max}, it is rather inefficient. The efficiency is $5/(5+39.7) =$ 11.2%. The efficiency increases to $5/(5+0.2)=96.2\%$ when V_C approaches its end value of V, but it takes 0.71τ to store 5% of E_{max} from 90% to 95%. This means that the capacitors should be charged (almost) to the source voltage V in order to have a high efficiency and the capacitance should be high, so that the voltage drop is minimal in the discharge period. Hence, only a fraction of the stored energy is active in the transfer process.

The frequency at which the energy is transferred or the amount of the energy that is transferred each cycle need to increase when the required output power of the charge pump is high. The combination of high efficiency and high power leads to very large capacitance values, because storing energy takes more time at the end of the charging process. These large capacitors cannot be integrated with the switches and control circuitry on a single IC in practice. External capacitors should be used instead. When efficiency is not a big issue and/or the load current is small, smaller capacitors can be used, which creates the possibility of integrating them.

7.3.3   Inductive voltage converters

The basic principle of using an inductive voltage converter is depicted in Figure 7.9. This type of converter is generally referred to as a DC/DC converter. As with capacitive converters, up-conversion and down-conversion of the battery voltage are possible. The time-discrete nature of the two converters can be clearly recognized by the switches.

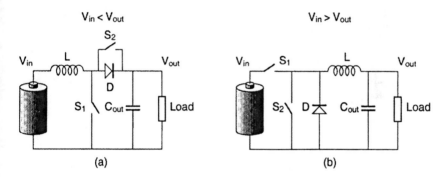

Figure 7.9: Basic schematic representation of an inductive voltage converter: (a) Voltage up-converter (b) Voltage down-converter

Switch S_1 is closed first and energy is stored in the inductor for both the up- and the down-converters. The inductor current ramps up linearly in the ideal case of zero series resistance. Switch S_1 is opened at a certain moment, depending on the desired ratio of V_{in} and V_{out} and the value of the load. The current will flow through the diode D, because the inductor current cannot change instantly, and the polarity of the voltage across the inductor changes, which leads to a decrease in the energy stored in the inductor and the current will ramp down accordingly. The energy that was stored in the inductor for the period when S_1 conducted will now be transferred to the output buffer capacitor C_{out}. Switch S_2 is not strictly necessary in either case.

When it conducts in the second phase, the efficiency will be higher than without the switch, since the voltage across diode D will be lower than without the bypass switch. Bypassing the diode by switch S_2 is commonly referred to as synchronous rectification.

The waveform of the current through the inductor is shown in Figure 7.10 for both the up- and the down-converter. For simplicity, it is assumed that the inductor current is never zero. This is called *continuous conduction mode*, as opposed to *discontinuous conduction mode*, in which the inductor current decreases to zero each switching cycle. The depicted waveform is valid for a high-efficiency converter, in which the voltage drop across the switches during current conduction can be neglected.

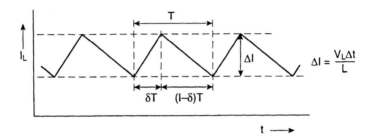

Figure 7.10: Inductor current I_L for the converters of Figure 7.9 in continuous conduction mode with ideal switches

The conduction period of S_1 is δT. The duty cycle δ is defined as the ratio of the time for which S_1 conducts and the total switching time T and has a value $0<\delta<1$. As a consequence, S_2 conducts for a period $(1-\delta)\cdot T$. In steady-state, in which case the input voltage, load current and duty cycle are fixed, all the energy that was stored in the inductor for the period δT will be transferred to C_{out} during $(1-\delta)\cdot T$. This means that the Volt·second product for the coil is equal in both periods, or $V_L(\delta T)\cdot\delta T=V_L((1-\delta)T)\cdot(1-\delta)\cdot T$. For the up-converter, this leads to:

$$V_{in}\delta T = \left(V_{out} - V_{in}\right)\!\left(1 - \delta\right)\!T \Rightarrow \frac{V_{out}}{V_{in}} = \frac{1}{1-\delta} \qquad (7.10)$$

whereas for the down-converter it leads to:

$$\left(V_{in} - V_{out}\right)\!\delta T = V_{out}\left(1 - \delta\right)\!T \Rightarrow \frac{V_{out}}{V_{in}} = \delta \qquad (7.11)$$

(7.10) and (7.11) illustrate that the ratio of V_{in} and V_{out} can be controlled by controlling the duty cycle δ. The ratio of the output and input voltages is greater than unity in the case of the up-converter and smaller than unity in the case of the down-converter, because $0<\delta<1$. An up/down-converter can be constructed by combining the configurations of Figure 7.9a and b. Four switches are necessary in that case and two conducting switches are always present in the current path. This leads to an efficiency that is poorer than that of single up- or down-converters, because then only one conducting switch is present in the current path.

The number of components needed to achieve a certain ratio of V_{in} and V_{out} is much smaller than in the case of capacitive converters. Only the inductor and diode are realized with external components in practice and the switches and the controller can be integrated. Another difference is that the ratio of V_{in} and V_{out} in inductive converters can be changed by controlling δ. In the case of a capacitive converter, the circuit topology has to be changed to achieve this or additional losses must be introduced in the form of an additional linear voltage regulator. Moreover, an essential difference between capacitive and inductive voltage converters is the efficiency of energy storage.

Efficiency considerations
Consider the storage of energy in an ideal inductor L from a voltage source V through a switch S with an on-resistance R_S. A voltage source is again to be considered, because this book deals with batteries as power sources and a battery can be regarded as a voltage source. This is shown in Figure 7.11.

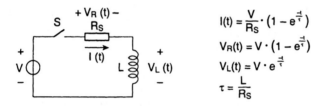

$$I(t) = \frac{V}{R_S} \cdot \left(1 - e^{\frac{-t}{\tau}}\right)$$

$$V_R(t) = V \cdot \left(1 - e^{\frac{-t}{\tau}}\right)$$

$$V_L(t) = V \cdot e^{\frac{-t}{\tau}}$$

$$\tau = \frac{L}{R_S}$$

Figure 7.11: Charging an inductor L by a voltage source V through a switch S with an on-resistance R_S

The energy $E_L(t)$ stored in L as a function of time t can be obtained from:

$$E_L(t) = \int_0^t V_L(t)I(t)dt = \frac{1}{2}LI^2 \cdot \left(e^{\frac{-2t}{\tau}} - 2e^{\frac{-t}{\tau}} + 1\right) \qquad (7.12)$$

where $I = V/R_S$. When t approaches infinity, the energy stored equals $E_{max} = \frac{1}{2}LI^2$. The energy $E_R(t)$ dissipated in R as a function of time t is obtained from:

$$E_R(t) = \int_0^t V_R(t)I(t)dt = I^2R_S \cdot \left(t - \frac{3\tau}{2} + 2\tau e^{\frac{-t}{\tau}} - \frac{\tau}{2}e^{\frac{-2t}{\tau}}\right) \qquad (7.13)$$

An important difference with respect to a capacitive converter is that $E_R(t)$ can be made zero when R_S is made zero. Hence, energy can be stored in an inductor from a voltage source with 100% efficiency in the ideal case when R_S is zero, as opposed to storing energy in a capacitor from a voltage source, in which case a certain amount of energy of $\frac{1}{2}CV^2$ is always lost, irrespective of the value of R_S, when the capacitor is charged from 0 V to a voltage V. This means that inductive voltage converters can achieve a 100% efficiency with ideal components. This is not the case for capacitive converters. This holds for a voltage source as input, which is the case in battery-powered equipment. Inductive voltage converters are for this reason found in many portable products.

Although an efficiency of 100% is theoretically possible with an inductive voltage converter, losses will occur in practice, resistive and capacitive losses. Resistive losses are caused by series resistance in the current path, for example an Equivalent Series Resistance (ESR) of the inductor and on-resistances of the switches. Moreover, each switch has a parasitic capacitance, which leads to capacitive switching losses in practice.

7.3.4   EMI problems involved in capacitive and inductive voltage converters

The time-discrete voltage converters can cause Electro-Magnetic Interference (EMI), for example, due to voltage ripple on the output buffer capacitor caused by the ESR, or due to current loops with varying currents that span a significant area. These problems can be avoided by carefully choosing the components, filtering and Printed-Circuit Board (PCB) layout. An output voltage control scheme with a fixed switching frequency outside the frequency bands of interest could also be considered. A linear regulator often follows an inductive voltage converter when the load is highly sensitive to EMI, as in RF circuitry in cellular phones. The voltage difference across the linear regulator should be kept low for efficiency reasons, as discussed earlier. However, the linear regulator has the advantage of offering additional filtering [9]. Moreover, the addition of a linear voltage regulator to the output of an inductive voltage up-converter implies the possibility of realizing a voltage up/down-converter [10]. Finally, one should realize that the output spectrum of a down-converter will be more favourable than that of an up-converter. The reason is the difference in shape and frequency content of the current that flows through the output capacitor for both converter types.

7.3.5   Inductive voltage conversion for efficiency control

The battery voltage has to be converted into a variable PA supply voltage for efficiency control; see section 7.2. An inductive voltage converter is most suitable for this from an efficiency point of view. Moreover, the ratio of V_{in} and V_{out} can be controlled relatively easily. This makes it easy to change from one converter output voltage to another. However, the influence of voltage ripple at the converter output on the PA's RF behaviour should be carefully investigated. This will be discussed in sections 7.6 and 7.7.

7.4   Simulation model derivation

Simple simulation models for a DC/DC down-converter and a PA will be derived in this section. Both models will be used in section 7.5 to compare different supply strategies of a PA by means of simulation.

7.4.1   DC/DC down-converter

The efficiency of a DC/DC converter is obtained from:

$$\eta_{DC/DC} = \frac{P_{out}}{P_{in,total}} = \frac{P_{out}}{P_{out} + P_{loss}} \tag{7.14}$$

The power loss P_{loss} includes ohmic loss due to parasitic series resistance R_{loss} and switching losses. The total switching losses are obtained from $f_{switch}C_{para}V^2_{switch}$, where f_{switch} is the switching frequency, C_{para} is the total parasitic capacitance that has to be charged and discharged each switching cycle, and V_{switch} is the switch drive voltage. V_{switch} should be as high as possible to minimize the switch resistance. V_{switch} is made equal to V_{in}, because this is the highest voltage available in a down-converter. This leads to the following expression for $\eta_{DC/DC}$:

$$\eta_{DC/DC} = \frac{V_{out}I_{out}}{V_{out}I_{out} + I_{out}^2 R_{loss} + f_{switch}C_{para}V_{in}^2} \tag{7.15}$$

The output current and power of a DC/DC down-converter are limited. Figure 7.9b illustrates that for a duty cycle $\delta=1$, switch S_1 is always conducting and the converter is nothing more than a connection from V_{in} to V_{out} through an always-conducting switch and a coil. This situation will eventually occur when V_{in} decreases towards V_{out}. The maximum current I_{out} that can be supplied at a certain difference between V_{in} and V_{out} is now obtained from:

$$I_{out} = \frac{V_{in} - V_{out}}{R_{loss}} \tag{7.16}$$

Maximum output power $P_{out,max}$ occurs when the load resistance R_{load} equals R_{loss}. The efficiency will be 50% in this case and $P_{out,max}$ is obtained from:

$$P_{out,max} = \frac{V_{in}^2}{4R_{loss}} \tag{7.17}$$

A conventional PA behaves as a current source and the value of this current I_{sup} drawn at the PA supply pin depends on the requested output power and the efficiency of the PA; see section 7.4.2. The battery voltage is the input of the converter and the supply voltage requested for the PA is the output when a DC/DC down-converter powers the PA. The converter will no longer be able to supply enough current at the requested supply voltage of the PA below a certain minimum voltage difference between the input and the output of the converter; see (7.16). As a consequence, the PA cannot supply the requested output power. The series resistance between the battery and the PA can be lowered in this case by connecting a low-ohmic bypass switch in parallel to the converter. The battery voltage may now drop a little further, since the bypass configuration has a lower resistance than R_{loss}. This will extend the talk time of a cellular phone. The PA supply voltage now varies with the battery voltage through:

$$V_{sup} = V_{bat} - I_{sup}(R_{bypass} \,/\!/\, R_{loss})$$ (7.18)

The simulation model for a DC/DC down-converter that will be used in section 7.5 is depicted in Figure 7.12.

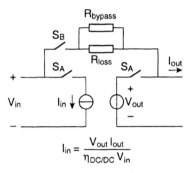

$$I_{in} = \frac{V_{out} I_{out}}{\eta_{DC/DC} V_{in}}$$

Figure 7.12: Simulation model for the DC/DC down-converter used in section 7.5

Efficiency $\eta_{DC/DC}$ in Figure 7.12 is given by (7.15). Switches S_A and S_B are assumed to be ideal. Switches S_A conduct when V_{in}-V_{out} is larger than $I_{out}R_{loss}$, where I_{out} is the current drawn at the output. Switch S_B does not conduct in this case and the output voltage is fixed by voltage source V_{out}. When V_{in}-V_{out} becomes smaller than $I_{out}R_{loss}$, only switch S_B conducts and the voltage at the output of the converter is given by (7.18), with $V_{bat}=V_{in}$ and $I_{sup}=I_{out}$.

7.4.2   Power Amplifier

The behaviour of the PA was described in section 7.2. The PA will be simply modelled as a current source in the simulations in section 7.5, having the value:

$$I_{sup} = \frac{P_{out}\delta_{PA}}{\eta_{PA}V_{sup}}$$ (7.19)

where P_{out} is the requested output power and δ_{PA} is the duty cycle of the PA, which is smaller than 1 in the case of TDMA systems. V_{sup} is the PA supply voltage, which is the voltage across the current source that models the PA, and η_{PA} is given by:

$$\eta_{PA} = \frac{V_{sup,opt}}{V_{sup}}\left(\eta_{max} - \frac{V_{nom} - V_{sup,opt}}{0.8}\right)$$ (7.20)

where the first term is found from (7.5) and (7.6). A modification is made to η_{max} in (7.20) when $V_{sup,opt}$ is lower than V_{nom}, hence when P_{out} is lower than $P_{out,max}$; see (7.7). The value of 0.8 was derived from measurements obtained with a practical PA applied in GSM phones [11]. This models the decrease in η_{max} with respect to the theoretical maximum efficiency $\eta_{max,theory}$ due to the increasing importance of saturation effects at lower supply voltages. The course of η_{PA} as a function of the supply voltage for maximum output power $P_1=P_{out,max}$ and a lower output power P_2 is shown in Figure 7.13.

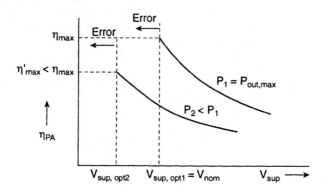

Figure 7.13: PA efficiency (η_{PA}) versus supply voltage (V_{sup}) for maximum output power P_1 and for $P_2<P_1$

An optimum supply voltage $V_{sup,opt}$ is shown for both output powers, where $V_{sup,opt}$ is lower at lower output powers, which follows from (7.7). The PA efficiency will decrease inversely proportionally to V_{sup} when V_{sup} is higher than $V_{sup,opt}$; see (7.20). The maximum attainable efficiency decreases with the output power in accordance with (7.20). The PA is no longer able to transmit at the requested output power when V_{sup} drops below $V_{sup,opt}$. This is indicated by the arrows labelled 'Error' in Figure 7.13. This can be taken into account in a PA simulation model by changing an error signal V_{error} from 0 ($V_{sup} \geq V_{sup,opt}$) to 1 ($V_{sup}<V_{sup,opt}$). The PA simulation model is

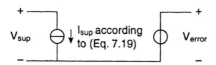

shown in Figure 7.14.

Figure 7.14: Simulation model for the PA used in section 7.5

7.5   Theoretical benefits of efficiency control

Different supply strategies of a PA will be compared by means of simulations in this section. In addition to the simulation models developed in the previous section, the Li-ion model of chapter 4 will be used to translate the efficiencies of the various supply strategies into operating times T_{oper}. The strategies differ in the application of DC/DC conversion. The situation where no DC/DC conversion is applied and a PA is directly powered by a battery will be taken as a reference. A DC/DC converter will be used for converting the battery voltage into the PA supply voltage in all other cases. The converters differ in efficiency, type, use of a bypass switch and control of the output voltage. The relative change in T_{oper} with respect to the reference situation will illustrate the possible benefits or drawbacks of the use of each of the DC/DC converters.

The efficiency of a converter, which is mainly determined by the on-resistance of the switches, of course influences T_{oper} to a large extent. Therefore, converters with different on-resistances of the switches will be compared. The desired supply voltage of the PA will be lower than the battery voltage in many practical cases. Therefore, most converters considered in the simulations will be down-converters. However, down-converters will not be able anymore to supply the desired PA power when the battery voltage approaches the desired PA supply voltage; see section 7.4.1. This situation does not occur when an up/down-converter is used. Therefore, this converter type will also be considered in one of the simulations. Alternatively, the battery voltage at which down-converters fail to supply the voltage requested by the PA can be decreased by applying a bypass switch; see section 7.4.1. The use of such a switch will be considered in some simulations to investigate the influence on T_{oper}. Of course, the on-resistance of the bypass switch determines its effectiveness. Different values for this on-resistance will therefore be considered. Finally, the output voltage of the converters will either be fixed or variable according to (7.7). The latter situation will illustrate the possible benefits of efficiency control. Seven supply strategies in total will be compared in simulations, including the reference situation.

7.5.1   Simulation set-up

The PA model was connected directly to the battery model in simulation 1. This is the reference situation. A DC/DC down-converter model was placed between the battery model and the PA model in simulations 2-6. A DC/DC up/down-converter was placed between the battery and the PA in simulation 7. All the simulations started with a full battery and ended either when the battery was empty, i.e. when the battery voltage had dropped below 2.7 V, or when V_{error} in the PA model had become 1, indicating that the PA was no longer able to transmit at the desired output power level. T_{oper} was determined for a range of output powers and for each of the supply strategies. The parameter values used in the PA model are listed in Table 7.4.

Table 7.4: Parameter values for the PA model of section 7.4 used in simulations 1-7

Parameter:	Value:	Unit:
P_{max}	35	dBm[1]
P_{out}	Variable	dBm[1]
V_{nom}	3.2	V
η_{max}	45	%
δ_{PA}	3/8	-

Note 1: P [dBm]=10·log(P [mW]), so 0 dBm = 1 mW

The specifications of 35 dBm for P_{max} and 3/8 for δ_{PA} were derived from the GSM standard, in which the future increase of δ_{PA} from 1/8 to 3/8 was anticipated. The maximum output power of the PA is 35 dBm, because a power loss of 2 dB between the PA and the antenna due to impedance matching and filtering was taken into account; see Figure 7.1. The values of V_{nom} and η_{max} are in agreement with present-day GSM PAs. The values of P_{out} increased from 21 dBm to 35 dBm in 2 dB increments. This increment size is used in the power control of PAs in GSM cellular phones, as will be described in more detail in section 7.7.

The supply strategies followed in simulations 2-6 differ in the parameters used in the DC/DC down-converter model. These differences concern:

- The value of the on-resistance of the switches (R_{switch}) and the ESR of the coil (R_{coil}), where $R_{switch}+R_{coil}=R_{loss}$. This influences the efficiency of the converter.
- The use of a bypass switch and, when used, its on-resistance (R_{bypass}).
- The control of the output voltage of the converter: V_{out} is fixed and equal to V_{nom} of the PA or V_{out} is variable and equal to $V_{sup,opt}$ for each P_{out} (efficiency control).

The parameter values for the DC/DC down-converter model shown in Figure 7.12 used in simulations 2-6 are listed in Table 7.5.

Table 7.5: Parameter values for the DC/DC down-converter model of Figure 7.12 used in simulations 2-6

Parameter:	Sim. 2	Sim. 3	Sim. 4	Sim. 5	Sim. 6	Unit:
V_{out}	3.2	3.2	3.2	3.2	Var.	V
R_{switch}	0.15	0.15	0.055	0.055	0.15	Ω
R_{coil}	0.05	0.05	0.022	0.022	0.05	Ω
R_{bypass}	10^9	0.1	10^9	0.05	10^9	Ω
C_{para}	20	20	20	20	20	pF
f_{switch}	400	400	400	400	400	kHz

Efficiency control was applied in simulation 6. Extra low-ohmic values for R_{switch} and R_{coil} were used in simulations 4 and 5. All values of R_{switch} and R_{coil} were obtained from several data sheets of DC/DC converters and coils [12],[13]. The use of a bypass switch with different on-resistances was considered in simulations 3 and 5. A value of 1 GΩ was used for R_{bypass} in the other simulations, which means that no bypass switch is used. The value of C_{para} was based on the values for practical converters, whereas f_{switch} was chosen in accordance with data sheets [12].

A fixed efficiency was used for the up/down-converter in simulation 7. No efficiency control was considered, and therefore an output voltage of 3.2 V was used. Simulations with design software for a DC/DC up/down-converter [14] were used to derive an efficiency of 85%. This is an average value for various input voltages, a switch resistance of 100 mΩ, a coil ESR of 20 mΩ, a switching frequency of 400 kHz and a load corresponding to a PA output power of 35 dBm.

7.5.2   Results and discussion

The results of simulations 1-7 are summarized in Table 7.6, which lists the relative change in T_{oper} with respect to the results of simulation 1. Therefore, the improvement with respect to this simulation realized in simulation 1 itself is 0% at all output powers. Several conclusions can be drawn from Table 7.6:

- A benefit of using a down-converter with a fixed output voltage of V_{nom}, instead of connecting the PA directly to the battery, is the positive effect on the operation time of the PA. Compare the results of simulations 2-5 with the result of simulation 1.

- A benefit of using a down-converter with lower-ohmic switches and a coil with a lower ESR is an increase in the operation time of the PA. Compare the results of simulations 4 and 5 with those of simulations 2 and 3.

- A benefit of using a bypass switch with the down-converter is a slight increase in operation time at maximum output power. The effect on T_{oper} is negligible at lower output powers, in comparison with the situation without a bypass switch. Compare the result of simulation 3 with that of simulation 2 and the result of simulation 5 with that of simulation 4 at maximum output power.

- Efficiency control yields a significant increase in T_{oper}, especially at lower output powers. Compare the result of simulation 6 with the results of all the other simulations.

- A drawback of using an up/down-converter is a decrease in T_{oper} at all output powers. Moreover, the silicon area of this converter is larger than that of a down-converter due to the two extra switches. Compare the result of simulation 7 with the results of all the other simulations.

Table 7.6: Simulation results obtained for different PA supply strategies. The table shows the relative changes in operation time T_{oper} of the PA in [%] with respect to the result of simulation 1, in which the PA was connected directly to the battery. The relative changes are shown for different values of the output power P_{out} of the PA

Simulation	Change in T_{oper} relative to simulation 1 [%]							
	P_{out} [dBm]							
	35	33	31	29	27	25	23	21
1: No DC/DC	0	0	0	0	0	0	0	0
2: DC/DC down, R_{switch}=0.15Ω, R_{coil}=0.05Ω, no bypass switch	+6.4	+10.4	+12.0	+13.3	+14.3	+15.3	+16.0	+16.6
3: DC/DC down (see 2) R_{bypass}=0.1Ω	+8.3	+10.4	+12.0	+13.3	+14.4	+15.3	+16.0	+16.6
4: DC/DC down, R_{switch}=0.055Ω, R_{coil}=0.022Ω, no bypass switch	+11.3	+12.6	+14.0	+14.8	+15.5	+16.3	+16.8	+17.2
5: DC/DC down (see 4) R_{bypass}=0.05Ω	+11.7	+12.7	+13.9	+14.8	+15.6	+16.2	+16.8	+17.2
6: DC/DC down (see 2) V_{out}=$V_{sup,opt}$ for all P_{out} (efficiency control)	+6.4	+41	+80.8	+133	+196	+274	+275	+275
7: DC/DC up/down, no bypass switch η=85%	-3.1	-4.2	-3.1	-2.5	-1.9	-1.3	-0.9	-0.5

Although not shown in Table 7.6, it was found in all the simulations that the operating time of the PA ended at maximum output power, because V_{error} became 1. This occurred at a battery voltage of 3.2 V in simulation 1. This is illustrated in Figure 7.13 at maximum output power, where V_{nom} equals 3.2 V. This occurred at a battery voltage higher than 3.2 V in simulations 2-6. It is tempting to say that T_{oper} will be lower when it ends at a higher battery voltage, but this is not the case. Inspection of the column for 35 dBm obtained in simulations 2-6 shows that T_{oper} with a down-converter is still higher than without a down-converter. The reason for this is the higher total efficiency of the solution with a down-converter in the discharge region where the battery voltage is still considerably higher than 3.2 V.

When the combination of a PA and a down-converter is regarded as a black box connected to a battery, the minimum battery voltage at which this black box can still transmit at maximum output power is higher than in the case of a bare PA. Any output power can be transmitted at any battery voltage when an up/down-converter is used in the black box. However, the simulations with an efficiency of 85% show that a drawback is the decrease in T_{oper} at all output powers. Alternatively, one could consider the use of a bypass switch that is switched on in the critical situation. This decreases the battery voltage at which the maximum output power can still be supplied by the black box, as described in section 7.4.1.

The PA model of Figure 7.14 draws average power when δ_{PA} is smaller than 1. For example, for $\delta_{PA}=3/8$, the model draws power continuously with a value of $3/8^{th}$ of the power that would be drawn during three time slots. This occurs in practice when a large buffer capacitor is used at the output of a DC/DC converter. The peak power during the transmit burst is then drawn from this buffer capacitor while the DC/DC converter supplies average output power. This is a sensible thing to do from an efficiency point of view. Smaller switches can be used in the converter, leading to smaller switching losses. At the same time, the large peak currents are drawn from the capacitor via a low-ohmic path, assuming that the buffer capacitor has a low ESR.

In order to obtain insight into what happens when no large buffer capacitor is used, some additional simulations were performed. A simple modification can be made to the PA model of Figure 7.14 to make it draw peak power instead of average power. The PA supply current is shown in Figure 7.15. An extra parameter T_{period} was added to the model. The PA draws peak power for the time $\delta_{PA}T_{period}$. Simulations 1 and 4 were performed once again using $\delta_{PA}=3/8$ and $T_{period}=1$ s. The change in T_{oper} observed in simulation 4 relative to the value in simulation 1 is shown in Table 7.7 for two output powers. Simulation 4 was used, because lower losses are especially important when peak power is drawn.

Table 7.7 shows that T_{oper} decreases at maximum output power when a DC/DC down-converter is used without a large buffer capacitor at the supply pin of the PA. However, it still makes sense to use a DC/DC down-converter at output powers lower than maximum, even without the application of efficiency control. T_{oper} is lower in absolute value when peak power is drawn from either the DC/DC down-converter or directly from the battery. Therefore it makes sense to use a buffer capacitor.

Table 7.7: Change in T_{oper} observed in simulation 4 (with a DC/DC down-converter) relative to the value in simulation 1 (no DC/DC converter) with a PA drawing peak instead of average power; see Figure 7.15

P_{out} [dBm]	Change in T_{oper} relative to the value in simulation 1 [%]
35	-22
33	+6

7.5.3   Conclusions

The operation time T_{oper} of a PA with the specifications listed in Table 7.4 powered by a down-converter is higher than that of a PA connected directly to a battery. Increasing the converter efficiency by using lower-ohmic switches has a positive effect on T_{oper} for all PA output powers. The use of a bypass switch only increases T_{oper} at maximum output power. The use of an up/down-converter with a realistic efficiency of 85% has a negative effect on T_{oper}. T_{oper} increases substantially relative to a fixed output voltage when the converter output is made variable with efficiency control. This is especially true at output powers lower than maximum. Therefore, the efficiency control supply strategy will be investigated in more detail by means of measurements in the remainder of this chapter.

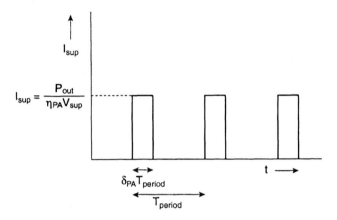

Figure 7.15: PA supply current I_{sup} of the modified PA model

7.6   Experimental results obtained with a CDMA PA

The benefit of efficiency control will be demonstrated in this section by means of measurements obtained with a CDMA PA with a variable supply voltage from a DC/DC down-converter [15]. The Li-ion battery model from chapter 4 will be used to estimate the actual talk time improvement realized in a cellular CDMA phone on the basis of the measured efficiencies. The PA was an EC 2068 CDMA PA from EiC Corporation and the converter was a TEA1210 DC/DC down-converter manufactured by Philips. A voltage source was used instead of a battery. For simplicity, this voltage source will be named battery in the remainder of this section.

7.6.1   Measurement set-up

The measurements have been split into two parts:

- In part 1, no DC/DC converter was used and the PA was connected directly to the battery and characterized; see section 7.6.2.
- In part 2, a DC/DC down-converter was used to power the PA. The results of part 1 of the measurements were used to determine the desired output voltage of the converter; see section 7.6.3.

The efficiency of the PA was measured in part 1 of the measurements for a range of battery voltages V_{bat} from 3.9 V to 2.7 V, which is the applicable voltage range for a Li-ion battery, and for a range of output powers $P_{out,RF}$. Sense lines were used to ensure the desired voltage at the supply pin of the PA. The set-up for part 1 of the measurements is illustrated in Figure 7.16.

The RF power meter was calibrated before use. $P_{out,RF}$ was controlled by means of the RF input power $P_{in,RF}$, supplied by a CDMA signal generator. For each value of V_{bat}, $P_{in,RF}$ was chosen so that $P_{out,RF}$ equalled the requested output power. Whether or not the CDMA specification was met was checked via the values of $ACPR_1$ and $ACPR_2$ with a spectrum analyzer. $ACPR_1$, which should be at least -44 dBc, was measured at 885 kHz from the centre frequency f_c, whereas $ACPR_2$ was measured at 1.98 MHz from f_c. $ACPR_2$ should be at least -60 dBc. The unit dBc expresses the suppression with respect to the carrier signal (c). This is an example of a specification in the frequency domain, as described in section 7.1. The minimum PA supply voltage $V_{sup,min}$ at which the ACPR specification could still be met was determined for each value of $P_{out,RF}$. The PA efficiency was determined for each value of V_{bat} larger than $V_{sup,min}$, as indicated in the upper left corner of Figure 7.16. Note that $V_{sup,min}$ can be higher than $V_{sup,opt}$. This will be explained in section 7.6.2.

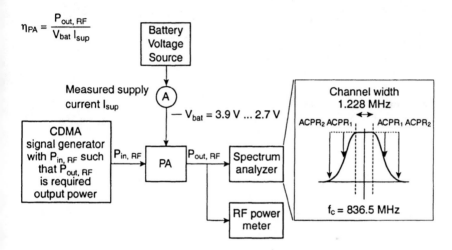

Figure 7.16: Set-up for part 1 of the measurements: no DC/DC converter

Figure 7.17 shows the set-up for part 2 of the measurements. In these measurements, the DC/DC down-converter powered the PA with $V_{sup,min}$ for each value of $P_{out,RF}$ found in part 1 of the measurements. Hence, efficiency control was applied. The battery voltage V_{bat} was applied at the input of the DC/DC converter using sense lines. The ACPR specification was checked for each value of V_{bat}. In addition, the efficiency η_{total} of the total solution including the DC/DC down-converter was determined, as indicated in the upper left corner of Figure 7.17.

7.6.2   Measurement results and discussion of part 1: no DC/DC converter

The results of part 1 of the measurements are summarized in Tables 7.8 and 7.9, in which the results are compared with calculations with the PA model in section 7.4.

Table 7.8: Results of part 1 of the measurements: PA connected directly to the battery

V_{bat} [V]	$P_{out.RF} = 28$ dBm $= 0.63$ W			$P_{out.RF} = 14$ dBm $= 25$ mW		
	I_{sup} [mA]	η_{PA} [%]	η_{PA} [%]: model[1]	I_{sup} [mA]	η_{PA} [%]	η_{PA} [%]: model[1]
3.9	544	29.7	29.6	161	4	5.2
3.5	556	32.4	33	161	4.5	5.9
2.7	-	-	-	160	5.8	7.6

Note 1: See section 7.4: Parameter values from PA (EC 2068) data sheet: V_{nom}=3.5 V, P_{max} = 0.63 W, η_{max}=33%

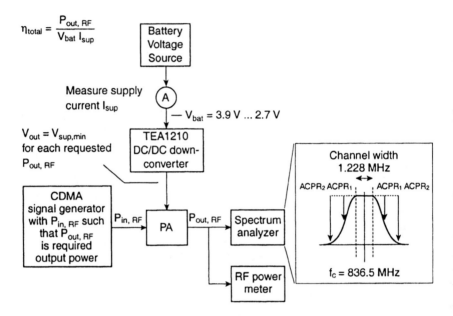

Figure 7.17: Set-up for part 2 of the measurements: Efficiency control was implemented with the aid of a DC/DC down-converter

Several conclusions can be drawn from Table 7.8:

- The PA indeed behaves as a current source, because I_{sup} is almost constant when V_{bat} changes.
- The PA can no longer supply the requested maximum output power of 28 dBm at a battery voltage of 2.7 V. This is in accordance with the results of the simulations discussed in section 7.5.
- The efficiency is exceptionally low at an output power of 14 dBm at all battery voltages. This is the average output power of a CDMA phone [16].
- The PA model predicts the PA efficiency quite reasonably. The predicted efficiency is higher than the measured efficiency at 14 dBm. This is because the CDMA system uses QPSK modulation. Therefore, the PA has to be linear, as was explained in section 7.1. The model assumes that the PA can have a voltage swing of V_{sup} in the output stage. However, a voltage headroom has to be taken

into account, because the PA has to be linear, which leads to a lower efficiency. This was discussed in section 7.2.2.

Several conclusions can be drawn from Table 7.9:

- $V_{sup,min}$ is lower at lower output powers, as was to be expected on the basis of the considerations outlined in section 7.4.
- The voltage headroom, explained above, is clearly observable in this table, in which $V_{sup,min}$ is higher than $V_{sup,opt}$ to meet the linearity specification of the considered CDMA system.
- The PA efficiency at $V_{sup,min}$, which is the maximum attainable efficiency, decreases as the output power decreases. This decrease is larger than that allowed for in the model which was fitted to a GSM PA; see (7.20).

Table 7.9: Measured $V_{sup,min}$ and calculated $V_{sup,opt}$ values

P_{out} [dBm]	$V_{sup,min}$ [V]: measured	$V_{sup,opt}$ [V]: model[1]	η_{PA} [%] at $V_{sup,min}$: measured
28	3.4	3.5	33.3
24	2.3	2.1	28.8
20	1.6	1.4	23.9
16	1.2	0.9	17.9
14	1.0	0.7	15.4

Note 1: See section 7.4: Parameter values from PA (EC 2068) data sheet: V_{nom}=3.5 V, P_{max} = 0.63 W, η_{max}=33%

7.6.3   Measurement results and discussion of part 2: with DC/DC converter

The DC/DC converter had a minimum output voltage of 1.3 V due to the design of the converter. Therefore, a value of 1.3 V was used instead of $V_{sup,min}$ for the output powers 16, 14 and 12 dBm, which unfortunately makes fair comparison difficult. The efficiency of the DC/DC converter was inferred from the measurements as follows:

$$\eta_{total} = \eta_{DC/DC} \ \eta_{PA,@V_{sup,min}} \Rightarrow \eta_{DC/DC} = \frac{\eta_{total}}{\eta_{PA,@V_{sup,min}}} \qquad (7.21)$$

where η_{PA}, valid at $V_{sup,min}$, from the last column of Table 7.9 was used. The efficiency η_{PA} at 1.3 V, instead of $V_{sup,min}$, had to be used in the calculations for output powers of 16 dBm and lower. This efficiency can be easily inferred from Table 7.9 by taking the current source behaviour of the PA into account.

Table 7.10 summarizes the results of part 2 of the measurements. The efficiencies of part 1 have been included in this table to enable comparison. This enables the calculation of the relative change in efficiency when efficiency control is applied. The measured DC/DC efficiencies are also compared with those calculated with the DC/DC converter model of section 7.4.

Several conclusions can be drawn from Table 7.10(a):

- Efficiency control yields a 5.4% improvement in efficiency when V_{bat} is 3.9 V.

• The PA is no longer able to transmit the requested output power of 28 dBm according to the CDMA specification at a battery voltage of 3.5 V. This was still possible in part 1 of the measurements, in which no DC/DC converter was used; see Table 7.8. This is in agreement with the conclusions presented in section 7.5. The combination of a DC/DC converter and a PA ceases to meet the specification at maximum output power at a higher battery voltage than without a DC/DC converter. This situation can be improved somewhat by using a bypass switch at maximum output power, as discussed in section 7.5.

• The measured $\eta_{DC/DC}$ is lower than the $\eta_{DC/DC}$ calculated with the model of section 7.4. This is because the TEA1210 used in the experiments was not optimized for use as a down-converter. The calculated value should be seen as an upper limit of the efficiency.

Table 7.10: Comparison of the efficiencies obtained in parts 1 (no DC/DC converter) and 2 (with DC/DC converter) of the measurements demonstrating the effect of efficiency control. The measured DC/DC converter efficiency $\eta_{DC/DC}$ is compared with the calculated value obtained with the simulation model of section 7.4

(a) $P_{out,RF}$ = **28 dBm**

V_{bat} [V]	$P_{out,RF}$ = 28 dBm = 0.63 W				
	η_{PA} [%] (no DC/DC, see Table 7.9)	η_{total} [%] (with DC/DC, eff. control)	Efficiency improvement [%] when eff. control is applied	$\eta_{DC/DC}$ [%] (measured)	$\eta_{DC/DC}$ [%] (calculated with the model of section 7.4)[1]
3.9	29.7	31.3	5.4	94	97.8
3.5	32.4	-	-	-	-
2.7	-	-	-	-	-

Note 1: See section 7.4: $V_{in}=V_{bat}$, $V_{out}=3.4$ V, $I_{out}=556$ mA, $R_{switch}=70$ mΩ, $R_{coil}=40$ mΩ, R_{bypass} = 1 GΩ (no bypass switch), $C_{para}=1$ nF, $f_{switch}=600$ kHz [12]

(b) $P_{out,RF}$ = **14 dBm**

V_{bat} [V]	$P_{out,RF}$ = 14 dBm = 25 mW				
	η_{PA} [%] (no DC/DC, see Table 7.9)	η_{total} [%] (with DC/DC, eff. control)	Efficiency improvement [%] when eff. control is applied	$\eta_{DC/DC}$ [%] (measured)[1]	$\eta_{DC/DC}$ [%] (calculated with the model of section 7.4)[2]
3.9	4	9.9	148	83.4	94.6
3.5	4.5	10.1	124	85.1	95.4
2.7	5.8	10.7	84	90.1	96.7

Note 1: With $\eta_{PA}=11.9\%$ at a supply voltage of 1.3 V, inferred from Table 7.9
Note 2: See section 7.4: $V_{in}=V_{bat}$, $V_{out}=1.3$ V, $I_{out}=162$ mA, $R_{switch}=70$ mΩ, $R_{coil}=40$ mΩ, R_{bypass} = 1 GΩ (no bypass switch), $C_{para}=1$ nF, $f_{switch}=600$ kHz [12]

Several conclusions can be drawn from Table 7.10(b):

• The higher the voltage difference between the input and the output of the DC/DC converter, the greater the gain in efficiency when efficiency control is applied. This is because the PA still operates at a reasonable efficiency when efficiency control is applied, whereas η_{PA} is quite low when the PA supply voltage is larger than $V_{sup,min}$. The maximum attainable efficiency is not 15.4%, as predicted by Table 7.9, but 11.9%, because the PA is operated at 1.3 V, which is higher than $V_{sup,min}$.

- The efficiency increases at all battery voltages because the output voltage of the DC/DC converter is 1.3 V, which is a lot lower than the battery voltage.
- The efficiency of the DC/DC converter calculated with the aid of the model of section 7.4 is too optimistic. The same explanation holds as for Table 7.10(a).

Figure 7.18 shows the PA efficiency as a function of $P_{out,RF}$ at $V_{bat}=3.5$ V with and without efficiency control. The gain in efficiency obtained with efficiency control increases when $P_{out,RF}$ decreases. This is in accordance with the results presented in section 7.5. It no longer holds at output powers of 16 dBm and lower, for which the η_{PA} and η_{total} curves move closer together again, because the DC/DC converter powered the PA with 1.3 V in all three cases, which is higher than $V_{sup,min}$ at these output powers. Although η_{total} has a value at 28 dBm in Figure 7.18, the CDMA specification was in fact not met at this output power with efficiency control at 3.5 V.

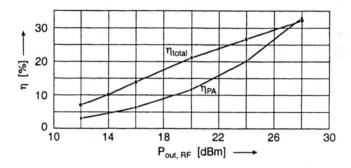

Figure 7.18: Efficiency without (η_{PA}, from part 1) and with (η_{total}, from part 2) a DC/DC down-converter at various output powers $P_{out,RF}$, $V_{bat}=3.5$V

7.6.4   Estimation of talk time increase in a complete CDMA cellular phone

Only the PA itself has been discussed so far. To predict the increase in talk time in an actual CDMA cellular phone with efficiency control, some additional information is needed. This information includes:

1. the power consumption of the rest of the phone;
2. the percentage of time for which the phone transmits at each of the possible output powers;
3. the battery voltage course during discharging.

The power consumption of the rest of the phone determines the output power up to which it still makes sense to improve efficiency. For example, when the rest of the phone consumes 1 W, a decrease in the power consumption of the PA from 2 W to 1 W will make a lot of difference in the phone's total power consumption. However, a decrease in PA power consumption from 2 mW to 1 mW hardly will influence the phone's total power consumption. The lower the power consumption of all the phone parts except the PA, the lower the output power for which efficiency control still makes sense.

The percentage of time for which the phone transmits a certain output power should be obtained from the network provider. Efficiency control does not lead to much improvement at maximum output power. Efficiency control is not very beneficial when the phone transmits at this output power a lot of the time. However, the general trend is towards transmitting at lower output powers most of the time, as described in section 7.1.

A battery discharge curve can be measured, which is time-consuming, or it can be simulated with the aid of the Li-ion battery model described in chapter 4, which takes only a limited amount of time. The efficiency improvement figures from Table 7.10 and Figure 7.18 will be translated into talk time improvements in simulations below. We will first consider the reduction in power consumption of a complete CDMA phone at a battery voltage of 3.5 V; see Figure 7.19. The power consumption of the rest of the phone was assumed to be 924 mW, which value was obtained from a practical CDMA phone. For the occurrence in time of the measured output powers, a rough estimate was used based on information obtained from the literature [16]. The average output power was taken to be 14 dBm.

Figure 7.19 illustrates that the total power consumption of the phone can still be decreased by 23% with efficiency control, even at a very low output power of 12 dBm. This can be understood as follows. Although the output power is only 16 mW, the PA efficiency both with and without a DC/DC down-converter is lower than 10 %. Therefore, the PA still consumes a significant amount of power and an improvement in this power consumption still contributes to a decrease in the phone's total power consumption.

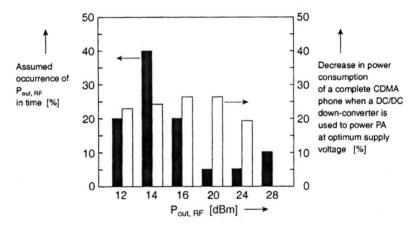

Figure 7.19: Reduction in power consumption of a complete CDMA phone at various output power levels $P_{out,RF}$ (in [dBm]) and the assumed occurrence in time of these output powers (in [%]). Battery voltage V_{bat}=3.5 V

The talk time simulations were performed using the set-up shown in Figure 7.20.

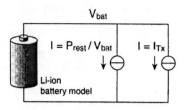

Figure 7.20: Set-up used for talk time simulations. The battery model was developed in chapter 4

Current I_{Tx} is either drawn by the PA when no DC/DC down-converter is present or is drawn by the combination of a PA and a DC/DC down-converter. Its value was chosen to be equal to I_{sup} in parts 1 and 2 of the measurements described earlier in this section. Table 7.8 showed that I_{Tx} is more or less constant at constant output power when no DC/DC down-converter is present. Table 7.10 illustrates that the total efficiency is more or less constant when a DC/DC down-converter is used for efficiency control. This means that the PA with DC/DC converter behaves as a constant power source with a value dependent on the PA's output power. The values of I_{Tx} at different output powers are summarized in Table 7.11. The PA output power is divided by the total efficiency of Figure 7.18 in the situation with a DC/DC converter. The value used for P_{rest} was 924 mW, as in Figure 7.19.

Table 7.11: I_{Tx} for various $P_{out,RF}$ values with and without a DC/DC down-converter

I_{Tx} value	$P_{out,RF} =$ 28 dBm	$P_{out,RF} =$ 24 dBm	$P_{out,RF} =$ 20 dBm	$P_{out,RF} =$ 16 dBm	$P_{out,RF} =$ 14 dBm	$P_{out,RF} =$ 12 dBm
With DC/DC	$\dfrac{P_{out}}{0.31} \cdot \dfrac{1}{V_{bat}}$	$\dfrac{P_{out}}{0.26} \cdot \dfrac{1}{V_{bat}}$	$\dfrac{P_{out}}{0.21} \cdot \dfrac{1}{V_{bat}}$	$\dfrac{P_{out}}{0.14} \cdot \dfrac{1}{V_{bat}}$	$\dfrac{P_{out}}{0.10} \cdot \dfrac{1}{V_{bat}}$	$\dfrac{P_{out}}{0.07} \cdot \dfrac{1}{V_{bat}}$
Without DC/DC	550 mA	365 mA	251 mA	183 mA	161 mA	144 mA

The occurrence percentage in time of $P_{out,RF}$ was taken from Figure 7.19, which led to the sequence for $P_{out,RF}$ in time illustrated in Figure 7.21. This was repeated every 100 s during the complete discharge of the battery.

According to the CDMA specification the combination of a DC/DC down-converter and a PA can no longer transmit the requested 28 dBm output power at battery voltages of 3.5 V and lower. The same holds for a separate PA at battery voltages of 3.4 V and lower. This effect was ignored in the simulations for simplicity. Hence, both with and without the DC/DC down-converter, the simulations proceeded from a full battery until the battery voltage dropped below 2.7 V.

The simulated increase in talk time with efficiency control was 24%. Besides on the assumed P_{rest} value, time distribution of output powers and a simulated battery discharge curve, this figure is based on measurement results obtained with a CDMA PA. Another assumption that has not yet been addressed involves the power control protocol. It was assumed that the supply voltage of the PA immediately had the right value as soon as an output power change occurred. In practice, the time that is available for this supply voltage change will strongly depend on the power control protocol. An example of including efficiency control in a complete GSM phone will be given in the next section. Information on the power control protocol is needed for this.

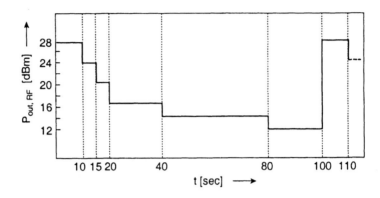

Figure 7.21: Sequence for $P_{out,RF}$ (in [dBm]) as a function of time (t [sec]) used in the talk time simulations

7.7   Application of efficiency control in a GSM cellular phone

The main aim of this section is to show how efficiency control can be implemented in an actual cellular phone, such as the GSM Spark phone manufactured by Philips [11]. First, some background information will be given on the GSM power control protocol. Then, the modifications of the phone needed to include efficiency control will be described. Finally, measurement results will be discussed and compared with results without efficiency control. The measurement results will also be compared with the results of calculations using the simulation models of section 7.4.

7.7.1   GSM power control protocol

Several classes exist in the GSM system. Each class is determined by, amongst other factors, a maximum power and a power control range. This range is from 5 dBm to 33 dBm at the antenna in the case of the Spark phone. The output power of the PA varies between 7dBm and 35 dBm, with an insertion loss of 2 dB between the PA and the antenna.

The base station controls the output power of the handset in steps of 2 dB. It communicates the required output power to the handset. Communication between a GSM handset and a base station is organized in so-called reporting periods, each reporting period consisting of 104 TDMA frames and lasting 480 ms [17]. Each TDMA frame lasts 4.616 ms and consists of 8 slots of 577 µs each. The structure of a reporting period is shown in Figure 7.22.

Each frame accommodates four Tx/Rx (Transmit/Receive) pulse combinations (the Tx and the Rx pulse each fit into a slot) for communication with four handsets, each Tx/Rx pair being separated by three slots. One pair is shown in Figure 7.22 as an illustration. The frame number (between 0 and 103), at which a reporting period starts, will depend on the TCH (Traffic CHannel) number, which may have eight values (TCH0..TCH7). For example, a reporting period lasts from frame 0 to 103 in the case of TCH0 and from frame 13 to frame 12 in the next group of 104 frames in the case of TCH1, *etc.*

The desired output power level is communicated in the Rx slots of the so-called SACCH frames, four of which are present in each reporting period. The frame numbers for the SACCH frames depend on the TCH number. For example, in the case of TCH0, the SACCH frames are located at frame numbers 12, 38, 64 and 90. All four SACCH frames have to be received to enable successful decoding of the information from the base station concerning the desired output power in the next reporting period. Hence, decoding this information can commence when TDMA frame 90 has been received in the case of TCH0. The change to the new output power in the next reporting period takes place at a rate of 2 dB every 60 ms (13 frames). The change commences in the first frame of the next reporting period. An example of the timing of a 4 dB increase in output power in a TCH0 channel is given in Figure 7.23.

Decoding the information in the four received SACCH frames takes four frames. Therefore, it becomes available in frame 95 in TCH0, which is 37 ms before the end of the reporting period. The actual change in output power starts in frame number 0 of the next reporting period. Therefore, the time between the Tx pulse in frame 103 and the Tx pulse in frame 0 in the next reporting period is available to change the output power by the first 2 dB. This time amounts to 4 ms.

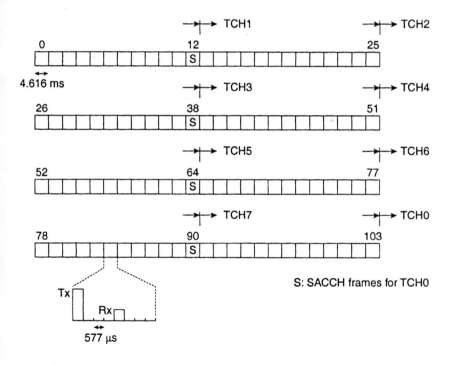

Figure 7.22: Structure of a reporting period in the GSM system

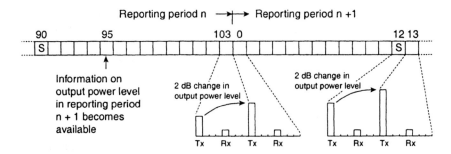

Figure 7.23: Example of a 4 dB output power increase in a TCH0 channel

Output power changes take place in small steps and relatively slowly in time, according to the GSM protocol. The DC/DC converter for efficiency control has to be able to change its output voltage within 4 ms. This needs to be done once every 60 ms at the most.

7.7.2 Modifications in the Spark GSM phone

The PA is usually connected directly to the battery in a Spark phone. The battery consists of four NiMH cells connected in series and has a nominal supply voltage of 4.8 V. The PA in the Spark phone is a BGY204 from Philips. This PA has a maximum efficiency η_{max} of 55%, a nominal supply voltage V_{nom} of 4.8 V and a maximum output power $P_{out,max}$ of 35 dBm. The PA connections in the Spark phone and the organization of power control are depicted in Figure 7.24. This figure is a more detailed extension of Figure 7.1. Impedance matching and filtering have been omitted for simplicity. Instead, an antenna switch for changing from Rx to Tx and *vice versa* is shown.

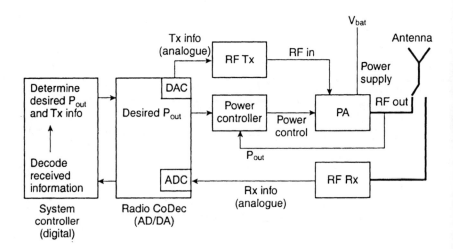

Figure 7.24: PA connections and power control organization in a Spark phone

The PA is connected to an RF input (*RF in*), which derives from the system controller through a Digital-to-Analogue Converter (DAC) and an RF Tx block, including a mixer. The PA amplifies this signal to a signal *RF out*, which is transmitted at the antenna. The PA has a power control pin, which is connected to a power controller. Power control is implemented as shown in Figure 7.4(a), because the BGY204 is a bipolar PA.

Data is received at the antenna during the Rx slots. The Rx info is digitized by an Analogue-to-Digital Converter (ADC) after being down-mixed. The digital information is interpreted in the system controller. The data that has to be transmitted in the next Tx slot is determined. Information on a new output power is received at certain designated times, as explained in section 7.7.1. When needed, the system controller programs the power controller with a new value immediately after the last transmit burst with the present output power.

The power controller provides the proper control voltage to the PA, starting immediately before the actual burst starts. It ramps up the power according to the predefined power-versus-time template; see section 7.1. The ramping time is 20 μs in the case of GSM. The power controller maintains the output power dictated by the base station after the ramping with a tolerance of ±1 dB. The signal *RF in* is amplified to the antenna during the burst. At the end of the transmit burst, the power controller ramps down the output power of the PA. Power control is performed in closed-loop mode, as discussed in section 7.1.

A DC/DC down-converter was included between the battery and the PA supply pin as a modification to the existing Spark phone. Existing hardware was used in the experiment. A TEA1204 DC/DC down-converter from Philips was used, whose output voltage was made controllable by adding some external hardware which consumed negligible power. Three possible output voltages could be generated, depending on the control voltage V_{con} applied to the external hardware:

1. the battery voltage could be passed through directly. Switch S_1 in Figure 7.9b conducted continuously in this situation;
2. a fixed output voltage of 3.7 V could be generated;
3. a fixed output voltage of 3.4 V could be generated.

The values of $V_{sup,opt}$ can be calculated from the PA parameters V_{nom} and P_{max} for each PA output power using (7.7). Table 7.12 shows the values of $V_{sup,opt}$ for the output powers at which the measurements were performed. The voltage applied with the DC/DC converter is shown in the fourth column. A PA supply voltage higher than $V_{sup,opt}$ was used at output powers of 29 dBm and lower, because only three voltages are possible due to restrictions of the employed hardware. Hence, no full benefit of efficiency control could be obtained in the experiments.

Table 7.12: Optimum supply voltages $V_{sup,opt}$ according to (7.7) and actual supply voltages obtained from the DC/DC down-converter for various output powers of the PA and at the antenna

P_{out} at antenna [dBm]	P_{out} at PA [dBm]	$V_{sup,opt}$ according to (7.7) [V]	Actual supply voltage using DC/DC down-converter [V]
33	35	4.8	4.8
31	33	3.8	3.7
29	31	3	3.4
27	29	2.4	3.4
25	27	1.9	3.4
23	25	1.5	3.4

The PA connections and power control scheme in the modified Spark phone with efficiency control are illustrated in Figure 7.25. The original power control loop was left intact. The supply voltages of all the other system blocks, except the PA, were not altered.

The control voltage for the DC/DC converter was supplied by a spare DAC in the Radio Coder-Decoder (CoDec). The software in the Spark phone was expanded slightly with a simple table, which translated the requested output power at the antenna into a control voltage V_{con} [18]. The system controller programmed the right V_{con} value in the spare DAC when the requested output power value was programmed into a register of the power controller. Hence, when needed, the DC/DC converter changed from one output voltage to another, immediately after the Tx burst with the previous output power. The employed DC/DC converter was not able to draw current from the output capacitor to make a step down in output voltage. Therefore, the PA drew current from the buffer capacitor at the PA supply pin in the first Tx bursts with a new output power. The voltage was kept at that value by the DC/DC converter as soon as the voltage had decreased to the voltage at which the DC/DC converter was installed.

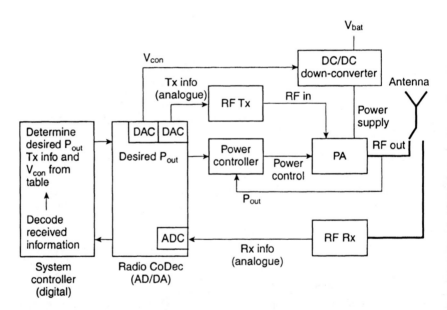

Figure 7.25: PA connections and power control scheme in modified Spark phone with efficiency control

Besides adding a DC/DC converter and adding a table to the software, an additional buffer capacitor was placed at the supply pin of the PA. A total capacitor value of 1 mF was obtained by adding capacitors to the 100 μF already present in the original phone. The reason for this has already been explained in section 7.5. The bursts (GSM is a TDMA system) should be drawn from a buffer capacitor with a low ESR value. The value of 1 mF was found to be a good trade-off between efficiency and capacitance value.

7.7.3 Measurement results and discussion

The original Spark phone shown in Figure 7.24 was compared with the modified Spark phone of Figure 7.25 in the experiments. A 4.8 V voltage source was used in both cases, instead of a battery. This voltage will be called V_{bat} in the remainder of this section. The phone was operated at different output powers, which were checked with a spectrum analyzer. The current consumption figures of the complete phone can be compared. Table 7.13 shows the measurement results. The phone was operated at channel 62, which is in the middle of the GSM Tx frequency band; see section 7.1.

Table 7.13 shows that the power consumption of the modified phone is 1.5% higher than that of the standard phone at maximum output power. This is because the DC/DC converter acts as a 'pass-through' for the battery voltage in the modified phone at maximum output power. Hence, an extra series resistance of S_1 and the ESR of the coil are present in the supply path; see Figure 7.9(b). The PA current consumption can be calculated at maximum output power with a supply voltage of 4.8 V and an efficiency of 55%, valid at 4.8 V. The difference when the values of P_{total} for the standard and modified phone are compared at maximum output power is attributable to an extra series resistance of 133 mΩ in the modified phone. The same currents flow into the PA in both cases. This value is in good agreement with the real value of 70 mΩ obtained for the switch resistance connected in series with 50 mΩ for the ESR of the coil. The series resistance can be lowered by applying a low-ohmic bypass switch, as described in section 7.5. The values of P_{total} given in Table 7.13 are average powers, which were obtained by dividing the power transmitted in the burst by eight.

Table 7.13: Power consumption of complete phone measured in a standard Spark phone and a modified Spark phone with efficiency control

P_{out} at antenna [dBm]	P_{out} at PA [dBm]	Standard Spark				Modified Spark (eff. control)			
		V_{bat} [V]	V_{PA} [V]	I_{bat} [mA]	P_{total} [mW]	V_{bat} [V]	V_{PA} [V]	I_{bat} [mA]	P_{total} [mW]
33	35	4.8	4.8	325	1560	4.8	4.8	330	1584
31	33	4.8	4.8	286	1373	4.8	3.7	272	1306
29	31	4.8	4.8	261	1253	4.8	3.4	240	1152
27	29	4.8	4.8	242	1162	4.8	3.4	224	1075
25	27	4.8	4.8	228	1094	4.8	3.4	214	1027
23	25	4.8	4.8	214	1027	4.8	3.4	207	994

Note 1: $P_{total}=V_{bat}I_{bat}$

P_{out} at antenna [dBm]	Change in P_{total} for modified Spark [%]
33	+1.5
31	-5.1
29	-8.8
27	-8.1
25	-6.5
23	-3.3

The decrease in power consumption as a function of the output power of the modified phone is in agreement with Figure 7.19 for a CDMA phone. Hence, efficiency control has the same effect on the power consumption of the complete phone, irrespective of the type of cellular system. No effect, or even a negative effect, as can be inferred from the 1.5% increase in total power consumption in

Table 7.13, was observed at maximum output power. The decrease in power consumption of the complete phone first becomes greater when the output power decreases. This is because the difference between the battery voltage and the optimum PA voltage $V_{sup,opt}$ increases. Hence, more can be gained from efficiency control. However, the benefit of efficiency control becomes less when the output power decreases further, because the power consumption of the rest of the phone remains constant. This is related to the ratio of power savings in the PA and the total power consumed by the phone. The decreases in power consumption were less in this experiment than in the experiments with the CDMA PA described in the previous section, because the V_{sup} of the PA in the GSM phone was in most cases higher than $V_{sup,opt}$, unlike in the CDMA PA experiment. Therefore, the full benefit of efficiency control was not realized in the experiments with the GSM phone.

The results shown in Table 7.13 can be verified with the aid of the PA model of section 7.4. The PA power consumption can be calculated with the PA model for both the standard and the modified phone and for each output power. A power consumption of 842 mW is found for the rest of the phone when this calculated PA power consumption is combined with the measured P_{total} value for the standard phone at maximum output power; see Table 7.13. This corresponds to a current of 175 mA at 4.8 V, which is in agreement with the results of the measurements. The changes in power consumption measured for the modified phone (see Table 7.13) can now be compared with the results of calculations in which the power consumption of the rest of the phone is assumed to be the same at all output powers. Moreover, the change in power consumption in the case of ideal efficiency control can be calculated. In that case, the DC/DC converter supplies the PA with $V_{sup,opt}$ at all output powers. The results of this exercise are shown in Table 7.14. A fixed efficiency of 90% was assumed for the DC/DC converter in all calculations. The PA duty cycle δ_{PA} was chosen to be 1/8, in accordance with the GSM system for which the Spark phone was built.

Table 7.14: Change in power consumption for a complete Spark phone relative to that of a standard Spark phone obtained in (a) measurements (see Table 7.13), (b) calculations based on the measurement results of Table 7.13 and the PA model of section 7.4, and (c) calculations of ideal application of efficiency control on the basis of the measurement results of Table 7.13 and the PA model of section 7.4

P_{out} at antenna [dBm]	P_{out} at PA [dBm]	(a) Change in P_{total} measured for modified Spark [%]	(b) Change in P_{total} calculated from (a) with the PA model [%]	(c) Change in P_{total} calculated for ideal efficiency control [%]
33	35	+1.5	+1.5	+1.5
31	33	-5.1	-5	-5
29	31	-8.8	-7.7	-11
27	29	-8.1	-6.7	-13.9
25	27	-6.5	-5.7	-15
23	25	-3.3	-4.8	-14.8

The calculated changes in column (b) of Table 7.14 are in good agreement with the measured changes in column (a). Hence, the benefits of efficiency control are again predictable with the models derived earlier in this chapter. The power consumption can moreover be decreased substantially by optimizing the DC/DC converter design and making it suitable for more output voltages. This is especially true at lower output powers, as can be seen in column (c). A similar exercise should be performed for real talk time calculations; see section 7.6. This exercise will not be repeated here.

Specific demands in both the frequency and the time domain exist for the signal transmitted at the antenna. For example, switching transient tests were performed using the modified Spark phone. The spectrum of the complete transmit burst was measured with a spectrum analyzer and had to be within a spectral mask in these tests; see section 7.1. The only violation occurred at an output power of 31 dBm and in channel 124, which is in the far right of the Tx frequency band. This violation was caused by PA saturation because the supply voltage of the PA was slightly lower than $V_{sup,opt}$ of 3.8 V. The supply voltage at which saturation occurs depends on the frequency of the transmitted signal. Moreover, the insertion loss between PA and antenna is frequency-dependent and the losses in these passive circuits are higher at the edge of the frequency band. Therefore, $V_{sup,opt}$ should also be made frequency-dependent. The violation disappeared when the PA supply voltage was increased a little.

Besides switching transient tests, many other tests have to be performed to determine the quality of the modulated signal at the antenna. All the necessary tests with the same PA and DC/DC converter as used in the experiments showed that the voltage output ripple of the DC/DC converter does not lead to violation of any specification [19]. The main reason for this is that the PA has a good power supply rejection ratio due to the high-bandwidth power control loop; see Figures 7.24 and 7.25. Placing buffer capacitors at the PA supply pin with a low ESR still makes sense, notably to reduce the ripple, but also to increase the system efficiency.

7.7.4   Conclusions of the experiments

The behaviour of the PA and the passive circuits between the PA and the antenna have to be characterized in more detail, for example as a function of frequency and temperature, in order to design a cellular phone with efficiency control. Then, an optimum PA supply voltage can be defined for a range of conditions, including output power, frequency and temperature. All frequency and time domain specifications should be met under these conditions. The output power and frequency are known quantities, while the temperature can be measured. The table included in the base-band part now becomes somewhat more complicated. However, the basic method of including efficiency control is still the same as described in this section.

7.8   Conclusions

Several trends in today's cellular telephony systems lead to a need for more efficient PAs. Simple calculations show that an optimum PA supply voltage $V_{sup,opt}$ exists for each output power. With efficiency control, the PA is powered with $V_{sup,opt}$ at each output power, irrespective of the battery voltage.

The brief description of DC/DC conversion principles given above has shown that inductive converters offer the best opportunity of delivering powers required by a PA in an efficient way. Any input battery voltage can be controlled to any output voltage value, irrespective of the load. Therefore, inductive DC/DC converters are the best candidates for implementing efficiency control in a cellular phone. The output voltage of the DC/DC converter is then controlled to $V_{sup,opt}$ at each output power.

Simulations and measurements have shown that efficiency control yields a significant reduction in PA power consumption, mainly at output powers lower than maximum. The PA power decreases only slightly at maximum output power, or even increases slightly, depending on the difference between the battery voltage and the PA supply voltage. The use of a bypass switch helps to decrease the battery voltage at which the PA can still supply maximum output power. The bypass switch does not yield any improvement at output powers lower than maximum. In general, the benefits of efficiency control are greater in the case of greater differences between the battery voltage and $V_{sup,opt}$. The results obtained with the PA simulation model in particular were found to be in good agreement with measurements.

Besides the realized reduction in PA power consumption, the actual benefit of efficiency control in a cellular phone depends on three other factors. First of all on the power consumption of the rest of the phone, secondly on the PA output power distribution in time and thirdly on the battery discharge curve. The power distribution in time is determined by the cellular system. The results presented in this chapter show that the battery models of chapter 4 can be used together with simulation models for a PA and a DC/DC converter to predict talk time improvements realizable with efficiency control in an actual cellular phone. These improvements can also be translated into battery size reductions when talk time improvement is not really an issue for a phone manufacturer. This leads to a smaller and lighter phone with the same talk time. Simulations with the battery model enable a system designer to quickly obtain an idea of what a battery management function, such as efficiency control, can offer.

Outlook
The implementation of efficiency control in a GSM phone shows that only minimal effort is required. A DC/DC converter has to be added between the battery and the PA. This DC/DC converter is already present in many cellular phones. The output voltage of especially Philips' inductive DC/DC converters, which have a digital controller, can easily be made controllable. The control signal can even be obtained directly from base-band in a digital format in a dedicated design. A table should be implemented in software to translate requested output powers into corresponding control voltages. The inclusion of frequency and temperature information in this table should be considered because the value of $V_{sup,opt}$ at which the output signal of the PA still meets all the cellular system specifications depends on the frequency and temperature. This was not considered in the simulations and measurements discussed in this chapter, because most measurements were performed at room temperature and in the middle of the cellular system frequency band. More PA measurements should be performed to determine the influence of the temperature and frequency on $V_{sup,opt}$. However, the basic principle of implementing efficiency control in a cellular phone remains the same as described in this chapter.

7.9   References

[1]   T.S. Rappaport, *Wireless Communications-Principles & Practice*, Prentice-Hall, Upper Saddle River, New Jersey, USA, 1996
[2]   J.D. Gibson (Ed.), *The Mobile Communications Network*, CRC press 8573, IEEE press PC5653, 1996
[3]   *Digital Modulation in Communication Systems- An Introduction*, Application Note 1298, Hewlett Packard, 1997
[4]   The name *efficiency control* originates from Ralf Burdenski

[5]   T.H. Lee, *The Design of CMOS Radio-Frequency Integrated Circuits*, Cambridge University Press, Cambridge, United Kingdom, 1998

[6]   S.C. Cripps, *RF Power Amplifiers for Wireless Communications*, Artech House, Norwood MA, USA, 1999

[7]   J. Smith, *Modern Communication Circuits*, McGraw-Hill Series in Electrical Engineering, Electronics and Electronic Circuits, New York, USA, 1986

[8]   F. Goodenough, "Low-Dropout regulators Get Application Specific", *Electronic Design*, pp. 65-77, May 13, 1996

[9]   B. Kerridge, "Mobile Phones Put the Squeeze on Battery Power", *EDN*, pp. 141-150, April 10, 1997

[10] L. Sherman, "Know the Peculiarities of Portable Power Supply Designs", *Electronic Design*, pp. 94-104, July 7, 1997

[11] H.J. Bergveld, *Increase of Efficiency of GSM Power Amplifiers by Varying the Supply Voltage: Measurement Results of a Demonstrator Built with Spark*, Nat.Lab. Technical Note 093/97, Philips Internal Report, April 1997

[12] Data sheets TEA12xx, *Philips Semiconductors*

[13] Data sheets from various coils, *Coilcraft*

[14] F.J. Sluijs, C.M. Hart, D.W.G. Groeneveld, S. Haag, "Integrated DC/DC Converter with Digital Controller", *International Symposium on Low Power Electronics and Design*, Monterey, USA, pp. 88-90, August 10-12, 1998

[15] H.J. Bergveld, A. van Bezooijen, A.J. Hoogstraate, *Efficiency Improvement of a CDMA PA Using a DC/DC Down-converter: Measurement Results*, Nat.Lab. Technical Note 185/99, Philips Internal Report, June 1999

[16] P. Asbeck, J. Mink, T. Itoh, G. Haddad, "Device and Circuit Approaches for Next-Generation Wireless Communications", Technical Feature, *Microwave Journal*, February 1999, Figure 8

[17] Final draft pr ETS 300 578:1994 (GSM 05.08 version 4.11.0)

[18] H.J. Bergveld, F.A.C.M. Schoofs, "A Communication System, Device and Method", *Patent US6298222, WO9931798, EP0960474*, filed 12 July 1998

[19] M. Simon, *RF-Amplifier + DC/DC-Converter: Test Report*, ABZ05-PA-0001, Philips Internal Report, Philips Technology Centre for Mobile Communication, January 1996

Chapter 8
General conclusions

The subject of Battery Management Systems (BMS) is very broad and involves many disciplines. The main focus in this book has been on simulations of BMS. As stated in chapter 1, simulations are a helpful means for obtaining an understanding of the behaviour of complex systems under a wide variety of conditions. Measurements of battery behaviour generally take a lot of time. With battery simulation models, simulations of battery behaviour in a portable device under development can be performed swiftly. The increased understanding of this behaviour can be applied to improve the BMS functionality. So far, no battery models were available for this purpose. Therefore, battery models with which improvements in BMS designs can indeed be realized have been described in this book.

The simulated behaviour should always be checked by comparing it with measurements. A great advantage in this respect is however that many situations can already be simulated in a relatively short time and the best possible situation can be selected before the start of the measurements. This can serve to give direction to the measurements, which significantly shortens the amount of time involved in the experiments. For example, an optimum charging algorithm can be selected in simulations. This algorithm can then be investigated in more detail by means of measurements. Tedious testing of different possibilities by means of measurements is prevented in this way.

The application of physical system dynamics to battery modelling is new. It turns out to be an elegant approach, in which simulation models of different battery types can be constructed on the basis of a single general set of building blocks. Once a mathematical description of the reactions and processes inside the battery has been derived, a model can be built for each type of battery 'brick by brick'. As a result, the models are based on physical and electrochemical theory and the various processes can easily be recognized in the model. The model serves as a 'transparent battery', in which the courses of the various reactions can be visualized. Because of the applied approach, a battery is represented by an equivalent network model. This enables simulation in the conventional circuit simulators commonly used by designers of portable devices. The battery models can therefore be simulated in both thermal and electrical interaction with their surroundings in the portable device. The best example of this was given in chapter 5, in which the NiCd model interacted both electrically and thermally with the surrounding system parts in a shaver.

The various examples given in this book illustrate that the desired complexity of a battery model depends largely on the application. The entire model was needed in the shaver example in chapter 5 because the focus not only was on battery voltage, but also on the combined effect of voltage (V), current (I), internal gas pressure (P) and battery temperature (T) in interaction with the other shaver parts. However, although possible, the model was not used as a 'transparent battery' in this example, because a shaver designer is not interested in what goes on inside the battery, but only in the SoC build-up and temperature development with the different charging algorithms.

Another example, in which a complete network model is definitely needed, is the charging simulation of Li-ion batteries in chapter 5. This example clearly shows that the developed models can readily be used by battery manufacturers to optimize the design of their batteries. Without the need to develop a wide range of prototypes and perform many time-consuming experiments, insight into the influence of certain design parameters on battery behaviour can be gained very quickly. Moreover, it is very easy to plot the simulated individual electrode potentials. In practice, measurement of these electrode potentials requires the insertion of a reference electrode inside the battery, which is not trivial. Therefore, the 'transparency' of the battery model, in the form of the easy accessibility of all the equilibrium potentials and overpotentials, is an advantage here.

In some cases the entire battery model will have to be used to gain insight in battery behaviour, after which the behaviour can be described in a simpler form. This was illustrated in chapter 6. The Electro-Motive Force (EMF) method was developed on the basis of the results of simulations and measurements. The overpotential description needed for 'time left' predictions was derived in simulations using the entire model. The resulting simpler battery description was used in a State-of-Charge (SoC) system. However, such a simple description can also be used in simulations in which a designer is only interested in battery voltage and does not care at all about what goes on inside the battery. A good example of such simulations is the talk-time simulation performed in chapter 7. In this case, a battery model was only needed to generate a discharge curve under various load conditions. A model based on an EMF curve and the overpotential description of chapter 6 will suffice in this case.

A good quantitative agreement between the results of simulations and measurements is important for increasing the usefulness of the models. It has been shown that good quantitative agreement can be obtained in a charging experiment at a 1 C-rate in the NiCd model with the aid of a dedicated mathematical method. This good agreement was obtained for the V, P and T curves simultaneously with a single parameter set. This inspires confidence in the applied modelling approach. Moreover, the outcome of the comparison of parameter values obtained with the mathematical method and available measured parameter values is promising, because most of the optimized parameter values were close to the measured values. However, parameter optimization alone is not enough, because quantitative optimization is an iterative process that also involves model modifications. With the model modification described in chapter 4 the quantitative agreement was improved for various charge currents simultaneously, although there was still room for improvement. However, the observed improvement was in the right direction, which again inspires confidence in the quality of the model. Various comparisons of simulation and measurement results obtained with the NiCd model discussed in other parts of this book have shown that usable predictions can already be made with the present NiCd model and parameter set. Examples are the self-discharge simulation discussed in chapter 4 and the charge simulations in chapter 5. Good quantitative agreement of charge and discharge curves simulated and measured under various conditions has also been demonstrated for the Li-ion model. However, temperature dependence will still have to be included in this model. Moreover, side reactions have not been implemented yet.

A large part of this book was devoted to the construction of the models themselves. The most important BMS improvements that have been realized with the constructed models are listed below:

- A cause of and solution for breakdown of cells in battery packs have been identified. The model enables simulations more complex than those presented in this book, in which the thermal behaviour of cells at different positions in the pack can also be taken into account.
- A new charging algorithm denoted as *thermostatic charging* has been found in simulations. The simulation results have been verified with measurements. This charging algorithm is very suitable for use in small portable devices with built-in chargers, because such devices often suffer from high internal temperatures during charging. The battery performance will degrade faster when the battery is exposed to high temperatures many times. High temperatures are prevented with *thermostatic charging*, whereas the same charged capacity can still be achieved in the same amount of time as in a standard charging regime with a constant current.
- A possible solution for capacity loss in Li-ion batteries during fast charging has been identified. Experiments will have to be performed to check whether the proposed solution indeed has the desired effect. If those experiments confirm the simulation results, fast charging with less capacity loss will become possible for Li-ion batteries after the capacity of the negative electrode has been increased by 10%.
- A new SoC indication method has been developed on the basis of the results of simulations. The results are promising, although many more experiments and improvements are still needed. The accuracy of the predicted remaining time of use is better than 5% for discharge rates of 1 C and lower. The predictions for large discharge currents at low temperatures are more reliable than those obtained with the existing bq2050 book-keeping system. However, the predictions obtained with the bq2050 at moderate currents and at room temperature are better than those obtained with the proposed system. Some suggestions for improvement of the proposed system were given in chapter 6.
- The battery models have been used as easy means for predicting talk time improvements realizable with *efficiency control*, together with the simulation models of a DC/DC converter and Power Amplifier (PA). A talk time improvement of 24% has been predicted for a Code-Division Multiple Access (CDMA) phone. This prediction is based partly on measurement results obtained with a CDMA PA. Measurements with a complete Global System for Mobile communication (GSM) phone have shown that reductions in phone power consumptions of up to 10% can be realized with *efficiency control*. This reduced phone power can also be translated into a longer talk time.

About the authors

Dr. Henk Jan Bergveld was born in Enschede, The Netherlands, in 1970. He studied electrical engineering at the University of Twente in Enschede from which he obtained his M. Sc. degree (*cum laude*) in 1994. He joined the Philips Research Laboratories in Eindhoven, The Netherlands, in 1994. His research interest was the design of Battery Management Systems and the modelling of rechargeable batteries to improve these designs. He worked on battery modelling with the co-authors of this book. He also worked with various product groups within the Philips organization to help improve the battery management functions in e.g. shavers and cellular phones. The results of these research activities resulted in his PhD degree, which he obtained (*cum laude*) from the University of Twente in 2001. He is currently a senior scientist in the Mixed-signal Circuits and Systems group focussing on the integration of Radio-Frequency transceivers in CMOS technology.

Dr. Wanda S. Kruijt was born in Badanah, Saudi-Arabia, in 1963. She studied Chemical Engineering in Amersfoort, The Netherlands, and subsequently Physical Chemistry at the University of Utrecht, The Netherlands, from which she obtained her M. Sc. Degree in 1989. She then started a research study on the initial stages of electrochemical phase formation. This work resulted in her PhD thesis entitled "Light Scattering from a Nucleating Surface" and her PhD degree from the University of Utrecht in 1993. She then joined the Philips Research Laboratories, where she continued her electrochemical specialism in the study and modelling of physical and (electro)chemical processes in rechargeable batteries. During this time she worked closely with both other authors of this book. She joined the Business Group Luminaires of Philips Lighting in 1998. She is currently working as a project manager at LumiTec, the pre-development group of BG Luminaires. She is responsible for the development and industrialization of large area LCD backlights.

Prof. dr. Peter H.L. Notten was born in Eindhoven, The Netherlands, in 1952. He was educated in analytical chemistry and joined the Philips Research Laboratories in 1975. While working at these laboratories on the electrochemistry of the etching of III-V semiconductors he received his PhD degree from the Eindhoven University of Technology in 1989. Since then his activities have been focussing on the research of hydride-forming electrode materials for application in rechargeable NiMH batteries and as switchable optical mirrors. At present he is responsible for the research programme "Portable Energy" within the Philips Research Laboratories. Since 2000 he has been appointed as part-time professor at the Eindhoven University of Technology where he is heading the group "Electrochemical Energy Storage". His main interests include modelling of various rechargeable battery systems and battery materials research.

Index

Printed in the United States
117896LV00003B/36/A

9 781402 008320